From gene to animal

From gene to animal
An introduction to
the molecular biology of
animal development

SECOND EDITION

DAVID DE POMERAI
Lecturer in the Department of Zoology
University of Nottingham

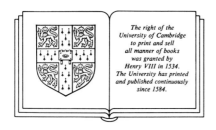

The right of the
University of Cambridge
to print and sell
all manner of books
was granted by
Henry VIII in 1534.
The University has printed
and published continuously
since 1584.

CAMBRIDGE UNIVERSITY PRESS
CAMBRIDGE

NEW YORK   PORT CHESTER
MELBOURNE   SYDNEY

Published by the Press Syndicate of the University of Cambridge
The Pitt Building, Trumpington Street, Cambridge CB2 1RP
40 West 20th Street, New York, NY 10011, USA
10 Stamford Road, Oakleigh, Melbourne 3166, Australia

First published 1985
Reprinted 1986

Printed in Great Britain at the University Press, Cambridge

British Library Cataloguing in publication data

De Pomerai, David
From gene to animal. – 2nd ed
1. Animals. Cells. Development. Molecular biology
I. Title
591.8761

Library of Congress cataloguing in publication data

De Pomerai, David, 1950–
From gene to animal: an introduction to the molecular biology of
animal development/David de Pomerai. --2nd ed.
    p.   cm.
Includes bibliographical references (p.   ).
1. Developmental genetics. 2. Developmental biology.
3. Molecular genetics. 4. Molecular biology. I. Title.
QH453.D4   1990
591.3--dc20   90–1468 CIP

ISBN 0 521 38192 4 hardback
ISBN 0 521 38856 2 paperback

In memory of Valerie and Odile,
who would have wished to understand;
and for Lesley,
who will have to try!

Contents

Preface to the second edition

This is virtually a new book. Dedicated palaeontologists will find skeletons of the fourth, fifth and sixth chapters of the first edition discernible in chapters 2 and 3 of the present version, while a few faint traces of the old chapter 7 appear in the new chapter 5. Chapter 4 has sprung *de novo* from scattered mentions in the old text, and likewise the bulk of chapter 5 is completely new. Much of the molecular biology of animal systems, dealt with at length in chapters 1–3 of the first edition, has now been jettisoned; a brief summary of this material now appears as chapter 1. The present book is already substantially longer than its predecessor, and to update the molecular material thoroughly would demand a great deal more space. In some ways this is a pity; several reviewers particularly praised the opening chapters of the first edition, while others liked the linking of molecular and developmental biology in a single text. However, yet others criticised the book on these same grounds, pointing out that molecular biology is now taught in foundation courses as well as in specialist options, and that many excellent texts cover this area. Moreover, molecular studies of development have so burgeoned and borne fruit over the past five years that they are now established as a distinctive speciality; this much is evinced by a new journal title – *Genes and Development* – which has become indispensable within three years of its appearance.

This second edition aims to explain the progress achieved towards a molecular understanding of developmental processes. Inevitably it is selective in its coverage; both *Caenorhabditis elegans* and *Drosophila* receive chapters (4 and 5) to themselves, while aspects of vertebrate development are covered in chapter 3 and at the end of chapter 5. This plan owes much to Adam Wilkins' text *The Genetic Analysis of Animal Development*. Likewise, the organisation and coverage of chapter 2 is based on the third edition of Eric Davidson's classic *Gene Activity in Early Development*. Chapter 4 is modelled on *The Nematode*

Caenorhabditis elegans edited by William Wood, which aims to become the bible of nematode molecular geneticists for many years to come. It is unfortunate that all three of these excellent books are priced way beyond the means of most students. As for *Drosophila*, brave indeed are they that venture into print concerning this creature, for surely their statements shall become outdated before they emerge from the printing press! I think this is a risk that simply has to be taken; chapter 7 in my first edition failed sadly on this score, since the whole subject began to change radically almost as soon as I had sent the manuscript to the printers, and only limited changes could be made at the proof stage. The unsatisfactory results of this were pointed out by a number of reviewers, and I can only hope that the present attempt withstands the test of time a little better. I make no apologies for devoting nearly half of this text to *Drosophila*, for it is here and in *C. elegans* that the cross-fertilisation between genetics, embryology and molecular biology has proved most fruitful. Some other topics have been excluded quite deliberately from this text, not because they are unimportant in animal development, but simply because the molecular information available is too limited. Thus embryonic fields and induction are mentioned only briefly; likewise the cell surface is given rather scant attention, except where genes of known developmental importance (e.g. *Notch, lin-12*) encode cell surface components. This is not to decry the recent molecular progress achieved in vertebrate systems involving the cadherins and CAM proteins, but we are still some way from understanding the genetic control of these molecules or their precise roles in development and differentiation.

Finally, I should like to thank a number of people who have contributed in various ways to this book. Firstly, I am grateful to the many reviewers of the first edition, both for their praise (which encouraged me to try again) and criticism (which showed me where improvements might be made next time). Several experts have provided detailed comments on parts of this second edition; it is a pleasure to thank John Sulston (chapter 4) and Mike Akam (chapter 5) for correcting my misapprehensions and broadening my knowledge of *C. elegans* and *Drosophila*, respectively, as well as Gary Morgan for comments on chapter 1. The errors that remain are my own; I hope they are not too numerous. I am very grateful to Mike Akam, Robb Krumlauf and Tom Kornberg for sending me material prior to publication. Special thanks are due to Mrs Marian Routledge for typing the enormous reference list with unfailing accuracy; the rest of the text was word processed on Amstrad 1512 and 1640 micros. Finally, I am indebted to my wife

Lesley for her patience and forbearance throughout the long rewriting process. I am grateful to the publishers for the opportunity to update parts of this text during the course of 1989, so as to keep abreast of recent developments in key areas.

DAVID DE POMERAI   *Nottingham, December 1989*

Abbreviations and definitions

rRNA   Ribosomal RNA.
tRNA   Transfer RNA.
mRNA   Messenger RNA.
pre-mRNA   Precursors of messenger RNA.
rDNA   Ribosomal DNA, i.e. DNA containing rRNA-coding
              genes.
S value   (As in 18S/5.8S/28S); a non-linear measurement of
              molecular size based on sedimentation rate.
kb   Kilobases (= 1000 bases); measurement of RNA chain
              length.
kbp   Kilobase pairs (= 1000 base pairs); measurement of gene
              or DNA-fragment length.
kd   Kilodalton (= 1000 daltons or molecular weight units);
              measurement of protein molecular size,
              e.g. 37 kd = 37 000 MW.
EM   Electron microscopy.
UV   Ultraviolet light.
pol I/II/III Nuclear RNA polymerases I (A), II (B) and III (C).
Animal pole   Region of oocyte containing nucleus, from which polar
              bodies are extruded during meiosis. Also that region of
              early embryo derived from same.
Vegetal pole Region of oocyte or early embryo opposite animal pole;
              often a yolk-rich part of the egg.
5', 3'   RNA chains are transcribed in a 5' to 3' direction. By
              convention, in DNA the 5' end of a gene denotes the
              transcriptional start (or cap) site, whereas the 3' end
              denotes the termination (or polyadenylation) site.
              Upstream or 5'-flanking sequences are those preceding
              the start of the gene, while downstream or 3'-flanking
              sequences lie beyond the end of the gene.

Gene
: This term is used throughout in a broad sense to describe any transcription unit (encoding an RNA or protein final product) plus its associated regulatory sequences (promoter, enhancers, etc.). Structural gene sequences denote the coding part of the gene as opposed to the regulatory sites, but in some cases these two may overlap. Although most genes encode a single protein or RNA product, there are instances where a single gene can encode several products via alternative splicing of multiple exons (see below) or cleavage of a large precursor protein/RNA.

Exon
: Those parts of a transcription unit which are represented in the messenger RNA. Not all of these are necessarily translated into protein (e.g. untranslated 5' leader or 3' trailer sequences in many mRNAs).

Intron
: Internal sequences of DNA within a transcription unit which are represented in the primary transcript (e.g. pre-mRNA) but which are absent from the cytoplasmic mRNA; intron transcripts are excised by splicing during RNA processing within the nucleus.

Genetic nomenclature: frequently the same set of initials is used to denote, on the one hand, mutations in a particular gene and the resultant mutant phenotype, and on the other, the wild-type gene and its RNA or protein products. To avoid confusion, the following conventions are adopted throughout this text (for a hypothetical gene *ex*):

ex
: the wild-type gene itself, including its regulatory DNA (promoter, etc.).

ex$^+$
: the wild-type RNA or protein products of that gene.

ex$^-$
: recessive null or deficient mutants lacking *ex*$^+$ function.

ex^D
: dominant mutants which express *ex*$^+$ products constitutively or inappropriately.

In some cases, where a particular mutation has not yet been assigned to the − or D classes, a phrase such as 'an *ex* mutant' will be used; it should be clear from the context that this does not refer to the wild-type gene or its promoter.

1

Gene organisation and control

The purpose of this introductory chapter is to present a brief overview of the organisation, packaging, transcription and regulation of genes in animal systems. In so doing, neither detailed explanations nor supporting evidence will be offered, although a few key references will be given where appropriate. It is assumed that much of the material summarised below will be familiar to readers from courses in molecular biology, and that further detail would be largely superfluous. Further explanation of particular points may be sought from any of several excellent texts covering the cell and molecular biology of both pro- and eucaryotes, including Lewin (1987), Watson (1987), Alberts *et al.* (1983) and Darnell *et al.* (1986).

Some knowledge of the following molecular techniques will be assumed (without further explanation) in the rest of the book: (i) DNA reannealing and RNA/DNA hybridisation for studying populations of nucleic acids; (ii) basic cloning technology for the isolation and bulk preparation of particular DNA sequences; (iii) the construction of chimaeric genes in which regulatory elements from one gene are fused to a foreign coding sequence (often a procaryotic reporter gene whose product can easily be detected); (iv) Southern blotting and restriction mapping with a cloned probe in order to study the genomic organisation of a gene; (v) basic DNA sequencing (Gilbert/Maxam and Sanger) techniques; (vi) Northern blotting to assess the size, tissue distribution and quantitative expression of transcripts from any gene for which a cloned probe is available; (vii) *in situ* hybridisation with a cloned probe to localise specific RNA transcripts; (viii) immuno-detection with poly- or monoclonal antibodies to localise particular proteins; (ix) Western blotting to identify and quantify specific proteins (after electrophoretic separation) using appropriate antibodies.

1.1 DNA organisation

All eucaroytes contain vastly more DNA per haploid genome than can be accounted for by the coding genes, of which there are between 4000 and about 100 000 different types in various animals. A variable proportion of the DNA is repetitive, consisting of multiple copies of the same or very similar sequences. Tandem arrays of short simple sequences repeated (inexactly) millions of times per haploid genome constitute the 'satellite' or simple-sequence DNA fraction, which reforms duplex structures rapidly during the reannealing of denatured DNA. Simple sequence DNA serves no coding function, and is transcribed only during oogenesis (see § § 2.3 and 2.4.2). Although such DNA may play some role in the structural organisation of chromosomes, much of it may be non-functional – sometimes described as 'junk' DNA or evolutionary debris. Other pointers towards this view of much of the eucaryotic genome include the frequent occurrence of transposable elements (see below) and of 'pseudogenes'; these latter are variant versions of *bona fide* genes which have become mutated in various ways so as to preclude their expression (e.g. non-functional promoters and/or premature stop codons: see § § 3.2.4).

By contrast, some coding genes are included in the 'middle-repetitive' DNA fraction present usually in a few hundred to a few thousand copies per haploid genome. These include the major ribosomal genes (encoding the 18S + 5.8S + 28S rRNA species), the 5S rRNA genes and the histone genes in, for example, sea urchins (α-type; see § 2.4.2) or *Drosophila*. These genes are arranged as tandem (head-to-tail) clusters each containing many copies of the same repeat unit, which normally includes the coding gene(s) plus 'spacer' sequences. Other repetitive genes show less regular organisation, with copies either loosely clustered (e.g. the tRNA genes, or histone genes in mammals) or dispersed to many different sites in the genome. Frequently, the different copies are variations rather than exact repeats, intergrading into families of related genes which have apparently diverged from a common ancestor. Examples of such genes include the actin genes (§ 3.1), the vertebrate globin genes (§ 3.2.4) and the vitellogenin genes (§ § 3.3.3 and 4.2). Some dispersed repetitive sequences are mobile and can transpose from one chromosomal site to another; there is evidence that some of these elements may be related to retroviruses. Mobile genetic elements often cause mutations at their sites of insertion, and this can be exploited in various ways by the experimenter (e.g. the Tc1 transposon in *Caenorhabditis elegans*; see § 4.2). Finally, transposable elements can be used as vectors for introducing novel genetic constructs into every cell of an embryo, as in the case of *Drosophila* P elements which only become transposed in the germ line (see § 5.1.3).

Sequences present in one or a few copies per haploid genome are described as unique or single copy. The vast bulk of protein-coding genes fall into this category. Because single-copy sequences reanneal slowly during DNA renaturation, it is possible to isolate them as that fraction which remains single-stranded after all repetitive sequences have reformed duplex structures; these latter bind to hydroxyapatite, whereas single-stranded DNA does not. Preparations of single-copy DNA (as a labelled tracer) are especially useful when comparing different populations of RNA by means of RNA-excess hybridisation (see § 2.3). If total DNA is used, hybrid formation between repetitive genes and their transcripts tends to obscure the much slower reaction between single-copy genes and their corresponding RNAs. In passing, we may note that most of the genes with which we shall be concerned in this book (i.e. protein-coding genes with important functions during animal development) fall into the single-copy category. However, this does not mean that repetitive DNA is unimportant; multicopy sequence elements located within coding or regulatory regions are often shared by functionally related groups of genes, and these have provided important clues about their modes of action and/or control. Two recurrent examples of this are the 'zinc-finger' and 'homeobox' elements which encode DNA-binding protein domains (see § 5.6).

Much of the genome in most eucaryotes is organised in a short-period interspersion pattern typified by the *Xenopus* genome, where short repetitive sequences (averaging 300 bp in length) alternate with longer single-copy sequences (averaging 750 bp). However, in *Drosophila* and certain other insects, the genome is organised in a long-period interspersion pattern, with extended blocks of repetitive DNA (averaging 5600 bp) separated by vast stretches of unique sequences (at least 13 000 bp in length). The significance of this contrast remains unexplained, since other insects including some dipterans (e.g. *Musca*) show *Xenopus*-type sequence interspersion, as do other organisms with very small genomes such as *Caenorhabditis elegans* (see § 4.2).

1.2 Chromatin and DNA methylation

In eucaryotic nuclei, the DNA is associated with a variety of nuclear histone and non-histone proteins in the form of **chromatin**. The fundamental repeating structure of chromatin is the nucleosome, which recurs at approximately 200 bp intervals along the length of the chromosomal DNA. Within each nucleosome, 145 bp of core DNA is wound around a histone octamer comprising two molecules each of the H2A, H2B, H3 and H4 core histones (this DNA–histone complex is termed the core particle). A single molecule of the fifth major histone, H1, is bound at the edge of the nucleosome to the 60 bp of linker DNA between adjacent core particles. A variety of nuclease enzymes cleave chromatin at approximately 200 bp intervals by making double-stranded cuts within the linker DNA. Incomplete digestion of chromatin with such enzymes (e.g. micrococcal nuclease) produces a 'nucleosome ladder' of DNA fragments with lengths of about 200 bp, 400 bp, 600 bp, etc., representing nucleosome monomers, dimers, trimers, etc. Under certain conditions, the ends of linker DNA flanking each cut site may become digested away, reducing the monomer fragments to 145 bp of core DNA protected by association with the histone octamer.

Nucleosomal chromatin is relatively resistant to digestion by the enzyme DNaseI; however, in the neighbourhood of active or potentially active genes, the chromatin is approximately 25-fold more sensitive to DNaseI attack. Thus the chick ovalbumin gene is DNaseI-resistant in brain chromatin where it is inactive, but much more sensitive in laying hen oviduct where it is highly active (see Garel & Axel, 1976 and § 3.3.2). EM spreading shows bulk chromatin as a characteristic 'string of beads' (nucleosomes evenly spaced along the DNA axis), but as a smooth fibre in highly active gene regions. These latter are also distinguished by the presence of side branches representing nascent RNA transcripts. However, nuclease digestion experiments indicate that a 200 bp periodicity is retained within DNaseI-sensitive smooth-fibre chromatin; moreover, all four core histones are present, as shown by immunological studies. These features suggest that DNaseI-sensitive 'active' chromatin represents a more extended conformation of the basic nucleosome structure. Whereas the extended DNA length is compacted by some 6–7 fold in nucleosomal chromatin (through being wound twice around each nucleosome), the degree of compaction is only about 2-fold in smooth-fibre chromatin. Another distinction between active and nucleosomal chromatin is the presence of two specific non-histone proteins (HMGs 14 and 17) bound to the linker DNA, possibly replacing histone H1 (see Weisbrod, 1982). Although most actively transcribed genes are in the smooth-fibre DNaseI-sensitive conformation, the converse is not necessarily the case. Thus some inactive genes are found to be DNaseI-sensitive, usually in tissues where they either have been or will become active (see § § 3.2.2 and 3.3.2). H1 may act as a general transcriptional repressor in nucleosomal chromatin, as well as being involved in the higher-order packing of chromatin into supercoils, etc.

Within broad regions of DNaseI-sensitive chromatin, certain limited sites are found to be especially susceptible to DNaseI attack. These DNaseI-hypersensitive (DHS) sites arise where particular protein factors bind to target sequences in the DNA (see e.g. Elgin, 1984). In many cases, DHS sites are found close to the 5′ ends of actively transcribed genes, but other DHS sites may occur well upstream from, within or downstream from such genes. The presence of DHS sites in a given gene region can be demonstrated by treating chromatin with extremely low concentrations of DNaseI and then digesting the DNA completely with an appropriate restriction enzyme. Among the resultant DNA fragments which hybridise with the corresponding gene probe will be a set of DNaseI-generated sub-bands as well as the standard restriction bands for that enzyme. The precise locations of DHS sites can then be mapped using indirect end-labelling methods. In cases where the same gene is active in different tissues and/or at different levels, alternative sets of DHS sites may characterise that gene region in each of its activity states (see Fritton *et al.*, 1984; § 3.3.2). Finally, there is evidence that certain DHS sites may 'mark out' a gene for future activation, and that such patterns of DHS sites may be inherited over many cell generations without actual expression of the gene (Groudine & Weintraub, 1982). Such a mechanism might perhaps underlie the phenomenon of determination, whereby sets of tissue-specific genes are selected for activation well before their products become expressed in detectable amounts.

Yet another possible explanation for determination and for the stability of the differentiated state (both heritable characteristics) involves the pattern of DNA methylation. Vertebrate DNA is typically methylated at a high proportion of C residues occurring in CG dinucleotides. Note, however, that there is no DNA methylation in many invertebrates, including *Drosophila* (Urieli-Shoval *et al.*, 1982). Methylated C residues are not incorporated directly into newly synthesised DNA strands. Rather, the methylation of C residues in vertebrate DNA is accomplished after replication by a methylase enzyme which is specific for (i) Cs within CG dinucleotides and (ii) hemimethylated DNA (Gruenbaum *et al.*, 1982). Because of semiconservative DNA replication, each daughter duplex will be hemimethylated, i.e. will contain one methylated parental strand and one newly synthesised non-methylated strand. The methylase will thus add methyl groups only to those Cs occurring in CG dinucleotides on the new DNA strand at positions where neighbouring Cs on the opposite parental strand are already methylated. In this way, the parental pattern of methylated and unmethylated sites will be passed on exactly from parent to both daughter cells, and so on. Methylation patterns in a particular gene region can be analysed by using two isoschizomer restriction enzymes which recognise the same target sequence containing a CG dinucleotide (e.g. CCGG, but which are respectively sensitive (*Hpa*II) and insensitive (*Msp*I) to methylation of the central Cs. Sites cut by *Msp*I but not by *Hpa*II identify some of the methylated Cs (only those contained in CCGG target sequences) within the DNA region mapped, whereas unmethylated target sequences will be cut by both enzymes.

There is considerable evidence that many active genes, and especially their 5'-flanking regions, are relatively undermethylated compared with bulk DNA, suggesting that demethylation might play some role in gene activation in vertebrates. However, this correlation may be coincidental, or else demethylation may be a consequence rather than a cause of transcription. The evidence on this point remains equivocal (Bird, 1984, 1986; Cedar, 1988). Demethylated CG-rich sequences are found near many housekeeping genes that are active at low levels in all cell types, but they are often absent from the neighbourhood of tissue-specific luxury genes (Bird, 1986). Thus demethylation is unlikely to play a general role in activating these latter, although it may mark out the former for constitutive activity. Methylation of DNA target sequences can prevent them from being bound by transcription factors (Iguchi-Ariga *et al.*, 1989), and there is also a mammalian protein which binds specifically to DNA containing methylated CGs (Meehan *et al.*, 1989).

Recently, differential methylation of maternal (low) and paternal (high) genomes during vertebrate gametogenesis has been implicated in genomic imprinting, whereby the two parental alleles of a gene may show differential activity during development of the zygote. However, the overall picture is complicated both by a loss of methylated sites during early development and later by *de novo* methylation of many new sites (except in the germ line, where previous imprinting may be erased during meiosis). Differential methylation may not be the sole mechanism involved in imprinting, and may only maintain the inactive state of non-expressed DNA (Monk, 1988). Heritable chromatin structures (e.g. patterns of DHS sites) may also be involved in the imprinting phenomenon, just as both may help to propagate active versus inactive chromatin states for particular genes through many cell generations.

Before leaving these general features of active chromatin, it is worth noting that active genes are found associated with the nucleoskeleton or nuclear cage (see Jackson *et al.*, 1981; Ciejek *et al.*, 1983; Jackson & Cook, 1985) at or near the periphery of the nucleus (Hutchison & Weintraub, 1985). Bulk nucleosomal chromatin is transcriptionally inactive (repressed) through association with histone H1. By contrast, the DNA in active gene regions appears to be under torsional strain (see review by Weintraub, 1985), and is associated with topoisomerase I (see e.g. Gilmour *et al.*, 1986) which may function in DNA supercoiling.

1.3 Transcriptional control

1.3.1 RNA polymerases and their products

There are three nuclear RNA polymerase enzymes in eucaryotes; these differ in their (complex) subunit compositions, their sensitivities to the fungal toxin α-amanitin, and in the sets of genes which they transcribe. Polymerase I (pol I) is resistant to α-amanitin and localised in the nucleolus. Its sole function is to transcribe the rDNA repeats into long (7–12 kb) rRNA precursors, each including one copy of the 18S, 5.8S and 28S units. These structural rRNA sequences become methylated and are cleaved out post-transcriptionally from the long precursor; the remaining (non-methylated) parts of this precursor are discarded during processing. All of this occurs within the nucleolus, where mature rRNAs are then assembled into ribosomes.

Pol III is inhibited by high but not low concentrations of α-amanitin; it is found throughout the nucleoplasm, and transcribes a limited set of small stable RNA species, including the 5S rRNAs, all tRNAs, and the U6 small nuclear RNA (snRNA; see below). Whereas 5S RNA requires only limited end-trimming prior to export, tRNAs are subjected to extensive post-transcriptional processing – including many unusual base modifications and often the removal of a small intron sequence adjacent to the anticodon loop. The process of intron removal from tRNA precursors follows a pathway distinct from that involved in pre-mRNA splicing (to be discussed in § 1.4 below). Some protozoans have intron-containing rRNA genes, and in such cases the intron sequences are removed from rRNA precursors by yet a third distinct mechanism (self-splicing, i.e. not requiring protein or other RNA cofactors; see Cech, 1983).

The remaining RNA polymerase, pol II, is also nucleoplasmic in location, but is sensitive to very low concentrations of α-amanitin. It transcribes all protein-coding mRNA precursors as well as most of the U-series snRNAs and a variety of transcripts with undefined functions. These last often include interspersed repetitive sequences, are apparently non-translatable, and are nucleus-confined in most tissues apart from oocytes (see also § § 2.3 and 2.4.2). Because the majority of protein-coding genes contain introns (with some exceptions, e.g. most histone genes), the removal of these sequences by the splicing of primary transcripts is an essential step in generating functional mRNAs for export to the cytoplasm. Nascent pol II transcripts become modified at their 5′ ends by capping, whereby a G residue (from GTP) is linked in reverse orientation through a triphosphate bridge onto the true 5′ terminus of the RNA (termed the cap site). This reversed G subsequently becomes methylated, and the resultant cap structure apparently protects the RNA from degradation. After the completion of transcription, a series of 50–200 successive adenosine residues [poly(A)] is added onto the 3′ terminus of most pre-mRNAs (again, the major histone mRNAs are exceptional on this count). Both splicing and polyadenylation will be covered in more detail in § 1.4 below. However, we may note in passing that the 3′ poly(A) tag provides a convenient means for purifying polyadenylated mRNAs and their precursors from the bulk of non-polyadenylated r- and tRNAs, since the former but not the latter will bind to an oligo(dT)-cellulose column.

1.3.2 Promoter sites and other transcriptional control elements

A number of techniques are available for identifying essential promoter elements and other sequences needed for optimal transcription, as well as for locating the sites where proteins bind to such DNA sequences. The latter are most often studied by variations on DNA footprinting or nuclease protection assays, whereby proteins bound tightly to sites within a DNA fragment will protect those sites from nuclease degradation; the sequences of both protected and non-protected regions are then determined (separately) to establish the nature and location(s) of the binding site(s). In the former category, DNA elements important for transcription can be identified by comparing the levels of gene expression obtainable from different lengths of regulatory sequence; usually this means the 5′-flanking regions of the gene, but sometimes includes intragenic and/or 3'-flanking sequences. Cloned genes plus their flanking regions may be transcribed *in vitro*, injected into *Xenopus* oocyte nuclei, or introduced by various means (microinjection, transfection or cotransformation) into heterologous cell types. Some of these approaches can be criticised on the grounds that tissue-specific transcription factors may be lacking in the recipient cell type. However, these should be present in nuclear extracts derived from the appropriate tissue, as used for *in vitro* transcription assays. If, on the other hand, cloned genes are introduced into homologous cells, background expression of the endogenous gene copies may pose problems. These are most simply circumvented by fusing regulatory regions from the gene of interest onto the coding region of a foreign 'reporter' gene. This last can encode any product which is never expressed by the host cells and for which there is a convenient assay. Typical choices for animal systems include the procaryotic genes encoding neomycin resistance (a selectable marker), chloramphenicol acetyltransferase (whose activity is easily quantitated), or β-galactosidase (whose protein product can be detected *in situ* by histochemical staining).

Various methods have been devised for introducing normal or fusion genes into every cell of a developing organism, including the germ line. Only thus is it possible to determine whether the flanking sequences used (from gene X) are sufficient to endow the introduced gene with the same pattern of temporal, spatial and quantitative regulation that characterises the expression of X during normal development. Techniques for achieving this in *C. elegans* are mentioned in § 4.2, and the pioneering approach using transposable P elements in *Drosophila* is discussed in § 5.1.3. Transgenic mice can be created by fertilising eggs with sperm *in vitro*, microinjecting gene constructs into the male pronuclei, and then transferring the early embryos into the wombs of pseudopregnant mothers; some of these embryos develop to term and express the introduced gene with varying degrees of specificity (see review by Palmiter & Brinster, 1985). At least in some cases, the introduced genes are expressed only in appropriate tissues in transgenic mice, for instance the rat elastase I gene in the exocrine acinar cells of the pancreas (Swift *et al.*, 1984), or the rat myosin light-chain 2 gene in skeletal muscle cells (Shani, 1985). Variations on this method can be used, for example, to target the expression of reporter genes by linking them to tissue-specific regulatory sequences; thus a rat elastase I promoter fused to a human growth hormone gene is specifically expressed in pancreatic acinar cells (Ornitz *et al.*, 1985). Similar constructs involving a diphtheria-toxin reporter gene destroy the developing pancreatic acinar cells, because only these cells express the toxin (Palmiter *et al.*, 1987). On occasion, a host gene may become disrupted fortuitously through the integration of introduced sequences, causing a novel mutant phenotype in that transgenic mouse; by cloning the DNA sequences flanking such inserts one can identify the wild-type gene (see Woychik *et al.*, 1985). Finally, it is possible to cure specific genetic defects by introducing wild-type copies of the appropriate gene into mouse zygotes of the mutant strain; thus complete myelin basic protein (MBP) genes can rescue the homozygous *shiverer* defect (involving partial deletions of both endogenous MBP genes: Readhead *et al.*, 1987). Such a cure extends even to the germ line (since *all* cells carry wild-type MBP genes) and is thus inherited by offspring of such transgenic mice.

The three eucaryotic RNA polymerases recognise different sequence elements in the promoters of the genes they transcribe. Other sequence features, often at some distance from the gene, may also participate in its transcriptional control. Most of these elements are recognised by other DNA-binding proteins generally termed 'transcription factors'. These interact in a modular fashion with the DNA control elements, and either stimulate or repress transcription. Some factors are ubiquitous, interacting with a wide range of genes in most cell types, while others are tissue-specific and may bind only to a few target genes in a single cell type. Factors binding to DNA sites at some distance from the promoter could affect transcription from that promoter in various ways. One possibility is DNA looping, such that a distantly bound transcription factor would also interact specifically with a target protein bound close to the transcriptional start site (Ptashne, 1986, 1988). In effect, the DNA loop would be cross-linked via interactions between two factors bound respectively at the proximal and distal sites. A protein bridge between two such sites is sufficient for activation even when the two are on separate DNA fragments (Muller-Storm *et al.*, 1989). Moreover, many transcription factors possess separable DNA-binding and protein-binding (activator) domains.

In pol I promoters, several 'nested' control elements are located upstream from the transcriptional state site, i.e. within the spacer region separating one precursor-coding region from the next. It used to be thought that these spacers were not transcribed, based largely on EM spreads of active rDNA repeat units (see e.g. fig. 2.1B). However, more recent data show that the spacer is indeed transcribed (into very unstable RNA), both as a result of readthrough originating from the major ribosomal promoter next upstream in the cluster, and also from initiation at a spacer promoter located some 2 kbp upstream from the start of the precursor-coding region. These spacer transcripts terminate at a specific sequence 200 bp upstream from the precursor initiation site, and this sequence is also an element in the major rDNA promoter (McStay & Reeder, 1986; Kuhn & Grummt, 1987; reviewed by Baker & Platt, 1986). However, spacer transcription does *not* serve to transfer engaged pol I enzymes from one ribosomal repeat unit to the next (Labhart & Reeder, 1989; Dunaway & Droge, 1989). At least two specific factors are required in additon to pol I for efficient initiation of transcription from the major ribosomal promoter in mammals; one of these binds directly to the promoter DNA, while the second binds to the preinitiation complex formed thereby (Learned *et al.*, 1986; Kato *et al.*, 1986; Tower *et al.*, 1986). Another protein binds specifically to DNA sequences downstream of the 28S gene which are required for transcriptional termination (Grummt *et al.*, 1986).

Promoter sequences recognised by pol III are typically (though not always) located internally within the gene. The prototype example is that of the *Xenopus* 5S genes, whose internal control region extends from +45 to +96 in the 120 bp gene (see § 2.4.2). Similar considerations apply in the case of tRNA genes, where the internal promoter is split into two functional domains (box A and box B) which must be a minimum distance apart (see e.g. Hofstetter *et al.*, 1981; Galli *et al.*, 1981; Sharp *et al.*, 1981). The first 11 bp of the internal control region in the *Xenopus* 5S gene are structurally and functionally homologous to the box A element of tRNA gene promoters (Ciliberto *et al.*, 1983), suggesting that common transcription factors might be shared by these two classes of pol III-transcribed gene. In fact, the major determinant of 5S expression is TFIIIA, a nuclear protein which binds specifically to 5S but not to tRNA genes (see § 2.4.2); however, two further transcription factors, TFIIIB and TFIIIC, are involved in the regulation of both tRNA and 5S rRNA genes (see Lassar *et al.*, 1983). Binding of TFIIIA/B/C to the centre of the *Xenopus* 5S gene apparently directs pol III to initiate transcription some 45 bp upstream from the 5′ border of the internal control region (Sakonju *et al.*, 1980), although the precise start site depends on sequences further upstream. In the *Drosophila* 5S gene, an upstream region from −39 to −26 is required for transcription (in addition to the internal control sequences), and this may function as a binding site for pol III (Garcia *et al.*, 1987). Even more surprisingly, certain other genes transcribed by pol III share several regulatory elements with pol II promoters (see below), including TATA boxes and distal enhancer sequences (reviewed by Sollner-Webb, 1988). One such enhancer upstream of the U6 snRNA gene (transcribed by pol III) is functionally interchangeable with the enhancer of the U2 snRNA gene (transcribed by pol II); an identical octamer sequence occurs within both of these enhancers and also in the U4 enhancer (again a pol II-transcribed gene: Bark *et al.*, 1987).

Many different DNA sequences and transcription factors participate in the regulation of transcription by pol II. Since specific examples of such control elements will be discussed in detail throughout this book, only general features will be mentioned here. The sheer diversity of protein-coding (pol II-transcribed) genes rules out any simple or universal pattern for their transcriptional regulation, although there are many features common to larger or smaller subgroups of these genes. It is here that the modular control of transcription comes into its own, involving multiple sequence elements which interact with tissue-specific and/or ubiquitous protein factors (Dynan, 1989). There is an important distinction between housekeeping and luxury genes in terms of their promoters. Whereas the latter contain a single transcriptional start site and characteristic upstream motifs such as the TATA and CCAAT boxes (see below), the former lack these features but often contain multiple start sites and an upstream GC-rich region (see e.g. Melton *et al.*, 1986; Mitchell *et al.*, 1986). An ubiquitous 'zinc finger' (see § 5.6) transcription factor termed Sp1 binds to GC-rich sequences (consensus GGGCGG; Dynan *et al.*, 1985) present both in housekeeping gene promoters (Dynan *et al.*, 1986) and in those of many tissue-specific genes. Recently, a second class of housekeeping genes has been identified, lacking both TATA box and GC-rich upstream elements. Basal expression from such promoters involves an initiator element (Inr), which is also recognisable in many TATA-containing pol II promoters. Since expression from an Inr can be boosted *in vitro* by adding a TATA box to fusion constructs, it is likely that the Inr sequence represents a minimal element for basal expression of the associated gene (Smale & Baltimore, 1989).

Turning to the better known genes encoding luxury products, 'promoter deletion' experiments identify the following sequence elements as being required for transcription in some/most cases:

(i)   The TATA box (varying from ATA to TATAAATA in different promoters) is located between -20 and -30 relative to the cap site ($+1$). This acts as a locator element to direct pol II to begin transcription at the cap site, since deletion of the TATA box results in multiple initiation sites as well as lower levels of transcription (Grosschedl & Birnstiel, 1980a). An ubiquitous factor termed

TFIID binds to the TATA box; subsequent interactions with further regulatory factors (ATF, TFIIB, TFIIE) and pol II generate a functional preinitiation complex. ATF interacts directly with TFIID and is required transiently for the assembly but not for the maintenance of this complex (Horikoshi *et al.*, 1988; Hai *et al.*, 1988a). The activator regions of proteins bound to upstream regulatory elements may also interact with TFIID (see Lillie & Green, 1989).

(ii) The (C)CAAT box, when present, is located further upstream, e.g. between −70 and −80 in mammalian β-globin genes (see § 3.2.3). This element is bound by the general transcription factor CTF (also called NF-1; Jones *et al.*, 1987), or rather by a family of related CTF factors (Santoro *et al.*, 1988; reviewed by Short, 1988). Distinct CTF species (Dorn *et al.*, 1987), or CTFs containing heterologous subunits (Chodosh *et al.*, 1988) may interact with the promoters of different genes. The overall picture here remains unclear, but evidently several transcription factors can bind to CCAAT boxes, perhaps in a gene-specific manner. Expression of a sperm-specific H2A histone gene in sea urchin is controlled via two octamer motifs (see below) and two upstream CCAAT boxes. Expression of this gene in embryos is prevented by a CCAAT displacement protein, which binds to sequences overlapping the proximal CCAAT box, so excluding the ubiquitous CCAAT-binding factor from activating this gene inappropriately (Barberis *et al.*, 1987).

(iii) Enhancers are regions of DNA able to activate transcription from a neighbouring promoter, and can do so independently of the position or orientation of the enhancer element relative to the coding sequence (reviewed by Serfling *et al.*, 1985). Thus enhancers have been identified upstream of the gene (as with the sea urchin α-type H2A gene; Grosschedl & Birnstiel, 1980b), as part of an intron sequence within the gene (e.g. in the rearranged immunoglobulin heavy- and κ-light-chain genes; see § 2.2.2), or downstream of the gene (as in several globin genes; see § 3.2.3). Enhancers interact with a number of transcription factors, some of which are ubiquitous while others seem to be tissue-specific and hence activate transcription only in cells of the appropriate type (see also § § 2.2.2 and 3.2.3). The enhancers associated with both globin and immunoglobulin (Ig) genes function optimally only in the appropriate cell-types (erythroid and lymphoid, respectively), by interacting with tissue-specific as well as general factors. Enhancers are effective over many kbp of DNA.

(iv) One element present in many enhancers (e.g. those of the Ig heavy-chain and histone H2B genes), and also in many promoters (including those of Ig heavy- and light-chain genes), is the octamer motif with a consensus sequence of ATGCAAAT. Among the octamer-binding factors, one (NF-A1 or Oct-1) is ubiquitous and involved *inter alia* in the activation of snRNA and histone H2B genes (Fletcher *et al.*, 1987), while another (NF-A2 or Oct-2) is lymphoid-specific and involved in the activation of rearranged immunoglobulin genes (Scheidereit *et al.*, 1987; see § 2.2). The ubiquitous Oct-1 factor can activate transcription from Ig promoters *in vitro*, though only the lymphoid-specific Oct-2 factor can achieve this *in vivo*, perhaps suggesting some dependence on the context of the octamer motif (LeBowitz *et al.*, 1988). Both of these protein factors share a bipartite DNA-binding POU domain which is partly encoded by a variant homeobox sequence; this domain is also present in the pituitary-specific transcription factor Pit-1 and in the *C. elegans unc-86* gene-product (see Robertson, 1988a; § § 4.4.1 and 5.6). The different target-gene specificities of the Oct-1 and Oct-2 factors (which bind to identical octamer sequences) may be mediated by parts of these proteins outside the POU domain (see Levine & Hoey, 1988; Herr *et al.*, 1988).

(v) The mammalian general transcription factor AP-1 binds certain enhancer elements and is related to the yeast transcriptional activator GCN4; both recognise very similar consensus sequences (GTGAGT^{C}AA). AP-1 activity in fact involves at least two distinct proteins (FOS and JUN), respectively related to the retroviral v-*fos* and v-*jun* oncogene products (see Curran & Franza, 1988). FOS and JUN proteins contain a conserved region with leucine residues recurring at 7 amino-acid intervals. These leucine side-chains would project from one side of an amphipathic α-helix, so that the leucines in two such helices lying side by side could interdigitate, resulting in protein dimerisation (the so-called 'leucine zipper': Landschulz *et al.*, 1988; Abel & Maniatis, 1989). This allows the possibility of forming heterodimers as well as homodimers. Thus the FOS/JUN heterodimer binds most effeciently to the AP-1 site (above), while JUN/JUN homodimers bind more weakly and FOS does not form homodimers nor bind alone. Several FOS and JUN-related proteins are known, resulting in a family of factors related to AP-1. One such is involved in activating genes whose transcription is stimulated in response to cyclic AMP (§ 3.3.1). This factor binds to a cAMP-response element (CRE) which includes a central C residue not present in the AP-1 binding site (Hai *et al.*, 1988b). A different DNA-binding/dimerisation motif is present in the MYC protein (related to the v-*myc* oncogene product), in the myogenic MyoD1 factor (see § 3.1), and in several *Drosphila* proteins (Murre *et al.*, 1989). Some polypeptides of this latter (helix-loop-helix) type also contain leucine zippers, increasing the range the possible interactions.

(vi) Steroid hormones become complexed with specific receptor proteins, and these then activate the transcription of target genes by binding to upstream enhancer-like elements (for further details, see § 3.3.1). Because these receptor proteins cannot activate transcription in the absence of hormone, they constitute a distinct class of ligand-activated transcription factors.

(vii) In addition to transcriptional enhancers, there are also transcriptional silencer sequences which have the converse effect of repressing transcription from a nearby gene (Laimins *et al.*, 1986), except when strong enhancer elements controlling the same gene are activated (see § 3.3.2 and Baniamad *et al.*, 1987).

Before leaving this topic, it is worth citing one example of modular control where several tissue-specific genes are regulated through the interactions of both tissue-restricted and ubiquitous transcription factors with a variety of promoter and enhancer elements. In the liver, DHS sites (§ 1.2) mark the positions where several proteins bind upstream from the cap site of the serum albumin gene (Babiss *et al.*, 1986). Sequences between −8.5 and −0.3 kbp upstream from this gene are required for tissue-specific expression in liver, but a distal enhancer element (located between −10.4 and −8.5 kbp) is also needed for optimal expression of fusion constructs in transgenic mice (Pinkert *et al.*, 1987). Similarly, the gene encoding the liver-specific protein transthyretin is regulated via a distal enhancer (−1.6 to −2.15 kbp) and two promoter sequences, all of which are tissue-specific (Costa, R. H. *et al.*, 1986). At least one liver-confined protein binds to the tissue-specific regulatory elements flanking the transthyretin and albumin genes, as well as to those near the α_1-antitrypsin gene (also liver-specific; Costa, R. H. *et al.*, 1988; Liu *et al.*, 1988). This factor (HNF-1 or LF-B1) is liver-specific, and its predicted sequence reveals a variant DNA-binding homeodomain (see § 6.6; Frain *et al.*, 1989). Since common enhancer and promoter sequences (e.g. the TATA box) are also present near these liver-specific genes, it follows that several ubiquitous transcription factors also participate in their transcriptional control.

1.4 RNA populations and pre-mRNA processing

The overall complexity of nuclear RNA (i.e. the range of different sequences represented therein) is at least 10-fold greater than that of the cytoplasmic RNA population derived from it. This figure applies equally to both undifferentiated and highly differentiated cell types, implying that a majority of the sequences transcribed within the nucleus are retained and/or turned over there, and are never exported to the cytoplasm. Part of this disparity arises from the presence of introns within most protein-coding genes, such that typical primary transcripts derived from these genes are 4–5-fold larger (8–10 kb) than the corresponding mRNAs (averaging 2 kb). Since intron removal by splicing occurs within the nucleus, it follows that 75–80% of each primary transcript never reaches the cytoplasm. In addition to this, a variety of pol II transcripts do not act as mRNA precursors, and may be utilised for purposes unknown within the nucleus. The snRNAs are also nucleus-confined, but like the other stable RNAs (r- and tRNAs) these make no significant contribution to the overall complexity of either nuclear or cytoplasmic RNA populations. This is simply because there are fewer than 100 different stable RNA species *in toto*, as compared with many thousands of different mRNAs in the cytoplasm, and a still greater diversity of pol II primary transcripts in the nucleus. In a typical mRNA population, kinetic hybridisation (with a complementary reverse-transcribed cDNA population) reveals several different abundance classes of mRNA sequences. A few highly abundant mRNA species (each present in many thousands of copies per cell) encode the luxury products characteristic of that cell type, e.g. globins in erythroid cells or ovalbumin in laying hen oviduct. A much wider range of moderately prevalent messengers (hundreds of copies of each per cell) variously encodes the rarer luxury products and/or the more prominent housekeeping products. Finally, there are many thousands of different rare-class mRNAs even in highly specialised cells, each present in only a few copies per cell (ranging down to figures suggesting perhaps one messenger molecule per several cells). Many suggestions have been made as to the functions of this great diversity of rare-class mRNAs, but a significant proportion of them are likely to code for rarer housekeeping functions. Where only a small amount of a stable protein is required per cell, the slow rate of ongoing synthesis needed to maintain this could be met by messengers present in very few copies per cell (see Galau *et al.*, 1977). However, not all rare-class mRNAs are necessarily translated into the corresponding proteins; in duck erythroblasts, for instance, only 200 out of 1400 rare-class messengers are present on the polysomes, while the remaining 1200 are non-polysomal and specifically excluded from translation (Imaizumi-Scherrer *et al.*, 1982).

The initial discovery of introns involved EM studies of R-loop structures formed by hybridising a cloned genomic gene to its corresponding mRNA. The resultant RNA/DNA hybrids contain multiple R-loops of single-stranded DNA (those parts of the gene represented in the mRNA, which displaces one strand of the DNA), but these are separated by double-stranded DNA loops representing the intron sequences (which are present in the gene but absent from the mRNA; see e.g. Tilghman *et al.*, 1978a). Similar hybrids between a cloned gene and the largest known precursor of its mRNA show only a single long R-loop spanning the entire gene; this indicates that the nuclear primary transcript is indeed colinear with its gene and includes the intron sequences (Tilghman *et al.*, 1978b). The discovery of introns resolved a paradox that had become apparent some years previously, namely that primary transcripts are much larger than their corresponding mRNAs, yet both the 5' cap and 3' poly (A) are conserved from the former to the latter. Rather than trimming primary transcripts from one end, the excess sequences are in fact removed from internal locations. The intervening sequences or introns are those present at internal sites within the gene and its primary transcript but absent from the corresponding messenger, whereas those sequences present in all three are termed exons. Note that messengers (composed only of exons) may include non-translated sequences at their 5' (leader) and/or 3' (trailer) ends, as well as the protein-coding regions lying between the translational initiation (AUG) and termination (stop) codons.

Splicing removes the intron sequences specifically from pol II primary transcripts acting as messenger precursors. This process must be highly accurate, since even a one-base error would result in a frame-shift mutation that would alter the entire amino-acid sequence encoded by the remainder of the resultant messenger. This is all the more impressive when one considers that some genes contain a multiplicity of introns, for instance 33 in the *Xenopus* vitellogenin genes (see fig. 3.11, later) or 50+ in certain vertebrate collagen genes. Sequence analysis of the intron/exon boundaries of innumerable protein-coding genes has revealed common consensus sequences present respectively at the 5' and 3' ends of all introns (the upstream and downstream splice sites), plus another within the intron at a site upstream from the 3' consensus sequence. The 5' and 3' consensus sequences do not permit self-complementary base pairing between the two ends of an intron in the pre-mRNA; rather they become base-paired (at least transiently) to a variety of small nuclear (sn) RNA molecules contained within specific ribonucleoprotein particles termed snRNPs or (more colloquially) snurps. The snurps involved in splicing pre-mRNAs are those containing the U1, U2, U5 and U4 + U6 snRNAs (see Maniatis & Reed, 1987). These snurps associate with the various intron consensus sequences in two stages to generate a multicomponent spliceosome complex (Grabowski *et al.*, 1985; Bindereif & Green, 1987; Konarska & Sharp, 1987) which can be visualised under EM (Reed *et al.*, 1988). The order of snurp binding during spliceosome assembly and even which snurp associates with which consensus sequence is still in dispute, although the weight of evidence favours U1 binding (transiently?) to the 5' splice site, U5 to the 3' site, and perhaps U2 to the consensus sequence upstream from the 3' site. This third sequence is now termed the branchpoint acceptor site or simply branch site. When splicing reactions were first studied *in vitro*, it became apparent that introns are released neither as linear nor circular molecules, but rather as lariats in which the 5' end of the intron is covalently linked to the 2' postion of an A residue (the branchpoint) some distance upstream from the 3' end (Ruskin *et al.*, 1984; Konarska *et al.*, 1985). During splicing, an initial cleavage is made at the 5' end of the intron; this part of the intron sequence is complementary to the branchpoint acceptor region, allowing limited base pairing which would serve to bring the 5' G residue close to the branchpoint A prior to their covalent linkage. At this point, the intron lariat still remains covalently attached to the 3' exon, whereas the 5' exon is not, though presumably it remains bound within the intact spliceosome. Finally, a second cleavage is made at the 3' end of the intron, so releasing this as a lariat, while the 5' exon becomes directly linked (spliced) to the 3' exon (see reviews by Weissmann, 1984 and Keller, 1984).

Several variations have evolved around this splicing theme. One of these is *trans* splicing, whereby segments of RNA are transferred from one molecule to another (see Sharp, 1987, and § 4.2); this feature is particularly notable in trypanosomes. A commoner variant is the use of alternative splicing. Single-copy multiexon genes can give rise to several protein isoforms by generating multiple mRNAs from different combinations of exons (e.g. in the case of the rat troponin T gene; Breitbart *et al.*, 1985). The proteins encoded by two alternatively spliced messengers will be related through exons common to both mRNAs, but will also have distinctive features arising from exons present in one

messenger but not the other. One consequence of this is that a mutation disrupting an exon of the latter type will selectively affect only one
(or a few) of the protein isoforms generated from that gene. One clear
example of this is given by a flightless strain of *Drosophila* in which a
large insertion of transposable DNA interrupts one of the optional
exons in the single tropomyosin gene (Karlik & Fyrberg, 1985). This
exon is normally utilised as part of the mRNA encoding an isoform
specific to the indirect flight muscles, which are rendered non-functional in the mutant because the appropriate type of tropomyosin cannot be synthesised. However, synthesis of a second tropomysin isoform
(in non-fibrillar muscles) can continue despite the large DNA insertion, because the exon containing this insert is absent from the mRNA
encoding the second isoform. Thus a sequence treated as an exon in
one tissue-specific splicing mode is treated as an intron in another. In
this and other examples, some exons appear to be obligatory while
others are optional (e.g. utilised in one tissue but not another). Presumably tissue-specific factors are involved in the choice of splice sites so
that one combination of exons is favoured rather than another. Alternative splicing patterns utilised in minority mRNA species may arise
because of the inefficient recognition of splice sites that are a relatively
poor match to the consensus sequences (hence they will tend to be passed over during splicing). Such optional splice sites frame a small 69
bp exon within a long intron in the mouse αA2 crystallin gene; primary transcripts from this gene can generate two mRNAs and hence
two protein products in the mouse lens, the minority species containing an insert of 23 amino acids at an internal site (representing the
small optional exon; King & Piatigorsky, 1983).

The addition of poly(A) to the 3′ end of pre-mRNA requires both a
conserved polyadenylation signal (AAUAAA) and sequences downstream of this (see Humphrey & Proudfoot, 1988), as well as a poly(A)
polymerase enzyme and possibly a 64 kd nuclear protein that binds to
the AAUAAA signal (Willusz & Schenk, 1988). However, pol II transcription often terminates some distance downstream from the
polyadenylation signal, whereas poly(A) is added onto an RNA terminus only 10–30 bases downstream from this sequence; this implies that
the 3′ end of the primary transcript is cleaved prior to polyadenylation.
The cleavage and polyadenylation processes both require multiple
protein factors, but whereas an RNA component (e.g. an snRNA?) may
be needed for 3′ cleavage in some cases (Gilmartin *et al.*, 1988), this is
not so for polyadenylation (McDevitt *et al.*, 1988). A definite role has
been established for the sea urchin U7 snurp in 3′ processing of αH3

histone mRNA (see Birnstiel *et al.*, 1985, and § 2.5.2). In some cases, two alternative polyadenylation sites may be combined with alternative splicing to generate multiple products from a single gene (e.g. the secreted and membrane-bound forms of IgM involving heavy chains differing only at their C termini; see § 2.2.2). As to the order of processing events for pol II primary transcripts within the nucleus, polyadenylation often precedes intron removal, in that poly(A)-containing primary transcripts and processing intermediates have repeatedly been observed. However, EM spreads of actively transcribing genes show frequent splicing of nascent transcripts (Beyer & Osheim, 1988), and in such cases introns are presumably removed on a first come, first served basis. Recent data indicate that several minor snurp species (e.g. those with U3, U8 and U13 snRNAs: Tyc & Steitz, 1989) are localised in the nucleolus, where they are presumably involved in the processing of ribosomal RNA precursors. Thus U3 snRNA can be cross-linked (by psoralen) to the 5′ region of the rRNA precursor, identifying a binding site which implicates U3 in the first cleavage of this precursor (Maser & Calvert, 1989). In summary, snurps participate in a multiplicity of intranuclear RNA processing events, including intron removal from pre-mRNAs, rRNA production in the nucleolus, and perhaps 3′ cleavage of pre-mRNAs.

A variety of post-transcriptional controls govern the export, stability, storage and translation of particular RNA species. Several specific cases are described later in this text, including 5S RNA storage in the form of complexes with TFIIIA in *Xenopus* oocytes, a sequence in TFIIIA mRNA which confers susceptibility to degradation in the embryo, the storage of various non-translated maternal mRNAs in the oocyte, and their selective mobilisation for translation after fertilisation (see § 2.4).

2

Molecular strategies in development

2.1 Introduction

Our picture of animal development is strongly coloured by the organisms most widely studied – namely sea urchins, *Drosophila, Xenopus*, chicken, rodents, and more recently the soil nematode *Caenorhabditis elegans*. Only in *Drosophila* and *C. elegans* (Brenner, 1974) is our developmental knowledge complemented by detailed genetics. Sea urchins are virtual non-starters in genetic terms, while among vertebrates the mouse has a clear advantage, with many inbred laboratory strains. Unfortunately, mutations in key genes acting early in mouse development are likely to produce embryonic lethal phenotypes which cannot easily be studied (but see also § 5.6). Of course, the absence of conventional genetics does not preclude molecular studies through gene cloning, RNA population analysis, etc., as for instance in sea urchins. However, such studies cannot easily focus on gene functions with defined effects during development. This option is widely available only in *C. elegans* and *Drosophila*, whose developmental patterns are in some respects unusual. For this reason, they are considered separately in chapters 4 and 5 respectively.

This chapter will deal with molecular aspects of several questions raised by classical studies of animal development. Is the genome identical in all cells of an organism, or are there irreversible changes in the nuclear DNA of particular cell types during their differentiation? To what extent are the RNA and protein populations of early embryos supplied from the fertilised egg? How is this developmental information built up during oogenesis, and for how long does it exert maternal control over embryonic development? When does the embryo's own genome become active, and when do signs of differential gene expression first distinguish one prospective cell type from another? Are regional differences built into the structure of the egg itself (as in mosaic development), or are they acquired during embryogenesis (as apparently occurs during regulative development)?

The discussion below is mainly confined to early development, where the following sequence of events is applicable in all animals:

Oogenesis $\xrightarrow{\text{fertilisation}}$ zygote $\xrightarrow{\text{cleavage}}$ blastula $\xrightarrow{\text{gastrulation}}$ gastrula $\xrightarrow{\text{organogenesis}}$ tissue differentiation

This whole field is reviewed much more fully by E. H. Davidson in his book *Gene Activity in Early Development* (3rd edition, 1986).

2.2 A constant genome?

A hallmark of metazoans is the specialisation of particular cell types to carry out different functions. Commonly this is reflected in the tissue-specific expression of abundant 'luxury' products, such as haemoglobin in vertebrate red blood cells, egg white proteins in the laying hen oviduct (both covered in chapter 3), or crystallins in the vertebrate eye lens. When the various cell types of a given animal are compared, their patterns of protein expression generally overlap in part but are otherwise distinctive (see Truman, 1974, 1982). This can arise at least partly from quantitative differences involving housekeeping functions common to many cell types (e.g. metabolic enzymes), as well as from the more familiar presence/absence of particular luxury gene-products. The process whereby cells acquire these distinctive patterns of gene expression is termed **differentiation**.

As to the molecular strategy of differentiation, three basic mechanisms can be envisaged: (i) **amplification** of genes whose products are required in high abundance for a particular differentiation pathway; (ii) **deletion** of genes whose products are *not* required (at any level) for such a pathway; or (iii) **differential gene expression** from a constant genome.

Although (iii) is the general rule in animal systems, examples of both amplification and deletion have been described in particular cases. A brief review of some salient evidence on genome constancy is given in the next three sub-sections.

2.2.1 Amplification

Until quite recently, the only clear example in this category was the amplification of ribosomal genes (rDNA) during oogenesis. Not all animals amplify these genes in their oocytes; in particular, those which undergo meroistic oogenesis (see § 2.4) are able to accumulate huge reserves of ribosomes by a different mechanism.

rDNA amplification occurs during the pachytene stage of meiotic prophase, and probably involves copying one or more chromosomal sets of ribosomal genes via a 'rolling-circle' DNA intermediate (see fig. 2.1*A*; Hourcade *et al.*, 1973). In each *Xenopus* oocyte, between 1.5 and 2.5 million copies of the ribosomal repeat unit are produced in this way (Perkowska *et al.*, 1968), compared with about 450 chromosomal copies per haploid genome. The amplified copies are in the form of free rDNA circles (nucleolar cores) each containing numerous repeat units. A single oocyte contains about 5000 such cores, and some 1500 **extrachromosomal nucleoli** are formed around them. These become engaged in active rRNA synthesis (see fig. 2.1*B*) during later oocyte growth. The amplified rDNA copies are gradually broken down after use and lost during early embryogenesis; unlike their ribosome products they are not utilised in the embryo. The egg-brooding frog *Gastrotheca* produces much larger and yolkier oocytes than *Xenopus*: notably, gastrulation in the former gives rise to an embryonic disc (cf. reptilian or avian embryos; del Pino & Elinson, 1983), rather than a *Xenopus*-type gastrula. Both the extent of rDNA amplification and the amount of rRNA produced are lower in *Gastrotheca* than in *Xenopus*, resulting in a lower rate of embryonic development (del Pino *et al.*, 1986).

Another case of gene amplification during normal development has been described in *Drosophila* egg chambers. The follicle cells secrete a proteinaceous chorion sheath around each egg prior to laying. Some 20 chorion proteins have been resolved (Waring & Mahowald, 1979), and their genes mapped to two major clusters in the genome. One such cluster contains four chorion protein (CP) genes spanning 6 kbp of DNA at site 66D on chromosome 3 (Griffin-Shea *et al.*, 1982). During the development of follicle cells, their chromosomes become **polytene**, i.e. several rounds of DNA synthesis occur without separation of the daughter duplexes, so that many DNA molecules come to lie side by side in register (see also § 5.2.1). But over and above this general increase in nuclear DNA content, the CP genes become specifically amplified during egg laying (Sprading & Mahowald, 1980). The amplification process in this case probably involves limited DNA replication in the region of the CP gene clusters (see fig. 2.2), since adjacent DNA regions spanning some 40 kbp are found to be amplified to a lesser extent (decreasing outwards from the 66D site; Spradling, 1981).

A

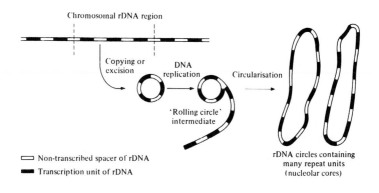

Chromosomal rDNA region

Copying or excision

DNA replication

Circularisation

'Rolling circle' intermediate

☐ Non-transcribed spacer of rDNA
▬ Transcription unit of rDNA

rDNA circles containing many repeat units (nucleolar cores)

B

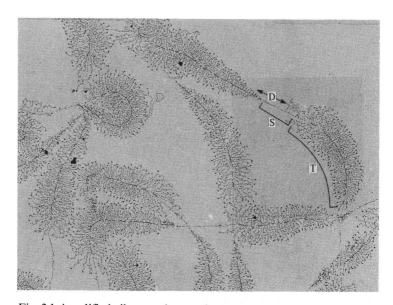

Fig. 2.1 Amplified ribosomal genes in *Xenopus*.

A Model for ribosomal gene amplification in young *Xenopus* oocytes.

B Active *Xenopus* ribosomal transcription units from oocyte nucleolar cores after spreading for electron microscopy. Each DNA axis (D) includes both transcribed regions (T) in the process of synthesising 40S ribosomal precursor RNA (seen as side branches of increasing length) and apparently silent spacer (S) regions, alternating in a regular tandem array. Magnification ×15 100. Electron micrograph kindly supplied by Prof. O. L. Miller Jr. (University of Virginia, Charlottesville) and reprinted with permission from the copyright holders. A. R. Liss Inc. from O. L. Miller Jr. & B. R. Beatty (1969) *J. Cell Physiol.* **74** suppl. 1, 225–32.

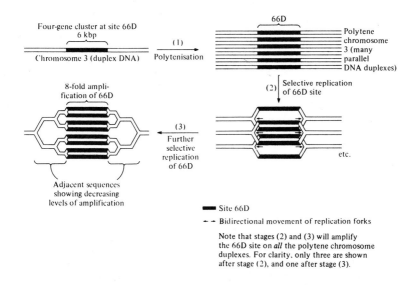

Fig. 2.2 Chorion protein gene amplification in *Drosophila* egg chamber cells (after Spradling, 1981).

Multifork replication structures representing active CP gene regions in *Drosophila* follicle cells can be visualised by EM spreading (Osheim & Miller, 1983). Interestingly, one of the genes at site 66D (but not the other three) is actively transcribed just prior to amplification, although the mRNA product is not at this stage translated into protein. Following amplification, all four genes in the cluster are actively transcribed, and their mRNAs are all translated into CPs (Thireos *et al.*, 1980; Griffin-Shea *et al.*, 1982). The other major cluster of CP genes lies on the X chromosome and contains six genes within an 18 kbp region. Again, DNA sequences on either side of this cluster become amplified to a lesser extent, but are not expressed during chorionogenesis (Parks *et al.*, 1986). This X cluster is split by the *ocelliless* mutation such that one half undergoes amplification but the other half does not. Several mutants are known which fail to produce sufficient CP material due to underamplification of one or both CP gene clusters; the mutations responsible mostly map at some distance form these clusters, suggesting the involvement of *trans*-acting factors (Orr *et al.*, 1984; Snyder *et al.*, 1986). Amplification also requires *cis*-acting DNA sequences within the CP gene clusters themselves (de Cicco & Spradling, 1984). One such essential region is located upstream from the s38 gene in the X-linked cluster, but it shares only 32 bp of homology with a region essential for amplification of the 66D cluster on chromosome 3 (Spradling *et al.*, 1987). This particular sequence is not required for transcription of the s38 gene, and may represent part of the replication origin needed for differential amplification. Amplification of specific gene regions also occurs in the 'DNA puffs' of salivary gland chromosomes in the insect *Rhynchosciara* (Glover *et al.*, 1982).

A variety of cell types are able to amplify particular DNA sequences when placed under selective pressure in culture (reviewed by Schimke, 1984). Chinese hamster ovary cells can metabolise low amounts of the cytotoxic drug methotrexate (MTX) by means of the enzyme dihydrofolate reductase (DHFR), which is encoded by a single-copy gene. Sudden exposure to high concentrations of MTX is rapidly lethal to all cells in such cultures. If the MTX concentrations are increased stepwise over an extended period in culture, however, a population of MTX-resistant cells is obtained. These cells contain levels of DHFR activity several hundred-fold higher than in normal (MTX-sensitive) cells, due to amplification of the DHFR gene by more than a hundred fold. In some cases, MTX-resistance is inherited stably over many cell generations, even in the absence of MTX itself. In other cases, MTX-resistance is rapidly lost from the progeny cells when MTX is withdrawn (unstable resistance). In the stably resistant cells, an expanded homogeneously staining region (HSR) of chromosome 2 has been identified, containing most of the amplified DHFR gene copies (Nunberg *et al.*, 1978). Moreover, a 135 kbp region of DNA becomes amplified in these cells; this is much larger than the DHFR gene itself, and may represent an entire **replicon** unit of DNA replication (Milbrandt *et al.*, 1981). The amplified sequences are probably in linear tandem arrays, which might arise from a multifork replication structure (cf. fig. 2.2) by recombination between repetitive sequences (see Roberts *et al.*, 1983). By contrast, in unstably resistant cells the amplified DHFR gene-copies are located on separate double-minute chromosomes, which are very small, lack centromeres and hence are easily lost during mitosis (Kaufman *et al.*, 1979).

One might expect such amplifications to be very rare events, detectable only because a single cell with the resistant (amplified) phenotype can survive the demise of many millions of sensitive (non-amplified) cells. However, there are data (Johnston *et al.*, 1983) to suggest that amplification may occur quite commonly in cultured cells even in the absence of selective agents; spontaneous increases and decreases in

DHFR gene number can be detected at relatively high frequencies (once per thousand divisions) in cultures never exposed to MTX.

Similarly, progressive increases in cadmium concentration can be countered by gene amplification in a cultured mouse cell line. The Cd^{2+}-resistant cells contain abnormally high amounts of metallothionein I (a protein which binds to and detoxifies heavy metal ions), due in part to amplification of the metallothionein I gene (Beach & Palmiter, 1981).

The ability to amplify DNA sequences *in vitro* does not necessarily imply that such mechanisms are used *in vivo* during normal differentiation. In many specific instances amplification clearly does *not* occur, even though the differentiated cell type may synthesise vast quantities of one or a few gene products. Thus globin genes are not amplified in erythroid cells (Bishop *et al.*, 1972), nor ovalbumin genes in chick oviduct cells, nor delta crystallin genes in chick lens (Zelenka & Piatigorsky, 1976). This was shown by mRNA or cDNA hybridisation to a vast excess of genomic DNA prepared from each of two tissues in which the gene product is respectively abundant and absent. Both give the same hybridisation curve, implying a constant number of gene copies per haploid genome.

2.2.2 Gene deletion and rearrangement

The same evidence which rules out gene amplification in specific instances, can equally be used to argue against a deletion model for differentiation. If globin gene sequences are present at the same copy number in erythroid cell DNA as in, say oviduct DNA, then the process of differentiation clearly cannot involve deleting the globin genes, which will never be expressed in oviduct cells. Of course, such examples do not provide a formal proof of genome identity in all cell types; the number of possible permutations is simply too vast. But even if deletions occur occasionally, these arguments do at least preclude the possibility that *all* unexpressed genes might be lost routinely.

There are indeed cases where DNA is deleted during development. As long ago as 1899, Boveri noted that chromosome ends are lost from all somatic nuclei during early cleavage in *Ascaris*, complete chromosomes being retained only in the primordial germ cell nucleus which will give rise to the gametes. The sequences lost from these chromosome ends are mostly satellite DNAs (Moritz & Roth, 1976), perhaps only required for chromosome pairing during meiosis. Even more dramatic instances of the same phenomenon occur in certain insects. In the gall-midge *Wachtliella*, somatic cells contain only four chromosomes, as against twenty in the germ line. Sixteen chromosomes are eliminated from all somatic nuclei during cleavage, a complete chromosome set being retained only in the primordial germ cells and their progeny (future gametes). Apparently these 16 chromosomes are largely concerned with oogenesis, which fails completely in their absence, though all somatic functions are normal; this was shown by ligaturing embryos in such a way that *all* nuclei undergo chromosome loss (Geyer-Duszynska, 1966). The somatic (S) chromosomes appear condensed and inactive in *Wachtliella* oocytes; by contrast, the E chromosomes required for oogenesis are dispersed and transcriptionally active in such cells (Kunz *et al.*, 1970).

Specific DNA rearrangements occur during the differentiation of antibody-producing cells (B lymphocytes) in the vertebrate immune system. These will be treated very briefly here, since their ramifications would justify a book in themselves! Basically each antibody is a Y-shaped molecule which contains four immunoglobulin (Ig) chains, two light and two heavy, linked together by disulphide bridges (fig. 2.3*A*). Each light or heavy chain comprises a variable (V) and a constant (C) region. Light chains are of two types designated κ and λ; in both the variable region of about 110 amino acids spans the N-terminal half of the molecule, while the constant region (of similar length) occupies the C-terminal half. Heavy chains are similar in structure, except that the constant region is much larger (some 300–400 amino acids long). Different antibody classes (1gA, 1gD, 1gM, 1gE or 1gG) are distinguished by their different heavy-chain constant regions, designated α, δ, μ, ε and γ respectively (the last of these being further subdivided into γ1, γ2a, γ2b and γ3). As the name implies, constant-region sequences are the same in all antibody molecules belonging to a given class. Variable regions, by contrast, vary widely in sequence between different antibodies within the same class. These V regions, as one might expect, are those principally involved in antigen binding. An enormous range of foreign proteins (antigens) entering the bloodstream can elicit production of highly specific antibodies by clones of B lymphocytes. How then is the requisite antibody diversity generated at the molecular level?

The number of different V-region coding sequences in the genome is large; about 300 for the mouse κ light chains, for instance, However, these V_κ genes encode only the first 95 amino acids of the κ chain, whose variable region is 108 animo acids long. The remaining 13 amino acids which link this fragment onto the C_κ sequence are in fact

encoded by a separate DNA sequence known as the J (joining) segment. About four different J-coding sequences occur in the mouse
genome, while the C_κ gene is unique. In embryonic DNA, the V_κ genes
are clustered together at a site distant from the J segment cluster, which
in turn is separated from the single C_κ gene by a long intron sequence
(fig. 2.3*B*); all three elements are located on the same chromosome.
During the differentiation of κ-chain-producing cells, this chromosome
is rearranged by somatic recombination, so that one of the V_κ genes
comes to lie immediately adjacent to one of the J sequences (Weigert *et
al.*, 1980). The sequences which originally separated these two sites are
thereby deleted from the functional κ gene created by this
rearrangement.

The site of DNA exchange between the end of the V_κ gene and the
start of the J segment is also variable, in that four possible base triplets
can result at the junction, encoding three different amino acids. Thus
the junction point itself becomes a source of heterogeneity in the variable
region, over and above the 1200 combinatorial possibilities arising
from 300 V_κ genes and 4 J segments. Overall, some 3000 different κ
chain genes can be generated by this rearrangement. Once a V_κ gene
has been positioned next to a J segment, transcription can take place
from a leader sequence 5′ to the V_κ gene through to a termination site
at the 3′ end of the C_κ gene (fig. 2.3*B*). Transcript sequences representing the long intron, together with any excess J-segment transcripts, are
removed from the precursor RNA by splicing. This generates a messenger which contains one V_κ sequence, one J sequence (that immediately following the junction point) and the one C_κ sequence (see fig.
2.3*B*). In the case of the mouse λ light chain system, there are two V_λ
genes located some distance away from a cluster of four C_λ genes, each
preceded by its own J_λ sequence plus an intron. Somatic recombination occurs so as to position one of the V_λ genes next to one of the J_λ
regions. The rearranged $V_\lambda/J_\lambda/C_\lambda$ unit is then transcribed, and the
intron separating the J_λ and C_λ sequences removed by splicing to produce a functional λ mRNA (see e.g. Bernard *et al.*, 1978). Overall, eight
different λ light chain variants are generated by this mechanism.

The situation is somewhat more complex in the case of heavy chain
expression. Here the variable region involves three distinct DNA
sequences (Early *et al.*, 1980a), i.e. one of the 100–200 different V_H genes
plus two separate junctional regions designated D and J_H (present
respectively in about 12 and 4 genomic versions). The J_H cluster is
again separated by an intron from the C_H region. Thus two somatic
recombination events are required in order to juxtapose a V_H gene next

A **Domains of an antibody molecule**

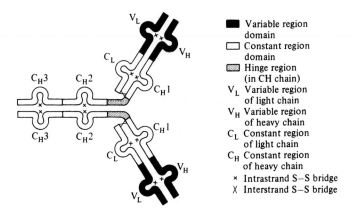

- ■ Variable region domain
- □ Constant region domain
- ▨ Hinge region (in CH chain)
- V_L Variable region of light chain
- V_H Variable region of heavy chain
- C_L Constant region of light chain
- C_H Constant region of heavy chain
- × Intrastrand S–S bridge
- ✗ Interstrand S–S bridge

B **Rearrangement pathway for κ light chain genes**

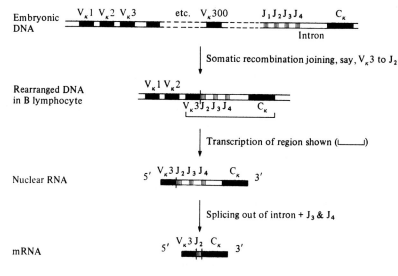

- V_κ Kappa chain variable region sequence
- C_κ Kappa chain constant region sequence
- J Joining segment sequence

C   **Rearrangement pathway for heavy chain genes (μ only)**

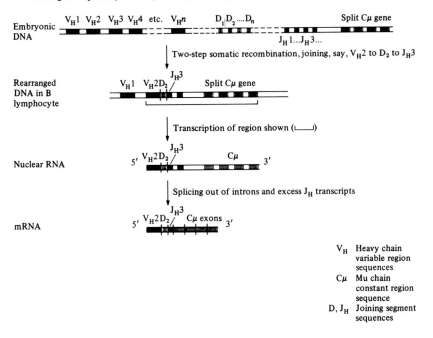

V_H Heavy chain variable region sequences

$C\mu$ Mu chain constant region sequence

D, J_H Joining segment sequences

D   **Irreversible heavy-chain class switching in humans** (split structure of C_H genes omitted for clarity)

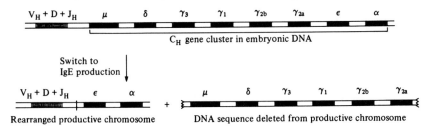

Fig. 2.3 Immunoglobulin gene rearrangements.

to a D segment and this in turn adjacent to a J_H segment. V_H, D and J_H sequences are all clustered separately in embryonic DNA, and are brought together by somatic recombination only during B lymphocyte differentiation (fig. 2.3C). These Ig gene rearrangements involve conserved nonamer (consensus ACAAAAACC) and heptamer (consensus CACAGTG) sequences. Both of these signal sequences precede all J and D genes (with the heptamer always proximal to the coding sequence), while their complements likewise follow all V and D genes. The proximal heptamers are separated from the distal nonamers by non-conserved spacer sequences which are either 12 or 23 bp long. The complementary heptamer and nonamer sequences probably become base-paired during recombination while the spacers form asymmetric unpaired loops; this would bring the coding V, (D) and J sequences close together for cutting and rejoining by recombination mechanisms that involve either deletion or inversion to remove the unwanted DNA. There is some imprecision at the junction sites between coding sequences, which allows one of several junctional codons to be formed (see above). However, this also leads to non-productive recombination events where the V and J segments are in different reading frames (hence no functional Ig chains can be produced). In the case of heavy-chain gene rearrangements, such null B cells can undergo a further recombination event which replaces the V_H segment originally chosen with a different one, so leading in some cases to a productive $V_H/D/J_H$ grouping (Reth *et al.*, 1986).

Transfection with DNA from pre-B cells can activate V(D)J recombination in a 3T3 fibroblast cell line (Schatz & Baltimore, 1988). Since the process can be repeated in a second round of transfection, this property is probably conferred by a single gene that is normally active only in lymphoid cells – presumably that encoding the V(D)J recombinase activity. Mice homozygous for the severe combined immune deficiency (*scid*) mutation are unable to rejoin coding regions into functional V(D)J units. Apparently the heptamer/nonamer recognition process can occur normally in *scid/scid* mice, but malfunction of the recombinase system leads to large DNA deletions rather than rearranged Ig genes (Malynn *et al.*, 1988; Lieber *at al.*, 1988). Note that rather similar DNA rearrangements also occur in the variable-region genes encoding the T-cell receptors (TCR) expressed by T lymphocytes. A common mechanism is implied, for example, by the fact that *scid/scid* mice lack functional T cells as well as B cells, being unable to produce either TCR or Ig variable-region coding units.

Several different C_H genes are available in both mouse and human genomes, linked in the order $5'$-μ-δ-γ_3-γ_1-γ_{2b}-γ_{2a}-ϵ-α-$3'$ on the same stretch of chromosome. Early in B lymphocyte differentiation, only μ heavy chains are expressed (IgM). Later, irreversible class switching may occur, e.g. to one of the IgG variants. When this happens, the DNA lying between the $5'$ end of the C_H cluster and the expressed gene (say γ_{2b}) becomes deleted by a recombination mechanism involving switch (S) sequences. A later class switch (say to IgA) would delete the γ_{2b}, γ_{2a} and ϵ genes, so removing all the C_H genes preceding α (Cory *et al.*, 1980; Burrows *et al.*, 1983). In this way, the same $V_H/D/J_H$ unit can be combined successively with different types of C_H chain. In some cases a single $V_H/D/J_H$ unit can be expressed simultaneously in combination with two types of C_H gene (e.g. μ and δ). This probably arises from alternative splicing of long transcripts spanning much of the C_H gene cluster (starting from the rearranged $V_H/D/J_H$ unit) without DNA deletion (Mather *et al.*, 1984; Shimizu & Honjo, 1984). Overall at least 5000 possible heavy-chain variants are encoded by the V_H, D, J_H and C_H genes (Molgaard, 1980; Marcu & Cooper, 1982). When multiplied by the available range of κ or λ light chain variants (see above), this gives well over 10^7 possible antibody molecules, encoded by a total of 10^3 or fewer gene sites. Additional diversity is generated by somatic mutation within the Ig genes (reviewed by Golub, 1987).

Each C_H gene is itself split by introns, usually into a long exon (encoding the CH1 domain), a short 'hinge' exon (H) and two further long exons (encoding the CH2 and CH3 domains; Sakano *et al.*, 1979). However, the μ gene is split into four long exons (Gough *et al.*, 1980), and its transcripts are subject to alternative processing pathways. IgM is initially expressed in a membrane-bound form, but later this switches to a secreted form. These two types of IgM differ only in the C termini of their μ heavy chains, designated μ_m and μ_s respectively. The $3'$ end of μ_m mRNA encodes a 41 amino acid hydrophobic region characteristic of transmembrane proteins, while the $3'$ end of μ_s mRNA encodes a different hydrophilic sequence of 20 amino acids. This distinction arises from alternative splicing and use of polyadenylation sites at the $3'$ end of the μ heavy-chain gene (Rogers *et al.*, 1980; Early *et al.*, 1980b). Early μ transcripts run on to a distal stop codon and polyadenylation signal which lie downstream of two small M exons encoding the transmembrane domain; during RNA processing, part of the fourth major exon (including an upstream stop codon and polyadenylation signal) is removed by splicing to give the μ_m mRNA. This includes both M exons but excludes the $3'$ end of the preceding exon 4. Later transcripts halt at an earlier point, after the upstream stop codon/polyadenylation signal but before the M exons; in this case RNA processing leaves the fourth exon intact, so generating the 3' end characteristic of μ_s mRNA (Danner & Leder, 1985). Although μ_m mRNAs remain present, there is preferential translation of μ_s mRNAs in later B cells, as well as an expansion of the Golgi apparatus relative to the endoplasmic reticulum (the respective sites of translation for μ_s and μ_m chains; Sitia *et al.*, 1987).

A lymphoid-specific enhancer is located in the intron between the J_H and C_H gene clusters in the heavy-chain locus (Gillies *et al.*, 1983; Banerji *et al.*, 1983). Another such enhancer occurs within the J-C_κ intron in the κ light chain gene-region (Queen & Baltimore, 1983; Picard & Schafner, 1984), though none has been found in the λ light-chain locus. Nevertheless, rearrangement of the κ gene locus coincides with (rather than precedes) the first onset of its transcription (Schlissel & Baltimore, 1989). Transcription of the κ gene is dependent on the presence of the κ enhancer sequence, as shown by transfecting rearranged human κ light-chain genes into mouse pre-B lymphocytes (Potter *et al.*, 1984). Similarly, optimal transcription of rearranged heavy-chain genes requires the heavy-chain enhancer; when this enhancer is linked to a heterologous promoter, *in vitro* transcription is stimulated much more effectively by B-cell extracts than by Hela-cell extracts (Augereau & Chambon, 1986; Wasylyk & Wasylyk, 1986). A single κ enhancer can activate transcription from two tandem κ promoters to equal extents, independent of the distance between these promoters (0.44 or 2.7 kbp) or their separation from the enhancer (1.7 or 7.7 kbp). This implies that the enhancer sequence can exert its influence over large DNA distances, and does not simply affect the nearest available promoter (Atchison & Perry, 1986).

However, the tissue-specificity of Ig gene transcription is not mediated solely by these enhancers, but also involves conserved sequence elements found in the V_L and V_H promoter regions (Falkner & Zachau, 1984; Bergman *et al.*, 1984; Grosschedl & Baltimore, 1985; Mizushima-Sugano & Roeder, 1986), some of which are related to sequences within the Ig enhancers. In particular, the octamer ATGCAAAT occurs upstream of all V_L and V_H genes as well as in the heavy-chain enhancer. These conserved sequence motifs interact in a modular fashion with an number of different transcription factors, only some of which need be tissue-specific (see e.g. Weinberger *et al.*, 1986b; Sen & Baltimore, 1986a). Thus the ATGCAAAT octamer can bind several different nuclear proteins, one of which (NF-A1 or Oct-1) is found in all cell types while another (NF-A2 or Oct 2) is largely confined to lymphoid cells (Singh *et al.*, 1986; Scheidereit *et al.*, 1987). Note that this octamer motif is not restricted to Ig promoter/enhancer sequences, but also participates in the regulation of several other genes, where the ubiquitous NF-A1/Oct-1 factor may play a role. Both NF-A1/Oct-1 and NF-A2/Oct-2 also bind to an unrelated heptamer sequence (consensus CTCATGA) present in Ig heavy-chain promoters; recent evidence suggests cooperative binding of NF-A2/Oct-2 to the octamer and heptamer elements (Poellinger *et al.*, 1989). A cloned cDNA (*oct-2*) specifying the lymphoid-specific NF-A2/Oct-2 factor includes an essential homeobox sequence which encodes a DNA-binding protein domain (see § 5.6; Ko *et al.*, 1988; Muller *et al.*, 1988b). This feature is also shared by NF-A1/Oct-1, and in both cases the homeodomain forms part of a larger POU domain (see § § 1.3.2, 4.4.1 and 5.6; Robertson, 1988a). The Ig heavy-chain enhancer contains seven protein-binding sites which interact with at least four different proteins (Petersen *et al.*, 1986); three of these proteins recognise other enhancers while one may be specific to the IgH enhancer. The κ enhancer is activated by the transcription factor NF–κB, which is present and active in nuclear extracts from κ-expressing B cells (Sen & Baltimore, 1986b). The NF-κB-binding site from the κ enhancer can act independently as a promoter element to confer lymphoid-specific expression on a heterologous gene (Wirth & Baltimore, 1988). However, the NF-κB protein is also present but inactive in the cytoplasm of many other cell types (Sen & Baltimore, 1986b; Bauerle & Baltimore, 1988a), where it is unable to bind to its target DNA. Post-translational activation (inducible by phorbol esters or bacterial lipopolysaccharide) releases NF-κB from this inhibition and promotes its translocation to the nucleus, so stimulating transcription from rearranged κ genes (Atchison & Perry, 1987; Bauerle & Baltimore, 1988a). In cells that do not express κ light chains, a specific inhibitor protein (IκB) interacts with NF-κB to form an inactive cytoplasmic complex, from which active NF-κB is released after phorbol ester treatment (Bauerle & Baltimore, 1988b). NF-κB is also used as an inducible transcription factor in several other cellular systems (Lenardo & Baltimore, 1989).

2.2.3 Nuclear equivalence

Cases of gene amplification or deletion seem limited to a few developmental systems *in vivo* (as discussed above). For most organisms, the nuclear DNA complement appears identical in all somatic tissues, as judged (crudely) by studies of DNA content and chromosomal karyotype. Of course, this is the outcome predicted by the conservative nature of DNA replication and mitosis, where each daughter cell normally receives an identical copy of the parental chromosome set. According to the differential gene expression model, most if not all nuclei in an organism contain the same set of DNA information, from which different **selections** of genes are expressed in the various cell types.

If most nuclei carry a full complement of genetic information, then in principle each of them should retain the same developmental capacities as the original zygote nucleus which initiates normal embryogenesis. This is the concept of **nuclear equivalence** or **totipotency**. In plants, individual cells from any tissue can be induced to regenerate a complete plant under appropriate culture conditions (Steward *et al.*, 1964). Thus cells in each plant tissue contain all the DNA information necessary to generate every other cell type; they are clearly totipotent. In higher animals, this regenerative ability is much more limited, and cell-type interconversions occur only among a limited range of tissues, e.g. in the vertebrate eye system. Moreover, regeneration in animals is usually restricted to embryonic systems, or else (in adults) involves reservoirs of undifferentiated cells. Perhaps the only exception on both counts is Wolffian regeneration, where a new lens is derived from fully differentiated dorsal iris cells in adult newts (reviewed by Yamada, 1977; Yamada & McDevitt, 1984). On this basis, animal cells do not appear totipotent; other approaches are needed in order to judge the equivalence of their nuclei.

One such approach involves nuclear transplantation into enucleated eggs (usually unfertilised). This method was pioneered in amphibian systems (e.g. Briggs & King, 1957; Gurdon, 1962; Gurdon & Uehlinger, 1966; Gurdon & Laskey, 1970), and subsequently extended to insects (Illmensee, 1972) and mammals (Illmensee & Hoppe, 1981). An unfertilised egg is first enucleated by removing its own nucleus (or a fertilised egg by removing both pronuclei), and is then microinjected with a single diploid nucleus from a somatic tissue cell. This nucleus initiates development of an embryo which, in genetic terms, will be a clonal 'copy' of the donor individual from which the somatic nucleus was taken. At least in some cases, partial or complete embryogenesis can ensue, and a fully functional adult may eventually develop. However, several interpretative difficulties arise, mainly concerning the low success rate of these experiments. In practice, complete development is obtained far more frequently from eggs injected with early embryonic donor nuclei; successful transplants involving late embryonic or larval donor nuclei are much rarer. This could mean that non-expressed DNA regions become permanently inactivated or deleted during differentiation, so that most of the later nuclei will be unable to initiate a full programme of development in the injected eggs (Briggs & King, 1957). Those few late nuclei which do give complete development might originate from a minority population of undifferentiated 'stem cells' in the donor tissue under study. None of the tissues used successfully to date is completely immune to this criticism.

However, an alternative explanation for these results has been forwarded by Gurdon and co-workers. Most differentiated tissue cells divide slowly if at all, whereas embryonic cells divide rapidly, particularly during early cleavage. Thus a differentiated cell nucleus will probably find it difficult to make the sudden transition from mitotic quiescence to rapid cleavage division following transplantation. This could well explain the various chromosomal abnormalities observed during mitosis in nuclear transplant embryos, usually leading to developmental arrest. Moreover, the success rate in these transplant experiments can be dramatically improved by starting with nuclei from blastula embryos which were themselves initiated by injecting differentiated cell nuclei into unfertilised eggs. Through several such cycles, the descendants of a differentiated cell nucleus become habituated to rapid embryonic division rates, and are then able to participate in normal development without such frequent mitotic abnormalities. Even so, the overall success rate remains low, although the sheer technical difficulty of these experiments provides a plausible explanation. Even the inactive nuclei taken from adult *Rana* erythrocytes can be reactivated by introducing them into enucleated oocytes rather than unfertilised eggs. A prolonged maturation period prior to parthenogenetic activation (by pricking: see § 2.5.1) apparently allows the introduced nucleus to prepare for rapid cleavage divisions (Di Berardino & Hoffner, 1983). Following nuclear transplantation from triploid juvenile erythrocytes into enucleated *Rana* oocytes, development can proceed in many cases as far as the feeding tadpole stage (Di Berardino *et al.*, 1986).

Those differentiated cell nuclei which do give rise to complete organisms are clearly totipotent; that is to say, they contain all the genetic information necessary for full development, and no essential DNA has been lost or permanently inactivated during their differentiation in the donor organism.

One final point should be made here. A culture of skin cells from a *Xenopus* tadpole will never give rise to a complete toad, yet a single nucleus from one of these cells can sometimes do so following nuclear transplantation into an unfertilised egg (Gurdon & Laskey, 1970). This points to a key role for egg cytoplasm in *eliciting* the full programme of development from a totipotent nucleus. Similar conclusions can be drawn from Spemann's 1928 hair-loop experiment on early newt embryos. If a fertilised newt egg is constricted symmetrically with a hair loop so that one half contains the nucleus, then development proceeds normally in the nucleated half, while the anucleate half remains uncleaved. Around the 16- or 32 cell stage, one of the nuclei (now smaller than initially) can escape through the narrow cytoplasmic bridge connecting the two halves. On entering the anucleate part, it can there initiate fully normal development of a second embryo. However, at this 16- or 32-cell stage no single blastomere will give rise to a complete embryo when allowed to develop in isolation. Again, the nuclei are equivalent and totipotent, but a substantial proportion of the original egg cytoplasm is necessary to elicit complete development. Oocyte cytoplasmic reserves are the topic of section 2.4.

2.3 RNA populations during development

The unfertilised egg is provided with vast stocks of RNA. Apart from ribosomes and tRNAs, these include a wide range of messenger RNAs and repetitive sequence transcripts. After fertilisation, many of these RNA species are used to direct or regulate key processes in early embryogenesis. In so doing, they exert *maternal* control over the initial stages of development (see § 2.5), since the RNA sequences responsible were encoded by the mother's genes and accumulated in the egg cell during oogenesis (§ 2.4). As a background to this, leading on from the theme of differential gene expression (§ 2.2), the present section deals with changes in overall RNA populations during sea urchin development (reviewed in Davidson *et al.*, 1982).

In 1976, Galau *et al.*, performed a detailed series by hybridisation experiments with mRNA populations derived from various sea urchin tissues and embryonic stages. Two types of tracer DNA were prepared from the total single-copy fraction of genomic DNA, by means of several cycles of mRNA–DNA hybridisation. One of these tracers, termed *mDNA*, comprised all those unique sequences which are represented in the cytoplasmic mRNA of gastrula embryos. The other tracer, termed *null DNA*, comprised all those unique sequences which are *not* represented in gastrula mRNA. Both mDNA and null DNA tracers were labelled and hybridised (separately) with an excess of mRNA prepared from each of the embryonic stages and adult tissues. From this the authors determined the complexity of mRNA sequences hybridising with each of the two probes. As an internal check, they also hybridised each mRNA population with total single-copy DNA, and showed that this overall complexity was equal to the sum of the complexities estimated separately with mDNA and null DNA tracers. (This confirms that mDNA and null DNA are mutually exclusive subsets of the original single-copy DNA fraction). In this way, they could assess the extent of both overlap and divergence between tissue mRNA populations, with respect to the gastrula messenger set (table 2.1).

The main conclusion which emerges from these findings is that distinct but partially overlapping gene-sets are expressed as mRNA at different developmental stages. As differentiation proceeds, there is a general reduction in the complexity of the messenger population, a feature apparent for whole embryos between the blastula and pluteus stages (table 2.1). This is true even though many new marker transcripts (characteristic of particular cell lineages: see § 2.7.2) are first expressed in the gastrula/pluteus embryo, but are absent from the egg and blastula RNA populations. Much lower RNA complexities characterise the various adult tissues tested (table 2.1), consistent with each becoming specialised for a particular set of luxury functions while retaining a basic set of housekeeping functions common to all cells. One apparent exception is adult ovary mRNA, which gives a high complexity estimate. However, this tissue will contain many immature oocytes in which high-complexity RNAs are being synthesised in preparation for the next generation. Similar results are obtained with an mDNA tracer complementary to oocyte total RNA (complexity 37×10^6 nucleotides). Some 73% of this 'maternal sequence set' is present in mRNA from 16-cell embryos, as compared to 56% in blastula mRNA and 53% in gastrula mRNA (Hough-Evans *et al.*, 1977). Some of this overlap may represent surviving maternal RNAs, but some will arise from new transcripts of the same type expressed by the embryo's own genes.

These overlapping but distinct mRNA populations contrast with the near-identity of blastula and pluteus nuclear RNA populations (Kleene & Humphries, 1977). Wold *et al.* (1978) found that most blastula mRNA sequences are also expressed in the nuclear RNA of adult intestine cells, but few of them are represented in the cytoplasmic mRNA of this tissue (as shown by using an mDNA preparation complementary to blastula mRNA). However, the complexity of nuclear RNA populations may conceal significant differences. Ernst *et al.*

Table 2.1. *Complexity of sea urchin mRNA populations using mDNA and null DNA tracers (from Galau* et al., *1976)*

Tissue source of mRNA	Complexity estimated with[a]		
	mDNA	null DNA	total unique DNA
Oocyte total RNA[b]	17×10^6	20×10^6	37×10^6
Blastula mRNA	12×10^6	15×10^6	27×10^6
Gastrula mRNA	17×10^6	0	17×10^6
Pluteus mRNA	14×10^6	0.6×10^6	14.6×10^6
Adult intestine mRNA	2.1×10^6	3.7×10^6	5.8×10^6
Adult coelomocyte mRNA	3.5×10^6	1.4×10^6	4.9×10^6
Adult tubefoot mRNA	2.7×10^6	0.4×10^6	3.1×10^6
Adult ovary mRNA	13×10^6	6.7×10^6	19.7×10^6

[a] All complexity estimates given in nucleotides of mRNA sequence.
[b] Note that total cellular RNA was used in this case.

(1979) used gastrula nuclear RNA to prepare a null fraction from total single-copy DNA, i.e. a population of unique sequences *not* represented in gastrula nuclear RNA. Adult intestine nuclear RNA reacts with some 3.6% of this null DNA tracer, but even this low proportion corresponds to a sequence complexity of $>10^7$ nucleotides. Thus sequences which remain silent in gastrula nuclei are transcribed in adult intestine nuclei. Since few of these differentially transcribed sequences are represented in adult intestine cytoplasmic mRNA, they are mostly nucleus-confined and may serve a regulatory rather than a coding function.

Most classes of repetitive DNA are represented at low or high levels in the cytoplasmic RNA of mature oocytes (Costantini *et al.*, 1978). These repetitive RNA sequences are apparently interspersed with single-copy sequences within long transcripts which resemble nuclear RNAs more closely than mRNAs (Posakony *et al.*, 1983). The repetitive elements are located in long 5′ or 3′ untranslated regions or else occur internally as unprocessed intron sequences. One representative example of a maternal interspersed repetitive transcript (ISp1) has now been cloned and sequenced (Calzone *et al.*, 1988a). The 3.7 kb ISp1 sequence contains a cluster of repetitive elements in its 5′ half together with single-copy sequences and a poly(A) tail in its 3′ half. ISp1 RNA is neither translatable itself, nor is it spliced to yield a translatable mRNA. A 620-base repetitive sequence found in the 5′ half of ISp1 is also present in many other interspersed maternal poly(A)+ RNAs, as well as in a prevalent set of polyadenylated RNAs (*c.* 600 bases long) which largely disappear by the time of gastrulation (Calzone *et al.*, 1988a). Similarly, interspersed repetitive sequences characterise many maternal poly(A)$^+$ transcripts in *Xenopus* oocytes (Anderson *et al.*, 1982).

2.4 Oogenesis

The cytoplasm of the unfertilised egg (mature oocyte) is a storehouse of developmental information (Raven, 1961). If this aspect were likened to a library or data bank, then other functions of egg cytoplasm would include a machinery depot, a food warehouse, a builder's yard and a power station! Typical oocytes contain vast stocks of the following: (i) maternal mRNAs and repetitive sequence transcripts: (ii) ribosomes, tRNAs and other protein synthetic machinery; (iii) food reserves, principally in the form of yolk (also glycogen, lipids, etc.); (iv) proteins required in quantity during cleavage, including histones, microtubule proteins, DNA and RNA polymerases; and (v) mitochondria. All of these reserves are utilised during early embryogenesis, and in the case of yolk the whole course of pre-hatching development depends upon egg supplies (except of course in mammals). The other functions are taken over by the embryo's own genes at different stages during embryogenesis (§ § 2.5 and 2.7.2 below), resulting in a progressive changeover from maternal to embryonic genome control.

Table 2.2 gives some indication of the vast storage capacity of a *Xenopus* oocyte as compared with a typical somatic cell. Although large, the mature oocyte in *Xenopus* is by no means as massive as that in birds, where the entire egg 'yolk' represents a single gigantic oocyte, nearly all of which is composed of yolk. Mammalian eggs are much smaller and less yolky, since the embryo no longer depends on internal food reserves after implantation (the nutrients required for later development being supplied via the maternal placenta). In other animal groups, the size and nutritional capacity of the egg reflect the time taken for that organism to reach hatching, after which it can utilise external food sources.

2.4.1 Meroistic versus panoistic oogenesis

In most animals oogenesis follows a **panoistic** pattern. That is, the oocyte develops as an individual cell, and therefore relies in large measure on its own synthetic capacity. As we shall see below, some materials (including yolk precursor proteins) are supplied from external sources. A coat of follicle cells surrounds the growing oocyte, form-

Table 2.2 *Comparison of oocyte reserves with those of an average somatic cell in* Xenopus

	Diameter (µm)	Ribosomes (pg)	5S + tRNAs (pg)	Nuclear DNA (pg)	rDNA (pg)	Mitochondrial DNA (pg)	Yolk (% dry weight)
Oocyte	1500	4×10^6	5×10^4	12(4C)	30 (amplified)	3×10^3	45%
Somatic cell	10–50	18	3	6(2C)	Barely detectable	0.06	Absent (except female liver)

ing close desmosome contacts with it; these do not, however, permit free passage of macromolecules into the egg. In the mollusc *Limnaea*, these contact regions become imprinted in the structure of the oocyte. Six subcortical patches of cytoplasm – distinguished by their staining properties from the surrounding yolk – are found in an asymmetrical band around the egg, marking points of close contact with the six follicle cells. Interestingly, this cytoplasmic organisation is found only in eggs after shedding, and is not apparent in oocytes still within the ovary (Raven, 1963). Presumably the pattern imprinted by the follicle-cell contacts remains implicit until shedding, when cytoplasmic reorganisation renders it explicit (see also § 2.6).

However, nucleic acid synthesis depends on the oocyte nucleus itself; thus the accumulation of large RNA stocks probably necessitates the lengthy growth period characteristic of oogenesis in panoistic systems (months or years in many vertebrates). Moreover, such specialised adaptations as ribosomal gene amplification (§ 2.2.1) and lampbrush chromosomes (§ 2.4.2) are far more prevalent in panoistic oocytes, apparently as devices for vastly increasing the rate of RNA production.

A different solution to the problem has evolved in many holometabolous insects (including *Drosophila*), whose pattern of oogenesis is termed **meroistic**. Here the oogonial stem cell undergoes a series of three or four incomplete divisions, resulting in a group of 7 or 15 **nurse cells** which remain connected to the future oocyte through open cytoplasmic channels known as **ring canals** (fig. 2.4). Each of the nurse cell nuclei becomes polyploid, which greatly magnifies the number of DNA duplexes engaged in RNA synthesis at active sites. The synthetic products of these nurse cells (both RNA and protein) are passed through the ring-canal system and accumulated in the growing oocyte (fig. 2.4; Bier, 1963), which consequently grows at their expense and may eventually engulf them. Throughout all this, the oocyte nucleus remains tetraploid (it is still in meiotic prophase) and seems relatively inactive in RNA synthesis. Thus large cytoplasmic reserves can be accumulated rapidly in meroistic ooctyes, without the need for an extended maturation period.

2.4.2 Accumulation of oocyte reserves

(a) Yolk

Yolk platelets are assembled in modified mitochondria in the oocyte cytoplasm. Each platelet is principally composed of phosvitin (33–34 kd) and lipovitellins I and II (110–120 kd and 30–34 kd). These elements are organised in a hexagonal lattice structure to form the crys-

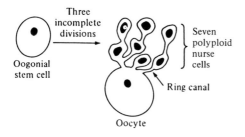

A **Ring canal system in *Hyalophora***
(diagrammatic; after King & Aggarwal, 1965)

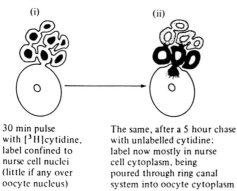

B **Incorporation of labelled cytidine in *Musca* oogenesis** (diagrammatic; after Bier, 1963)

(i)

(ii)

30 min pulse with [^{3}H]cytidine, label confined to nurse cell nuclei (little if any over oocyte nucleus)

The same, after a 5 hour chase with unlabelled cytidine: label now mostly in nurse cell cytoplasm, being poured through ring canal system into oocyte cytoplasm

Fig. 2.4 Meroistic oogenesis.

talline array characteristic of yolk platelets. The yolk proteins are not synthesised in the oocyte, but instead are processed from longer precursor proteins supplied exogenously. In vertebrates, the precursors known as vitellogenins (approx. 170 kd) are actually synthesised in the liver, transported in dimeric form to the ovary via the bloodstream, and are taken up into the growing ooctyes by pinocytosis. Typically, a single vitellogenin molecule is cleaved to yield one copy each of the phosvitin and lipovitellin I and II units for assembly into platelets (see also § 3.3.3). Thus the massive stores of yolk required in the oocytes of non-mammalian species (particularly lower vertebrates) originate from vitellogenin mRNAs and proteins synthesised in the liver under hormonal control (§ 3.3.3). In *Drosophila*, vitellogenins originate from the fat body, and in *C. elegans* from the intestine (§ 4.2).

(b) *Protein synthetic machinery*

Ribosomes are also stored in vast numbers (table 2.2) in the oocyte cytoplasm. In many panoistic systems (though not in sea urchins), the rDNA repeats comprising 18S + 5.8S + 28S genes become amplified in the young oocyte, facilitating rapid synthesis of ribosomal RNA during later growth (§ 2.2.1). In meroistic oocytes, a large fraction of the RNA transferred from nurse cells to the oocyte is known to be ribosomal. The 5S rRNA genes, by contrast, are not specifically amplified in oocytes. In some species a large reserve of genomic 5S genes is pressed into service only during oogenesis. For instance, in *Xenopus laevis* some 24 000 oocyte-type 5S genes (per haploid genome) are transcribed very actively by pol III in young oocytes. This large burst of 5S RNA synthesis precedes the onset of active ribosomal RNA transcription (from amplified rDNA copies), though some new production of 5S RNA occurs through the rest of oogenesis. The 24 000 oocyte-type 5S genes are not utilised in somatic cells, which rely instead on a much smaller set of somatic-type 5S genes (450 copies per haploid genome); these differ from the oocyte-type 5S genes in 6 out of 120 base-pairs. Whether oocyte-type as well as somatic-type 5S genes are transcribed, or only the latter, depends on the availability of a transcription factor, TFIIIA, which binds to the centre of the 5S gene between positions +45 and +96 (Sakonju *et al.*, 1980; Bogenhagen *et al.*, 1980; Engelke *et al.*, 1980). This binding (to either type of 5S gene) may be sustained through many rounds of pol III transcription by means of multiple DNA-binding 'zinc-finger' motifs in the TFIIIA protein (see § 5.6; Enver, 1985). The single-copy TFIIIA gene is expressed at high levels during early oogenesis (maximally 10^{12} molecules of TFIIIA per oocyte), but at much lower levels in all somatic cells (Shastry *et al.*, 1984; Ginsberg *et al.*, 1984; Taylor *et al.*, 1986). Selective expression of somatic-type 5S genes is correlated with limited TFIIIA availability in somatic cells, whereas the oocyte-type 5S genes are expressed efficiently only when excess TFIIIA is present. This has been demonstrated by injecting the two types of 5S gene into cleaving embryos with or without co-injection of exogenous TFIIIA (Brown & Schlissel, 1985). Also. the introduction of TFIIIA mRNA into fertilised eggs can transiently reactivate the oocyte-type 5S genes via increased TFIIIA production (Andrews & Brown, 1987). However, *in vitro* binding experiments reveal little difference in the affinity of TFIIIA for the two types of 5S gene (McConkey & Bogenhagen, 1988), suggesting that other factors may be involved. Stable transcription complexes (Brown, 1984; Darby *et al.*, 1988) are formed by the sequential binding of three distant transcription factors to the central control region of the 5S gene. TFIIIA binds first, followed by TFIIIC and TFIIIB; only then can pol III associate with this preinitiation complex to begin transcription (Lassar *et al.*, 1983; Bieker *et al.*, 1985). There is evidence *in vitro* that TFIIIC stabilises the binding of TFIIIA to somatic-type 5S genes much more effectively than to oocyte-type 5S genes (Wolffe, 1988). This difference could account for the preferential transcription of somatic-type 5S genes when TFIIIA is limiting (Wolffe & Brown, 1988). Finally, TFIIIA also binds to the 5S RNA product, forming a stable 7S storage complex; its binding to both gene and product may be mediated via a common site (Huber & Wool, 1986). This feature may limit the period of abundant 5S-gene transcription, since rising levels of 5S RNA will tend to sequester the available TFIIIA into 7S storage complexes, so reducing the amount available for binding to the ooctye-type 5S genes. Later in oogenesis, stored 5S RNAs become incorporated into ribosomes, along with new 18/5.8/28S rRNAs synthesised from amplified rDNA (see § 2.2.1). tRNA production follows a similar schedule to that for 5S RNA synthesis, with very active transcription by pol III in young oocytes; this involves both TFIIIB and TFIIIC, but not TFIIIA (Lassar *et al.*, 1983). Ribosomal proteins are also required in vast amounts. Even so, their genes are not amplified, but are expressed abundantly during *Xenopus* oogenesis (Pierandrei-Amaldi *et al.*, 1982). Two cloned ribosomal protein genes (L1 and L14) are differentially expressed when injected into *Xenopus* oocyte nuclei; whereas the L14 protein is accumulated in vast excess, the L1 product is not – apparently because of a splicing block which prevents maturation of the L1 (but not L14) transcripts and confines them to the nucleus (Bozzoni *et al.*, 1984).

(c) *Maternal proteins and messenger RNAs*

The maternal control of early cleavage development (§ 2.5) implies that proteins and messenger RNAs synthesised during oogenesis can be stored in the egg and utilised in the cleaving embryo. This has led to the 'masked messenger' hypothesis (see Spirin, 1966); 'masked' because the messenger-containing ribonucleoprotein (mRNP) particles found abundantly in many animal eggs are not used for protein synthesis until after fertilisation.

In *Xenopus*, only 2% of the putative mRNA present in early oocytes is polysomal, rising to about 20% in late oocytes (see Chapter V in Davidson, 1986). This implies that mRNAs are stored in a non-translatable form in the early oocyte, probably through association with a set of poly(A)-RNA-binding proteins whose levels decrease as oogenesis proceeds (Richter & Smith, 1983). These proteins can be extracted from early oocytes and reversibly repress the translation of test mRNAs *in vitro* (Richter & Smith, 1984). In later *Xenopus* oocytes, injected mRNAs compete with the endogenous messengers for transla-tion, suggesting the limited availability of some maternal translation component(s). In the case of mRNAs which are normally translated by membrane-bound polysomes, the rate-limiting component is an initia-tion factor associated with rough endoplasmic reticulum (Richter *et al.*, 1983). In sea urchins, there is also some evidence that the translational capacity of the oocyte is limited (Colin & Hille, 1986). In an egg-derived cell-free translation system, protein synthesis can be stimulated two-fold by the addition of purified elongation factor eIF2 (Winkler *et al.*, 1985). *In vivo*, the distribution of eIF2 – though not its total amount – changes markedly between the egg and early embryo; this may in part account for the activation of translation after fertilisation (Yablonka-Revueni & Hille, 1983). The egg extracts studied by Winkler *et al.* (1985) gave efficient translation of protein-free mRNAs purified from sea urchin eggs, implying that mRNA-binding (masking) pro-teins might normally prevent the translation of most messengers in the egg. By contrast, Moon *et al.* (1982) found that mRNP particles from unfertilised eggs can act as templates for protein synthesis *in vitro*, and indeed are translated just as efficiently as protein-free mRNAs pre-pared from them. In prefertilisation eggs, some of the factors which inhibit translation might affect ribosome function (see e.g. Danilchik & Hille, 1981) rather than the messengers as such.

Among the mRNP particles stored in unfertilised sea urchin eggs, messengers coding for the histones are prominent (Skoultchi & Gross, 1973). No detectable size changes occur in any of the five major histone mRNAs between the unfertilised egg stage when they are not translated, and cleavage stages when they are engaged in active histone synthesis (Lifton & Kedes, 1976). The histone composition of sea urchin embryos changes sequentially from one set of subtypes to another during early development. The histones incorporated into the chromatin of the earliest cleavage cells are special cleavage-stage (CS) variants of H1, H2A, H2B and H3, plus a non-CS H4 (Poccia *et al.*, 1981). These CS histones are translated in the unfertilised egg from mRNAs newly synthesised by the female pronucleus (Brandhorst, 1980), and their production continues up to the 16-cell stage. The unfertilised egg also contains much larger stocks of maternal mRNAs encoding the α-type histones; these proteins become incorporated into chromatin later during cleavage. The α-histone mRNAs are not synthesised until late in oogenesis, during maturation of the egg (Angerer *et al.*, 1984), and they are not translated until well after fertilisation (Wells *et al.*, 1981). In fact, these α-histone mRNAs remain sequestered within the pronucleus of the egg, and are only released into the cytoplasm when the nuclear membrane breaks down prior to first cleavage (De Leon *et al.*, 1983). Later shifts in the histone composition of sea urchin embryos involve embryo-coded 'late' histone mRNAs. The α-histone mRNAs are atypical in respect of their nuclear localisation and release, since most other maternal mRNAs are stored in the egg cytoplasm (Angerer & Angerer, 1981).

There are several different patterns for storing maternal gene products as mRNAs and/or as proteins. For instance, enormous stores of calmodulin protein are synthesised during oogenesis, whereas the corresponding mRNA is relatively rare in unfertilised sea urchin eggs (Floyd *et al.*, 1986). The small subunit (41 kd) of the enzyme ribonucleotide reductase is encoded by a prominent maternal mRNA, which is abundant in the sea urchin egg and translated efficiently after fertilisation; by contrast, the large subunit of this enzyme is apparently accumulated in protein form during oogenesis (Standart *et al.*, 1985). Stored maternal mRNAs encoding the sea urchin α- and β-tubulins are not translated maximally until cleavage is under way (Alexandraki & Ruderman, 1985). Conversely, maternal cyclin messengers are translated immediately after fertilisation; the cyclin proteins are synthesised prior to, and selectively degraded during, each of the first few cleavages in sea urchin embryos (Evans *et al.*, 1983).

In *Xenopus*, some maternal mRNAs are differentially distributed in particular regions of the oocyte, being concentrated for instance at the animal or vegetal pole (King & Barklis, 1985). A number of these localised mRNAs have been cloned (Rebagliati *et al.*, 1985); they include the Vg1 mRNA which encodes a protein related to mammalian transforming growth factor-β (Weeks & Melton, 1987). Notably, Vg1 mRNA is distributed throughout the cytoplasm of immature oocytes, but later becomes localised in a vegetal-pole crescent (Melton, 1987). Since micro-injected Vg1 mRNA becomes similarly localised, this process must depend on features of the Vg1 transcript itself (see Yisraeli & Melton, 1988). Maternal histone mRNAs accumulate during early oogenesis in *Xenopus*, and this stored pool is maintained by new synthesis balancing degradation [see (*d*) below]. Maternal histone mRNAs are relatively stable in the oocyte, perhaps through interactions with RNA-binding proteins and/or because 50–75% of them are polyadenylated (reviewed by Woodland *et al.*, 1983). These histone mRNAs become selectively deadenylated during hormonal maturation of the egg; this may destabilise them and so ensure that they are degraded during cleavage, to be replaced by new embryo-coded histone mRNAs (Woodland *et al.*, 1983). RNA stability in the oocyte is increased by the presence of long poly(A) tails, but certain other sequences (e.g. one present in TFIIIA mRNA) confer a marked susceptibility to degradation during early development (Harland & Misher, 1988). Histone proteins are synthesised from some of the maternal histone mRNAs during oogenesis, at an average accumulation rate of 50 pg/hour for each core histone in immature *Xenopus* oocytes. Histone H1A is accumulated at a slower rate, mainly during early oogenesis (van Dongen *et al.*, 1983). However, during oocyte maturation the rate of histone synthesis increases by 50-fold, due mainly to selective mobilisation of stored histone mRNAs (Ruderman *et al.*, 1979). There is no evidence in *Xenopus* for shifts of histone subtype, as observed during early sea urchin development (see above). Other stored maternal proteins in *Xenopus* include tubulins (Pestell, 1975), certain actins (Merriam & Clark, 1978), the transcription factor TFIIIA (which becomes sequestered by binding to 5S RNA during oogenesis; see (*b*) above), both DNA and RNA polymerases, and vimentin (notably associated with the vegetal pole germ-plasm; Tang *et al.*, 1988).

The overall sequence complexity of the maternal RNA population increases during sea urchin oogenesis. According to Hough-Evans *et al.* (1979), the RNA population of immature (previtellogenic) oocytes contains only half the range of RNA sequences present in mature (vitellogenic) oocytes. Thus new RNA species are added sequentially to the pool of maternal transcripts during oogenesis. Much of the maternal RNA population in sea urchin eggs comprises interspersed repetitive-sequence transcripts that serve no apparent protein-coding function (Davidson *et al.*, 1982; see also § 2.3. above). Similar inter-spersed repetitive transcripts are probably derived from the lampbrush chromosomes in *Xenopus* oocytes (Anderson *et al.*, 1982), yet these structures appear to be absent from the oocytes of most sea urchin species.

(d) *Lampbrush RNAs*

During the long diplotene phase of meiotic prophase (often lasting weeks, months or even years), the paired chromosomes of the tetraploid nucleus become extremely decondensed in the oocytes of many animal groups. Characteristically, the duplex DNA becomes looped out from the chromosomal axis at many sites. Both the number of loops and their sizes vary widely between animal groups. Those in amphibians are particularly well developed; in the newt *Triturus* each haploid chromosome set contains some 5000 loops (fig 2.5*B*), each ranging in length from 50 to 200 μm (Callan, 1963). Organisms using the meroistic pattern of oogenesis show little if any development of lampbrush chromosome structures in their oocyte nuclei. Moreover, the small panoistic oocytes of both mammals and sea urchins also lack lampbrush chromosomes (see Chapter V in Davidson, 1986). These facts suggest that lampbrush chromosomes are a device for vastly increasing the rate of RNA production (see below) in large panoistic oocytes which are unable to take advantage of nurse-cell assistance.

Apparently each loop represents a 'domain' of chromosomal DNA held in place at its base by the chromomere matrix (more condensed). Two DNA duplexes (chromatids) lie side by side at this stage in meio-sis; these are closely associated in the chromomeres and chromosomal filament (fig. 2.5*A*), but not in the loops themselves, which instead form symmetrically disposed pairs, each loop comprising a single DNA duplex.

These DNA loops are the sites of intensive RNA synthesis, as shown for example by autoradiography after labelling with radioactive uridine. Sometimes the entire loop appears to be transcribed, but in other cases clearly delimited transcription units can be resolved by EM spreading techniques (fig. 2.5*C*). Individual loops may contain two or more transcription units separated by non-transcribed spacer DNA, and sometimes these lie in opposite orientations (Angelier & Lacroix, 1975). The nascent RNA transcripts become complexed with a variety of nuclear proteins, some of which are specific to the transcripts from only a few defined loops (Scott & Sommerville, 1974).

The size and complexity of these lampbrush loop transcripts suggests

A

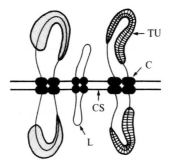

C   Condensed chromomere region at base of loop
CS   Less condensed chromosomal filament
     (two DNA duplexes side by side)
L   Decondensed loop of duplex DNA
TU   Transcription unit on loop, giving polarised
     accumulation of RNA products and
     associated proteins (as RNP)

Fig. 2.5 Lampbrush chromosomes.

A Semidiagrammatic representation of lampbrush chromosome loops in *Triturus* (after Callan, 1963).

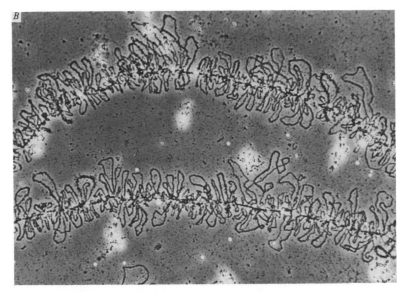

B Parts of two homologous chromosome pairs, showing numerous lampbrush loops, from an oocyte of the newt *Notophthalmus*.
Viewed under phase contrast in a centrifuged preparation. Photograph reproduced by kind permission of Prof. J. G. Gall (Carnegie Institution, Baltimore). Magnification ×434.

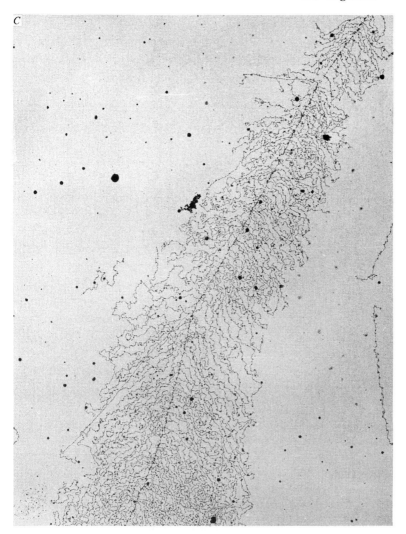

C Part of a large transcription unit in a newt lampbrush
chromosome loop. Electron micrograph of a spread preparation
showing central DNA axis and lateral branches representing
primary transcripts. Magnification ×19 000. Photograph kindly
supplied by Prof. O. L. Miller Jr. (University of Virginia,
Charlottesville) and reprinted with permission from the editor of
Acta Endocrinologica. From O. L. Miller Jr. & A. H. Bakken (1972)
Acta Endocrinol. Suppl. **168**, 155–73.

that they are typical nuclear RNAs. It must be stressed that transcription rates during the lampbrush phase are extremely rapid, perhaps 100-fold faster than in normal somatic nuclei. This has been shown by direct measurements of uridine incorporation rate, and is also implied by the dense packing of polymerase molecules and their RNA products along the length of transcribed regions (fig. 2.5C). This rapid transcription rate, combined with the large numbers of loops, the sheer length of many transcription units (up to 100 kbp), and the long duration of the lampbrush phase, together imply that enormous amounts of lampbrush RNA must be synthesised.

A large fraction of this RNA turns over within the nucleus, but about 11% of it is stable and accumulates in the cytoplasm of growing (stage 3) oocytes, dropping to 5% in late (stage 6) oocytes (Anderson & Smith, 1978). Since transcription is very rapid (see above), this implies that stable RNAs are exported to the cytoplasm at up to 1000 times the average somatic rate during the protracted lampbrush phase of oogenesis (see Chapter V in Davidson, 1986). In *Xenopus*, most of the prevalent maternal mRNAs are already represented in the cytoplasm of early stage 1–2 oocytes (Rosbash & Ford, 1974; Golden *et al.*, 1980), though lampbrush chromosomes do not become visible under the light microscope until stage 3 and later. However, numerous large transcription units are being actively transcribed even in young stage 1–2 oocytes, as shown by EM spreading studies (Hill & MacGregor, 1980). The development of visible lampbrush chromosomes in stage 3 oocytes is accompanied by an increase in the density of polymerase packing on such transcription units (about 8-fold denser at stage 3 than at stages 1–2). This implies a higher rate rather than a different pattern of transcription; thus a similar range of RNA species may be synthesised both before and during the lampbrush stage. It has also been shown that the steady-state amounts of prevalent non-mitochondrial maternal mRNAs do not increase significantly during later oogenesis (Rosbash & Ford, 1974; Golden *et al.*, 1980). This may mean that the rate of new mRNA synthesis (processed from at least some lampbrush transcripts) is balanced by degradation, i.e. dynamic rather than static maintenance of the maternal mRNA population. Although 5S RNAs are stored stably in the oocyte cytoplasm (as 7S RNPs; see (*b*) above), this is not necessarily the case for mRNAs. Overall, the maternal stocks of mRNA in the *Xenopus* oocyte are most probably derived from lampbrush transcript precursors. However, many such transcripts are not processed to messengers, though alternative functions for these remain as yet speculative.

A large fraction of lampbrush RNA consists of transcripts from repetitive DNA sequences, and even satellite DNAs are transcribed during the lampbrush phase (e.g. Varley *et al.*, 1980). In one specific instance, Diaz *et al.* (1981) have shown that long stretches of satellite DNA sequences separating the histone gene clusters are transcribed from lampbrush chromosome loops in the newt *Notophthalmus*. The histone genes in this species are organised into two large clusters, containing between them some 800 copies of a 9 kbp repeat unit. Each such unit comprises one copy of each of the five major histone genes, and is separated from the next repeat by 50 kbp or more of satellite I DNA (Stephenson *et al.*, 1981). The synthesis of satellite transcripts from these histone gene regions probably represents 'readthrough' transcription beyond the termination signals normally observed in somatic tissues. Both strands of the satellite DNA are transcribed, probably because one of the histone genes (H2B) is present in the opposite orientation to the other four (H4, H2A, H3 and H1) within each cluster (Stephenson *et al.*, 1981). Thus the two strands of satellite I DNA are represented in transcripts from *different* transcription units of opposite polarity – initiated either from the H1/H4/H2A/H3 promoters on one strand *or* from the H2B promoter on the other (Diaz & Gall, 1985). There is no evidence that promoter sites within the satellite regions are used to initiate transcription during oogenesis. The readthrough transcripts are probably processed into functional histone messengers, since no satellite I transcripts are detected outside the nucleus (Diaz & Gall, 1985).

Many of the poly(A)$^+$ interspersed repetitive transcripts found in amphibian eggs (Anderson *et al.*, 1982) may originate from inefficient processing of similar readthrough transcripts. An RNA processing block has been demonstrated in the case of transcripts from exogenous genes encoding the L1 ribosomal protein, following their introduction into *Xenopus* oocyte nuclei (Bozzoni *et al.*, 1984). This finding (and others using heterologous genes) suggests that inefficient or incomplete RNA processing may be a feature of later oogenesis. It remains unclear whether unprocessed (or partially processed) RNAs are later converted into functional mRNAs for use during early development. However, there may also be a large fraction of interspersed repetitive transcripts which cannot be spliced to messengers and which serve no protein-coding function, as is the case in sea urchin eggs (Posakony *et al.*, 1983; Calzone *et al.*, 1988a).

2.5 Fertilisation and cleavage

2.5.1 Maturation and fertilisation

During fertilisation two haploid pronuclei, from sperm and egg respectively, fuse to form the diploid zygote nucleus whose mitotic descendants will produce a complete new organism. Only a single sperm can fertilise each egg, and at least two distinct mechanisms act to prevent polyspermy (see review by Wassarman, 1987). In mammals, there are species-specific sperm receptors (the ZP3 glycoprotein in mouse) on the surface of the zona pellucida surrounding the egg. Once sperm cells of the same species have been recognised and bound by these receptors, the acrosome reaction is initiated. This involves fusion and vesiculation of the outer membrane of the sperm-head acrosome, so exposing the inner membrane and its associated enzymes. In mouse the acrosome reaction is induced by the sperm receptor (ZP3) itself; *inter alia*, an acrosomal protease (acrosin) becomes activated, and this may assist the sperm to penetrate the zona pellucida. In sea urchins, a polysaccharide constituent of the jelly coat induces the acrosome reaction, while a species-specific glycoprotein receptor on the vitelline envelope recognises an acrosomal component (bindin) of the sperm head. (Note that both of these functions are combined by mouse ZP3.) When the first sperm reaches the surface of the egg, fusion occurs between their respective plasma membranes. At least in sea urchins, this causes a transient depolarisation of the egg membrane, which temporarily prevents any further sperm–egg fusions (fast block to polyspermy). Stored cortical granules underlying the egg plasma membrane are then released into the perivitelline space which separates this membrane from the zona pellucida (cortical reaction). These granules contain proteolytic and other enzymes which modify various zona constituents and inactivate the sperm receptors, so providing a slow but lasting block to polyspermy (zona reaction). Thus ZP3 extracted from the zonae of fertilised mouse eggs is altered such that it can no longer bind sperm nor promote the acrosome reaction *in vitro* (reviewed by Wassarman, 1987). In this way, a single species-specific fertilisation event is ensured, even though many sperm may bind initially.

Apart from its genetic contribution, the sperm cell adds little in terms of cytoplasm to the huge unfertilised egg (but see § 4.3.2). It does, however, serve a key triggering function to initiate embryonic development. For instance, protein synthesis in activated massively following fertilisation in many invertebrate systems. At least in sea urchins, this involves a decrease in ribosome transit time, so that mRNA chains of the same length are translated over twice as fast in the zygote as in the unfertilised egg (Brandis & Raff, 1978). As discussed earlier (§ 2.4.2c), one or more translation factors appear to be rate-limiting in the unfertilised egg. A change in the distribution of eIF2 after fertilisation may contribute towards the increased rate of protein synthesis (Yablonka-Revueni & Hille, 1983). So too may the activation of cap trimethylation after fertilisation, whereby trimethyl caps are added to the 5' ends of some 10^7 RNA molecules (many of them histone messengers; Caldwell & Emerson, 1985). Most of the increase in protein synthesis, however, results from the rapid assembly of new polysomes using preexisting maternal mRNAs and ribosomes.

In vertebrate systems such as *Xenopus*, a major increase in the rate of protein synthesis is induced hormonally prior to fertilisation, during final **maturation** of the oocyte. By no means all maternal messengers in *Xenopus* oocytes become polysomal during maturation; some specific mRNAs are held back and are not incorporated into polysomes until after fertilisation (Dworkin *et al.*, 1985).

Hormonal stimulation also activates meiosis in *Xenopus*, whereby the 4C oocyte gives rise to a haploid egg plus two small polar bodies; note that the second meiotic division becomes arrested at metaphase and is only completed after fertilisation. Maturation-promoting factor (MPF) activity appears and disappears in phase with the meiotic cycle in *Xenopus* eggs; i.e. it rises prior to the first division, then falls on its completion, and rises again before the second division – remaining high in unfertilised eggs (Gerhart *et al.*, 1984). Both the appearance and disappearance of MPF activity are regulated post-translationally. MPF is also involved in the mitotic cycle, participating in a general cell-cycle oscillator mechanism (Newport & Kirschner, 1984). MPF-related proteins promote the transition from G2 into M phase in many eucaryotic cell-types, so regulating entry into mitosis. The oscillator becomes arrested during second meiotic metaphase in the unfertilised egg (so stabilising MPF); it can be restarted by fertilisation or Ca^{2+} ions, both leading to a rapid loss of MPF activity (Newport & Kirschner, 1984). MPF is composed of two major proteins, one of which (*c.* 34 kd) has protein kinase activity (acting *inter alia* on histone H1; Arion *et al.*, 1988). This MPF component (pp34[cdc2]) is related to the *cdc2* gene-product which regulates the onset of mitosis in the fission yeast *Schizosaccharomyces pombe* (Dunphy *et al.*, 1988; Gautier *et al.*, 1988). The other major component is cyclin (see § 2.4.2c), which interacts with pp34[cdc2] in MPF (Draetta *et al*, 1989) to control entry into the cell division cycle (see Murray, 1987, 1988, 1989; Dunphy & Newport, 1988). Cyclins are related to the *cdc13* gene product of *S. pombe*, which acts in conjunction with the *cdc2* product in the control of mitosis. The regulation of MPF activity is undoubtedly complex, involving multiple kinases and phosphatases acting separately on both pp34^{cdc2} and cyclin, together with the latter's degradation and resynthesis which drive the cell cycle (Murray & Kirschner, 1989). For instance, MPF activation prior to mitosis requires both dephosphorylation of pp34^{cdc2} and phosphorylation of cyclin (see Murray, 1989); a phosphatase (INH) which reverses this latter step may thus inhibit MPF activation (Cyert & Kirschner, 1988). Autocatalytic features include the ability of MPF to activate its own (inactive) precursors and to phosphorylate cyclin.

Fertilisation may also trigger a radical reorganisation of the egg cytoplasm, resulting in several distinct regions or *ooplasms*. Examples of this include the subequatorial band of red pigment granules established in zygotes of the sea urchin *Paracentrotus*, and five distinct cytoplasmic sectors formed in fertilised eggs of the sea squirt *Styela* (see § 2.6.1).

In *Xenopus*, the outer cortex of the egg is a gelated layer of cytoplasm under the plasma membrane, containing embedded pigment granules in the animal hemisphere. Immediately after fertilisation this cortical layer contracts asymmetrically towards the site of sperm entry, a movement initiated by the sperm centriole which reorganises the microtubules of the egg-cortex cytoskeleton (Ubbels *et al.*, 1983). This cortical contraction can be induced by Ca^{2+} ions or by the calcium ionophore A23187; these agents apparently loosen or dissolve the structural connections between the dense contractile cortex and the cytoskeleton of the inner cytoplasm, so permitting the former to move relative to the latter (Merriam & Sauterer, 1983). Note the key role played by calcium in this and other aspects of the fertilisation process, a function mediated via calmodulin and other calcium-binding proteins.

The whole process of fertilisation can be mimicked in some cases by parthenogenetic activation. Thus *Xenopus* or sea urchin unfertilised eggs can initiate cleavage after pricking with a rusty needle or exposure to abnormal pH or ionic conditions. In such cases, the cleavage nuclei are of course haploid, and development usually ceases by the time of gastrulation. A few cases of natural parthenogenesis are also known, giving complete development in the absence of fertilisation. Examples include some insects (e.g. aphids) and rotifers during particular seasons of the year, and certain all-female populations of parthenogenetic lizards found near the edge of their species range.

2.5.2 Cleavage

This first stage of embryonic development is characterised by very rapid cell division in most animal groups (with mammals as the main exception). Basically, cleavage subdivides the huge zygote cell into many smaller **blastomeres**, with a corresponding increase in the volume-ratio of nucleus to cytoplasm (approaching average somatic values in the blastula). Where yolk is concentrated towards the vegetal pole of the egg, cleavage produces large vegetal **macromeres** and smaller animal **micromeres**. During this process, any distinctive ooplasms in the zygote will become **partitioned** among particular groups of cells (see § 2.6). In most animal groups the blastula is a hollow sphere of cells, the internal cavity being termed the blastocoel. In those groups with very yolky eggs, cleavage is confined to the animal pole region, producing a bilayered blastodisc capping the uncleaved mass of yolk (e.g. in birds and reptiles).

The rapid mitotic rate of cleaving embryos engenders enormous demands for several specific proteins, among them histones (needed to associate with the newly replicated DNA to form chromatin), tubulins (for mitotic spindle formation), and DNA and RNA polymerases. These demands are met from large maternal stocks of proteins accumulated during oogenesis, supplemented in some cases by new protein synthesis from stored maternal mRNAs (see § 2.4.2c above). In the case of histones, even this device is insufficient and new embryo-coded histone mRNAs are pressed into service from the cleavage (sea urchin) or midblastula (*Xenopus*) stage onwards.

A high rate of protein synthesis is also required in order to sustain the rapid production of histones, etc., in cleaving embryos. The maternal stock of ribosomes is so vast in most animals that little new synthesis of rRNA occurs during cleavage. In *Xenopus*, anucleolate σ_{nu}/σ_{nu} homozygotes with little rDNA of their own (Steele *et al.*, 1984) can survive to the swimming tadpole stage using only these maternal ribosomes; even in normal embryos, rRNA synthesis remains at a relatively low level until this stage. Apparently a few of the rDNA repeat units are active during embryogenesis, while the majority remain repressed. Embryo-coded messengers for the ribosomal proteins normally appear from gastrulation onwards, well before the onset of active rRNA synthesis in *Xenopus* (Pierandrei-Amaldi *et al.*, 1982). In homozygous anucleolate mutants, the two ribosomal proteins investigated are synthesised in normal amounts up to the tailbud embryo stage; thereafter both the proteins and their mRNAs become unstable and are degraded, apparently because of the absence of active rRNA synthesis (Pierandrei-Amaldi *et al.*, 1985). In sea urchins, rRNA synthesis is not fully activated until the larval feeding-pluteus stage. In mammals, however, active rRNA synthesis is initiated during cleavage, probably because the maternal ribosome stocks are much smaller. Activation of 5S and tRNA synthesis usually occurs around the midblastula stage (see below).

Pre-mRNA synthesis begins during cleavage in most types of animal embryo. At least in some cases, messenger RNAs derived from these transcripts are utilised during normal cleavage development (e.g. new histone mRNAs in sea urchin and *Xenopus*, as mentioned above). In *Xenopus*, embryonic gene activity is almost undetectable during the first 12 cycles of rapid and synchronous cleavage divisions; i.e. these early cleavages depend largely if not wholly on reserves of maternal mRNAs and proteins. After completion of the eleventh cycle, when the rate of cell division becomes slower and asynchronous, the transcription of many embryonic genes is activated. This phenomenon is termed the midblastula transition (MBT; Newport & Kirschner, 1982a). The timing of the MBT is not influenced by the number of cells nor by the time elapsed since fertilisation; rather it depends on reaching a critical ratio of nuclei to cytoplasm. Plausibly, the egg cytoplasm might contain limited stores of an inhibitory factor which becomes titrated out by the increasing number of nuclei and is thus depleted by the twelfth cleavage cycle (hence transcriptional activation at the MBT). Evidence for this model has been obtained by injecting cloned yeast tRNA genes into early (pre-MBT) *Xenopus* embryos. Normally such genes are expressed transiently and then shut down until the MBT. However, co-injection of competing DNA can overcome this pre-MBT suppression of transcription, provided that a critical quantity ($\geqslant$ 24 ng) of DNA is used; this is the amount of DNA present at the twelfth cleavage in normal embryos (Newport & Kirschner, 1982b). Thus large amounts of injected DNA appear to titrate out the putative cytoplasmic inhibitors. The activation of transcription and of cell motility at the MBT is in fact controlled via the lengthening of the cell cycle which takes place after the eleventh cleavage (Kimelman *et al.*, 1987). Specifically, embryonic gene transcription becomes activated if DNA synthesis is interrupted during the first eleven cleavages, whereas the blastomeres become motile if cleavage is interrupted. Both features can be induced precociously in pre-MBT embryos by treatment with cycloheximide, while inhibitors of DNA synthesis or of cytokinesis induce only embryonic gene activation or cell motility respectively (Kimelman *et al.*, 1987). This implies that the rapid cleavage cycle in pre-MBT embryos suppresses both cell motility and embryo gene transcription; the timing of the MBT is thus controlled via the mitotic cell-cycle oscillator (involving MPF etc; see § 2.5.1). Similar patterns of transcriptional activation at the midblastula stage have been reported in both *C. elegans* (§ 4.3.1) and *Drosophila* (§ 5.1.2), though some zygotic genes are expressed before the MBT at least in the latter species. In the case of one *Xenopus* gene (GS17) whose transcription begins at the MBT, this activation is dependent on a 74 bp enhancer-like sequence some 700 bp upstream from GS17. This sequence can confer MBT activation when linked to a heterologous gene (Krieg & Melton, 1987), but it remains to be seen whether other genes are activated at the MBT via similar control elements. Figure 2.6 shows a schematic summary of the time courses for synthesis of different RNA classes from the embryonic genome.

Although many embryo-coded messengers are transcribed (and some translated) during cleavage, these do not appear essential for the cleavage process itself, which proceeds largely under *maternal* control. The drug actinomycin D has been used to block transcription during early embryonic development in a wide range of animal species. In most cases, cleavage occurs normally, but gastrulation is blocked. At face value, this would suggest that the mRNAs synthesised during cleavage are a *preparation* for later developmental stages, rather than

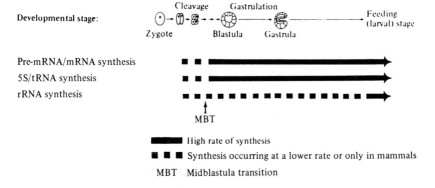

Fig. 2.6 Time courses for RNA synthesis during early development.

for immediate use. Unfortunately, actinomycin has several toxic side effects, does not effectively block synthesis of all types of RNA if given at low dose levels, and may also fail to penetrate the embryo properly. Even where appropriate controls have been performed, the results of actinomycin experiments remain ambiguous.

However, a variety of other experiments point to the same conclusion; namely, that cleavage processes are predominantly under maternal control, and can continue in the absence of any contribution from the embryonic genome. The most dramatic evidence for this is provided by the partial blastulae obtained when enucleated eggs from sea urchins (Harvey, 1936) or *Rana* (Briggs *et al.*, 1951) are activated parthenogenetically. Cell walls subdivide the egg cytoplasm and amphiaster figures are formed within the 'cells', despite the total absence of chromosomes. Many specific features of cleavage development are also under direct maternal control. One example from classical embryology concerns the number of primary mesenchyme cells produced in late sea urchin blastulae. In cases where viable hybrids can form between two sea urchin species, each characterised by different numbers of these cells, the hybrid embryos always follow the maternal pattern (table 2.3; Driesch, 1898). Another example is the direction of spiral displacement between different tiers of cleavage blastomeres in gastropod molluscs, a feature which is later reflected in left- or right-handed coiling of the shell. In some normally right-handed (dextral) species such as *Limnaea*, left-handed (sinistral) coiling is inherited as a simple one-gene recessive character; however, phenotypic expression of the coiling genotype always lags one generation behind the standard Mendelian pattern (Boycott *et al.*, 1930). Pre-

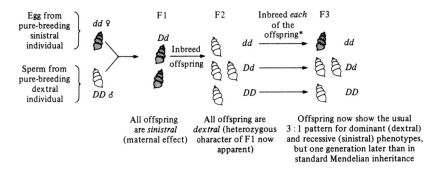

Fig. 2.7 Maternal inheritance of shell coiling pattern in the gastropod *Limnaea peregra*: D. dominant allele for dextral coiling; d. recessive allele for sinistral coiling. *N.B. These molluscs are hermaphrodite.

Table 2.3. *Numbers of primary mesenchyme cells in sea urchin interspecies hybrids (after Driesch, 1898)*

♀ species	♂ species	Average number of primary mesenchyme cells
Sphaerechinus	× *Sphaerechinus*	33 ± 4
Echinus	× *Echinus*	55 ± 4
Sphaerechinus	× *Echinus*	35 ± 5

sumably, it is the *maternal* genes, acting through factors accumulated in the egg cytoplasm, which determine the direction of shell coiling in all of the offspring (fig. 2.7). Cytoplasm taken from dextral eggs and transplanted into sinistral eggs confers a dextral cleavage pattern on the recipients, whereas the converse transplant has no effect (Freeman & Lundelius, 1982). This implies that sinistral cleavage is the ground state or default option, which is modified by the wild-type dextral gene product when available.

The duration of maternal control during early development varies considerably between different animal groups, and also according to the parameter examined (some functions being taken over by embryo genes sooner than others; see e.g. fig. 2.6). In sea urchin interspecies hybrids, paternal-type histone mRNAs (encoded by the embryo's own genes) can be detected as early as the 2-cell stage (Maxson & Egrie, 1980). During normal sea urchin development, the embryo-coded histone mRNAs synthesised during cleavage are of the same α-type as the majority of maternal histone mRNAs (see § 2.4.2c above). Indeed, soon after fertilisation, the sperm histones in the male pronucleus are replaced by maternal CS histones and new α-type histone transcripts become detectable. Embryo-coded α-histone mRNAs account for the bulk of histone synthesis during cleavage from the 16-cell stage onwards. The rate of histone synthesis declines sharply after the midblastula stage, concomitant with a lengthening of the cell cycle; this is linked to a decrease in the half-life of α-histone mRNAs. The α-histone genes are repeated about 400-fold in most sea-urchin haploid genomes, and are organised in tandem clusters of a 7 kbp repeat unit. This comprises one copy of each of the five major histone genes, separated by non-coding spacer sequences, and all in the same transcriptional orientation. Nevertheless, each histone gene is transcribed monocistronically rather than as part of a polycistronic precursor (Mauron *et al.*, 1981). An enhancer sequence is located upstream (−111 to −165) of the *Psammechinus miliaris* αH2A gene (Grosschedl *et al.*,

1983), and there are also two internal sites in this gene which interact with chromatin proteins that stimulate its transcription (Mous *et al.*, 1985). Also in *P. miliaris*, the correct 3' processing of the αH3 mRNA requires interaction with a *trans*-acting snRNP complex containing U7 snRNA (Galli *et al.*, 1983; Birnstiel *et al.*, 1985); this probably involves specific base-pairing between U7 and the 3' sequences of αH3 pre-mRNA, allowing specific cleavage of the latter at its correct 3' terminus (Strub *et al.*, 1984). Transcription of the α-histone genes ceases in the late blastula, possibly through the depletion of maternal *trans*-acting transcription factors interacting with the 5' regulatory sequences of these genes (Vitelli *et al.*, 1988; DiLiberto *et al.*, 1989). Thereafter, histone production depends on a set of late-variant histone genes, present in only a few copies each and not arranged in tandem clusters (Kedes & Maxson, 1981; Maxson *et al.*, 1983). Note that these so-called late histones are in fact expressed from the beginning of development, both in oocytes and in early cleavage embryos (Mohun *et al.*, 1985); their products, however, are quantitatively outweighed by the vast amounts of α-histones produced during cleavage. These late histone genes can be regarded as the somatic set, since only they are utilised during larval and adult stages; their expression is subject to cell-cycle regulation both in late embryonic and adult cell types. By contrast, the tandemly repeated α-histone gene set appears to be a special device for boosting the rate of histone production during cleavage. This is not found in *Xenopus* or *Drosophila*, though the latter has tandemly repeated histone genes which are used throughout development (Lifton *et al.*, 1977). The changeover from early to late histone variants in sea urchins is achieved largely through the cessation of α-histone expression, and this takes place smoothly and synchronously in all blastomeres (Angerer *et al.*, 1985), apart from those which withdraw from the mitotic cycle and hence express neither.

In *Xenopus*, paternal H1 histone genes are activated at the midblastula stage, while stored maternal H1 messengers disappear in the early gastrula, as shown by Woodland *et al.* (1979) using 'interspecies hybrids' between *X. laevis* enucleated eggs (containing maternal cytoplasmic reserves but no chromosomal DNA) and *X. borealis* sperm (providing paternal H1 genes of a different type). In both *Xenopus* and sea urchins, the new embryo transcripts first appear while maternal messengers of the same type are still available. Indeed, most of the prevalent RNA species expressed by the early embryo (zygotic transcript set) overlap with those represented in the maternal transcript set accumulated during oogenesis (see § 2.3 above; Chpater III in Davidson, 1986). This probably explains why cleavage can proceed in embryos treated with actinomycin D (to block new transcription), although gastrulation is prevented.

In mammals the onset of embryo genome control occurs during the earliest stages of cleavage, with many genes being activated at the 2-cell stage. A case in point is that of β_2 microglobulin in mouse development; a mutant form of this protein encoded by the paternal gene (hence absent from the maternal reserves) can be detected in embryos as early as the late 2-cell stage (Sawicki *et al.*, 1981). A rather more involved experiment leading to similar conclusions was performed by Monk & Harper (1978); in early mouse embryos, the amount of an X-linked enzyme, hypoxanthine phosphoribosyl transferase (HPRT), shows maternal genome influence (XX or XO) only up to the 8-cell stage, after which the embryo's own genotype (several possible permutations) assumes control over HPRT levels. Maternal HPRT mRNA is gradually replaced by embryo-coded messengers during early cleavage (Harper & Monk, 1983); thus HPRT minigene constructs are expressed in mouse 2-cell embryos, but this is negated by HPRT antisense DNA (Ao *et al.*, 1988). Although considerable stores of maternal mRNA are present in mouse eggs, they direct developmental events only up to the 2-cell stage. Thus injections of α-amanitin (blocking new mRNA synthesis) have little effect on the pattern of protein synthesis in mouse zygotes, but cause marked alterations in 2-cell and later embryos. The levels of most maternal mRNAs decline sharply after first cleavage, and they are replaced sequentially by new embryo-coded transcripts. Two-cell embryos contain only about 10% of the maternal histone and actin mRNAs that were present in the unfertilised egg, but by the 8-cell stage these mRNAs are beginnng to accumulate once again as a result of embryonic gene transcription (Giebelhaus *et al.*, 1983). Many features of early mammalian development up to the 8-cell stage depend upon mRNAs transcribed from the embryo's own genes prior to the early 4-cell stage (Kidder & McLachlin, 1985).

However, this difference between mammals and other animal groups as regards the timing of embryo genome activity, is more apparent than real. Strictly it applies only to developmental stage and not to 'real' time, because mammalian embryos cleave much more slowly than those of other animals. By the time a mammalian zygote has cleaved once or twice, a *Xenopus* or sea urchin embryo will have reached the blastula stage or beyond. Thus maternal influence may persist for a similar number of hours in both situations, though not for the same number of cleavage divisions.

2.6 Mosaic versus regulative development

2.6.1 Mosaic systems and cytoplasmic determinants

One problem which we have skirted so far is how groups of embryonic cells are directed to follow particular developmental pathways. We have already seen that maternal influences extend well into cleavage development (and in some cases beyond). If this maternal information were unequally distributed between different parts of the egg cytoplasm, it would become partitioned during cleavage into discrete groups of blastomeres, and could then direct each group towards a particular fate. In essence, this is the pattern of *mosaic* development found in several invertebrate groups, including many molluscs and ascideans (sea squirts). In such systems the zygote cytoplasm is highly organised, and may be subdivided into regions distinguished by different pigmentation or yolk density. Such ooplasms may have distinct boundaries, or may intergrade into each other, e.g. through gradients.

(a) Ascideans

A classic case of this type was described in the sea squirt *Styela* by Conklin (1905). Here, sperm entry triggers a process of cytoplasmic streaming in the egg, from which five distinct sectors emerge: clear cytoplasm, pale grey and dark grey yolky regions, plus pale and dark yellow crescents on one side of the egg. The emergence of these cytoplasmic sectors from an apparently homogeneous unfertilised egg can be interpreted in terms of three independent and sequentially acting gradient influences within the egg (Catalano *et al.*, 1979). By the 64-cell stage (late cleavage) each cytoplasmic sector has become segregated into a different group of blastomeres. More importantly, each of these blastomere types has a distinct development fate. Thus cells containing the clear cytoplasm develop into ectoderm, those with pale grey material into notochord, those with dark grey yolky cytoplasm into endoderm, those with pale yellow crescent material into coelomic mesoderm, and those with dark yellow crescent material into the larval tail muscles. Moreover, these cell fates are for the most part irrevocably fixed. If the cells containing one type of egg cytoplasm are destroyed (blastomere deletion), then the embryo develops without the structures normally formed by those cells, and cannot 'regulate' to make good the loss. If instead such blastomeres are removed and allowed to develop in isolation, they form only the appropriate tissue type and no others. Thus embryos of this type develop as *mosaics* of cells whose future fates are largely mapped out by their inheritance of cytoplasmic substances (termed *morphogenetic determinants*) from different regions of the fertilised egg.

Partial embryos can develop from fragments of fertilised ascidean eggs, both from those fragments which contain the zygote nucleus and also from those anucleate fragments which have received a transplanted embryonic nucleus (Tung *et al.*, 1977). Notably, the range of tissues which develops in the latter case is determined by the cytoplasmic constituents of the anucleate egg fragment and not by the tissue source of the transplanted nucleus.

If early ascidean embryos are treated with cytochalasin B, they continue DNA synthesis but further cleavage is blocked. Marker enzymes that would normally appear later in development in certain groups of cells (e.g. future muscle or gut), still appear on schedule in such cleavage-arrested embryos, but only in the precursors of the appropriate cell groups. Thus even a cleavage-arrested zygote can go on to form a multinucleate single cell which may express differentiated ultrastructural characteristics of several cell lineages, suggesting that multiple gene-activating factors are localised in different cytoplasmic regions of the unfertilised egg (Crowther & Whittaker, 1986). Cleavage-arrested embryos can be used to trace the ancestry of particular cell types; thus descendants of the B4.1 blastomere pair (which inherit the yellow crescent 'myoplasm') express the muscle-specific marker enzyme acetylcholinesterase (AChE; Meedel & Whittaker, 1984). If *Styela* embryos are compressed at the time of third cleavage, then the plane of this cleavage becomes meridional rather than equatorial, giving an abnormal 8-cell embryo in which the myoplasm is present in four cells rather than two. If such embryos are then cleavage-arrested, the AChE muscle marker appears on schedule in three or four of the eight cells (rather than the two muscle precursors identified in normal 8-cell embryos following cleavage arrest). These extra cells in which AChE appears after compression would normally contribute to the ectoderm, suggesting that the diversion of myoplasm into these precursors has altered their cell fate (Whittaker, 1980). Note that muscle markers such as AChE and myosin ATPase are mainly encoded by embryonic transcripts which are synthesised from about gastrulation onwards, and not by pre-synthesised maternal mRNAs (Meedel & Whittaker, 1983; Meedel, 1983). This suggests that the timing of their transcription is dependent on some internal clock (cf. the MBT in *Xenopus*), since they still appear on schedule in cleavage-arrested embryos. This clock mechanism requires a definite number of DNA replication cycles, since the DNA-synthesis inhibitor aphidicolin blocks AChE appearance in seventh but not eighth or ninth generation cells (see Satoh, 1979; Satoh & Ikegami, 1981).

A different approach to analysing ascidean development utilises a panel of monoclonal antibodies (MAb's), each recognising a tissue-specific protein expressed in one particular cell lineage (see e.g. Nishikata *et al.*, 1987a; Mita-Miyazawa *et al.*, 1987). Muscle differentiation can be followed using one such MAb (Mu2, which recognises a 220 kd muscle-specific protein), confirming *inter alia* that this protein is embryo-coded and can only be expressed after the completion of a certain number of DNA replication cycles (Nishikata *et al.*, 1987b). A different MAb (HG6B2), recognising a component of the zygote myoplasm, can prevent muscle differentiation and the appearance of AChE when injected into fertilised eggs (Nishikata *et al.*, 1987c).

The later fates of particular blastomeres can be followed by intracellular injection of a tracer enzyme (see e.g. Nishida, 1987). Thus at the 64-cell stage, 44 of these cells go on to generate a single cell-type, while at the 110-cell stage this number has risen to 94. Using this approach, an almost complete embryonic cell lineage has been derived, confirming the invariant cell fates adopted by the vast majority of blastomeres in ascideans. Only two candidate 'equivalence groups' have been identified. where cell:cell interactions rather than cell ancestry may specify cell fate (Nishida, 1987). One such example involves the two melanocyte precursors, one of which will become ocellus (primary fate) while the other becomes otolith (secondary fate) (Nishida & Satoh, 1989). Complete and similarly invariant cell lineages have been derived for the nematode *Caenorhabditis elegans*, using a laser microbeam to ablate particular blastomeres (see § 4.3; Sulston *et al.*, 1983). Here too there are instances of equivalence groups where the adoption of particular cell fates depends on intercellular interactions (reviewed in chapter 4 below).

The molecular nature and mode of action of the localised maternal determinants is obviously of key importance in ascidean embryos. Most studies to date have focussed on the yellow crescent myoplasm of *Styela*, from which the muscle lineages (described above) are derived. There is a notable concentration of maternal actin mRNA within the myoplasm of the fertilised egg; this region contains 45% of the actin mRNA but only 5% of the total poly(A)$^+$ RNA (Jeffery *et al.*, 1983), as revealed by *in situ* hybridisation. However, this localised maternal actin mRNA mainly encodes a cytoplasmic isoform and not the muscle-specific isoform that characterises later muscle cells. In fact, the muscle-specific actin is specified by a 2 kb mRNA which is not expressed at significant levels until the gastrula stage. This expression is spatially restricted to muscle lineage cells, i,.e. primarily in descendants of the B4.1 blastomere pair (Tomlinson, C.R. *et al.*, 1987a), but later also in the secondary muscle cells derived from b4.2 and A4.1 (see also Nishikata *et al.*, 1987b). However, some of the muscle-specific actin is apparently translated from maternal actin transcripts detectable in the egg, as well as a major contribution from embryo-coded mRNAs (Tomlinson, C.R. *et al.*, 1987b). It remains to be seen whether this maternal muscle-specific mRNA is confined partly or wholly to the myoplasm. The myoplasmic region of the zygote has a characteristic cytoskeletal structure (part of the egg cortex) which permits its isolation by cell fractionation methods. Some of the proteins associated with this myoplasmic cytoskeletal domain (MCD) are unique or at any rate rare elsewhere in the zygote, whereas no unique maternal mRNAs are associated with the MCD (see Jeffery, 1985a, b).

(b) Molluscs

An interesting example of mosaic development is provided by certain gastropod (*Ilyanassa*) and scaphopod (*Dentalium*) molluscs. Here the egg contains regions of clear cytoplasm both at the animal pole (containing the nucleus) and at the opposite vegetal pole (polar plasm), separated by a wide band of coloured yolky material. Prior to first cleavage, the vegetal polar plasm becomes extruded as a **polar lobe** connected to the rest of the zygote by a narrow cytoplasmic bridge. During first cleavage the polar lobe is passed to only one of the two daughter blastomeres, into which it is resorbed on completion of the division. The lobe-carrying cell is designated CD, its lobeless companion AB. At second cleavage the same thing happens, so that the polar lobe is transferred to only one cell (D) out of the four (C, like A and B, is lobeless). This is shown diagrammatically in fig. 2.8 (see Wilson, 1904a,b).

If this polar lobe is cut off prior to either first or second cleavage, then the embryo which develops is deficient in mesodermal tissues such as muscle etc. Thus the polar lobe presumably contains mesodermal determinants, directing those cells which inherit them to develop towards a mesodermal fate. Although the pattern of polar lobe formation breaks down during later cleavage, the mesodermal determinants continue to be shunted into particular daughter cells through several further divisions, as shown by blastomere deletion experiments in *Ilyanassa* (Clement, 1962). After sixth cleavage, the 4d cell (primary

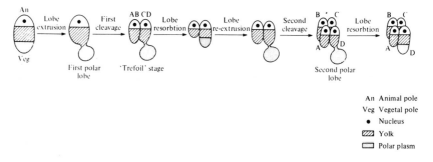

An Animal pole
Veg Vegetal pole
• Nucleus
▨ Yolk
▭ Polar plasm

Fig. 2.8 Early cleavage in *Dentalium* or *Ilyanassa* (diagrammatic).

mesentoblast) contains most of the mesodermal determinants, and from it are descended the primary mesodermal cells of the later embryo (but see Clement, 1976).

The role of the polar lobe is in fact rather more complex than this, since the dorso-ventral (D/V) polarity of the embryo as well as mesoderm formation depends on its presence. If polar lobe formation in *Dentalium* zygotes is inhibited with cytochalasin B, then the first cleavage is symmetrical, i.e. both daughters receive polar lobe material (rather than just the CD cell). Second cleavage then generates CDCD or CCDD 4-cell embryos which later develop duplications of many lobe-dependent structures – such as apical tuft, mesodermal bands and post-trochal structures (Guerrier *et al.*, 1978). These findings suggest that the D/V axis is not preformed in the unfertilised egg, but rather arises epigenetically through the fusion of the polar lobe with one of the first two blastomeres (CD). This explains the duplication of lobe-dependent structures when lobe material is partitioned into both of these blastomeres. In gastropods which do not form a polar lobe (e.g. *Patella*), the 4-cell stage is equal; D/V polarity first becomes apparent between the fifth and sixth cleavages, when one of the vegetal macromeres (designated 3D) makes contact with animal micromeres and then divides to give the 4d primary mesentoblast as one of its daughters. If the 3D cell is removed, one of the other macromeres can substitute for it and generate a primary mesentoblast, but this cell is not formed if contact between macromeres and micromeres is prevented (van den Biggelaar & Guerrier, 1979). The same is true in *Limnaea*, another equally cleaving spiralian gastropod (Martindale *et al.*, 1985); in this case, cytochalasin treatment results in radially symmetrical development by preventing contact between one of the macromeres (that which normally specifies the D quadrant) and the overlying micromeres. Thus the D/V axis in molluscs can be determined either early in cleavage by means of the polar lobe (when present), or else later through interactions between blastomeres.

The small polar lobe of the gastropod *Bithynia* contains an RNA-rich vegetal body which is passed to the CD cell at first cleavage and then divided between the C and D cells at second cleavage. Indeed, the C cell in *Bithynia* receives much of the material that goes to the D cell in *Ilyanassa*; thus lobe-dependent structures develop better in ABC than in ABD ¾ embryos (Cather *et al.*, 1976). In the polychaete worm *Sabellaria*, the first polar lobe (passed to CD) is required both for apical tuft formation and for post-trochal development, whereas the second polar lobe (smaller and passed only to D) is needed for the latter but not for the former. If this second polar lobe is removed, then both C and D derivatives give rise to apical tuft (normally formed only from descendants of C). If an equal second cleavage is induced by SDS treatment (second polar lobe material passed to C as well as D), then neither C nor D derivatives form apical tuft. This suggests that the second polar lobe contains an inhibitor of apical tuft formation which is shunted into the D cell at second cleavage. By contrast, determinants stimulating apical tuft formation are segregated into the CD cell at first cleavage, and hence are inherited by both C and D cells. Since the inhibitor is normally passed only to the D cell from the second polar lobe, this means that only C derivatives can form apical tuft during normal development (Render, 1983). However, the nature of both the determinant(s) and inhibitor(s) of apical tuft formation remain unknown.

One obvious line of investigation is to compare the protein synthetic profiles in normal and delobed embryos, since any lobe-confined messengers should be absent from the latter, leading to some depletion in the range of proteins synthesised. At least in *Ilyanassa* embryos, most of the newly synthesised proteins detected on 2D gels are ubiquitous; they are not confined to particular organ systems such as the mesodermal derivatives produced in normal but not lobeless embryos (Collier, 1983). However, any differences involving proteins synthesised from rare lobe-specific mRNAs would not have been detected by this approach – which can only pick up fairly prevalent proteins. Several changes in protein synthetic pattern occur even in isolated polar lobes (which have no nuclei); these presumably involve the selective translation of stored maternal mRNAs (Brandhorst & Newrock, 1981). Removal of the polar lobe in *Ilyanassa* delays a switch of histone H1 subtypes, retarding both the disappearance of early H1 variants and the appearance of later subtypes (Flenniken & Newrock, 1987). Factors controlling these H1 switches appear to reside in the D cell lineage (which receives the polar lobe material) in normal embryos.

Note that the actual cytoplasmic contents of the polar lobe during the first or second cleavage divisions are not necessarily crucial to its function. If *Dentalium* zygotes are centrifuged gently, the yolky material can be displaced to the vegetal pole. Nevertheless, a polar lobe still forms at its normal vegetal site, even though it is now filled with yolk instead of clear polar plasm. Such an embryo is able to develop normally, and moreover removal of the polar lobe prior to first or second cleavage still results in a mesoderm-deficient embryo (Verdonk,1968). So morphogenetic determinants must still be localised in the polar lobe, despite its altered cytoplasmic contents after centrifugation. One likely explanation is that these determinants occur embedded in the **cortex** of the polar lobe region, rather than free in the polar plasm. However, their nature remains to be determined (cf. in *Styela* above).

(c) Other examples

Cytoplasmic localisation plays a key role in germ-cell determination in many animal groups. Clear examples of this are afforded by those organisms where the somatic nuclei undergo chromosome diminution during cleavage (e.g. *Ascaris, Wachtliella*; see § 2.2.2). Only one or a few nuclei escape this process, namely those destined to found the germ-cell line. The primordial germ-cell nuclei are always located at one pole of the embryo (posterior pole in insects), in a region of cytoplasm distinguished by granular inclusions termed polar granules. If nuclei are not permitted to enter this region (e.g. by ligaturing the embryo; Geyer-Duszynska, 1966) or if the same region is cauterised by narrow-beam UV irradiation before any nuclei have migrated into it, then germ cells fail to develop. In cases of chromosome diminution, this means that *no* nuclei retain their full chromosomal complement. UV irradiation of the polar region has similar effects in many organisms which do not undergo chromosome diminution (e.g. *Drosophila*), i.e. no germ cells are formed, and the adults developing from such embryos are sterile (no gametes). The effects of UV-cauterisation can be reversed by subsequently microinjecting cytoplasm from the corresponding polar region of an unirradiated egg; germ cells will then develop normally from any nuclei migrating into the injected region. This cytoplasm must therefore include germ-cell determinants, possibly associated with the characteristic polar granules (see § 5.1.2).

The molecular identities of morphogenetic determinants remain in most cases obscure, as do their mechanisms of action. However, in

Drosophila several maternal determinants controlling aspects of the overall body pattern have been characterised in both genetic and molecular terms (see § 5.3.1 and § 5.4.2). In particular, the anterior determinant encoded by the *bicoid* (*bcd*) gene forms a protein gradient declining steeply away from the anterior pole (Driever & Nusslein-Volhard, 1988a) during early embryogenesis. This in turn depends on the sharp localisation of *bcd⁺* mRNA at the anterior pole of the egg, where it is bound via the products of two other maternally acting genes, *exuperantia* and *swallow* (Belerth *et al.*, 1988; see § 5.4.2). The *bcd⁺* protein contains a DNA-binding homeodomain and has been directly implicated in the transcriptional regulation of zygotic genes which become active in anterior parts of the embryo. This general strategy of action is clearly plausible, but remains as yet unproven, for many of the other examples of maternal determinants discussed above.

2.6.2 Regulative systems

In embryology, the term **regulation** means that an embryonic system is able to make good deficiences caused by removing some part of that system. Whereas early cleavage blastomeres from mosaic embryos show restricted developmental fates, those from **regulative** embryos are able to develop a complete set of tissues. In the starfish *Asterina*, all eight blastomeres from an 8-cell embryo can regulate to form complete ⅛-sized larvae (Dan-Sohkawa & Satoh, 1978). In sea urchins, regulative ability persists up to the 4-cell stage but is later lost.

Nevertheless, there is good evidence in sea urchin systems for cytoplasmic localisation, which may sometimes be visibly apparent (e.g. the subequatorial pigment band in *Paracentrotus*; § 2.5.1). From about the 8-cell stage onwards, sea urchin blastomeres show restricted developmental fates (fig. 2.9). Thus when tiers of blastomeres from 16-cell embryos are separated and grown in isolation, the three cell types develop differently (fig. 2.9). The mesomeres (a ring of eight cells at the animal pole) form **animalised** embryos, consisting mainly of ectodermal derivatives but deficient in other tissues. The macromeres (four larger cells in the vegetal half of the egg) form **vegetalised** embryos composed mainly of endoderm (gut), but lacking a normal complement of ectoderm. The micromeres (four small cells from the vegetal pole) form primary mesenchyme derivatives such as skeletal spicules when cultured *en masse* (Okazaki, 1975; Harkey & Whiteley, 1983).

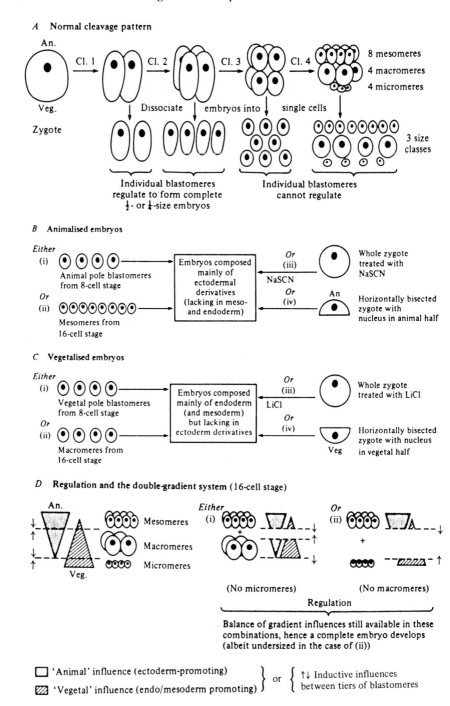

Fig. 2.9 Early sea urchin development.

However, these distinct developmental fates are already latent in the fertilised sea urchin egg. If a zygote is bisected equatorially between the animal and vegetal poles, then in cases where the animal half received the nucleus, an animalised (mainly ectoderm) embryo develops, which fails to gastrulate or form primary mesenchyme. In cases where the vegetal half received the nucleus, the embryo which develops is often vegetalised (deficient in ectoderm) but is able to gastrulate and form primary mesenchyme. This was first demonstrated in *Paracentrotus* where the subequatorial pigment band identifies the vegetal half of the zygote, but is equally true in species lacking this feature – such as *Hemicentrotus* where the jelly canal provides a stable landmark for the animal pole (Maruyama *et al.*, 1985). The conclusion is the same in both cases: determinants for gastrulation and primary mesenchyme formation are prelocalised in the vegetal half of the uncleaved zygotes. Similar effects can be obtained by treating whole embryos with chemical agents; lithium chloride causes embryos to become vegetalised (inhibiting ectoderm formation), while sodium thiocyanate causes them to become animalised (inhibiting endoderm/mesoderm formation).

These and other experiments led to the double-gradient hypothesis of Runnström (1928) and Hörstadius (1928). In essence this postulates an ectoderm-promoting influence centred on the animal pole and decreasing towards the vegetal pole, together with an endoderm/mesoderm-promoting influence centred on the vegetal pole and declining towards the animal pole (fig. 2.9). Normal development would depend on a balance between these two gradients. For instance, combinations of macromeres (high vegetal influence) with mesomeres (high animal influence) will regulate, i.e. develop into complete embryos, despite the absence of micromere material. Perhaps more surprisingly, complete embryos are also formed from combinations of mesomeres with micromeres (highest level of vegetal influence). In this case the resulting embryos are dwarf, since they lack the macromere material representing nearly half the volume of a 16-cell embryo. However, such micromere/mesomere combinations do *not* develop into an endoderm-deficient embryo composed mainly of mesoderm and ectoderm, as would be expected in a typical mosaic system. Intercellular interactions are clearly required in order to redistribute the gradient influences across the remaining embryonic material, such that a normal complement of endoderm also forms. Indeed, these findings are also interpretable (perhaps more simply) in terms of inductive interactions between different tiers of blastomeres. Thus cell fates may become altered whenever spatially separated tiers are juxtaposed by the experimenter; for instance, endodermal tissue would be induced where micromeres abut on mesomeres, and so on (see Hörstadius, 1973, and chapter VI of Davidson, 1986; Davidson, 1989). A corollary of either view is that each cell must be aware of its position in the embryo as a whole (e.g. relative to the animal and vegetal poles) and can adjust its fate in accordance with altered positional information when parts of the system are deleted or added, such that the entire system regulates to form a normal structure. This is an important feature of regulative systems and pattern formation in general (see Wolpert, 1969), but is a topic beyond the scope of the present text.

As mentioned above, individual sea urchin blastomeres from both 2- and 4-cell stages can regulate to form complete embryos. Yet the system of morphogenetic influences appears to be established before first cleavage, as revealed by the zygote bisection experiments (fig 2.9). This apparent paradox is simply resolved; the first two cleavage divisions are vertical (meridonal), and do not cut across the animal–vegetal axis (see fig. 2.9). In other words, each 2- or 4-cell-stage blastomere contains the same balance of animal and vegetal influences as does the original zygote. Only when third cleavage cuts across the animal–vegetal axis in an equatorial (horizontal) plane do the blastomeres lose their equivalence. Each individual blastomere from the 8-cell stage will contain a preponderance of either the animal or vegetal influence, and is thus no longer able to give rise to a normal balance of embryonic tissues.

This relationship between cleavage planes and the segregation of morphogenetic determinants can be used to explain the apparently regulative or mosaic nature of most types of embryo. In mosaic embryos, even the first cleavage results in an unequal segregation of these determinants (obvious in the case of *Dentalium*; fig. 2.8), so that neither daughter cell has the same balance of developmental potentialities as the zygote. In regulative embryos, on the other hand, each early cleavage blastomere receives a complete and balanced set of determinants, and it is only during later cleavage divisions that some determinants are apportioned preferentially into one group of cells rather than another. At this stage, the cells will cease to be equivalent in their developmental capacities. Even so, such cells do not follow a completely mosaic fate thereafter, as discussed above. One simple way to account for this supposes that morphogenetic factors can act inductively between cells in regulative embryos, but not in mosaic embryos. In fact, many intergradings between these patterns are possible, and most mosaic embryos show evidence of intercellular communication during their later development. As we have seen, the D/V axis is established early in some mollusc species via maternal localisation in the polar lobe, but in other species this feature arises later in cleavage through inductive contacts between the 3D macromere and the overlying micrometers (see § 2.6.1b). Although *C. elegans* develops largely as a mosaic embryo, some cell fates are dependent upon early intercellular interactions (Priess & Thomson, 1987; see § 4.3). Further subtleties are introduced by the fact that cytoplasmic localisation is sometimes established *progressively* during the early cleavage divisions, rather than peformed in the egg or zygote. A clear example of progressive localisation is provided by the segregation of two sets of determinants during cleavage in the ctenophore *Mnemiopsis* (Freeman & Reynolds, 1973; Freeman, 1976).

In mammalian embryos, cell position appears to exert an overriding influence on future development. Thus cells located on the outside of the 16–32 cell *morula* become extra-embryonic trophectoderm derivatives (e.g. placenta) following cavitation, while those on the inside develop into the embryo proper (inner cell mass or ICM). Up to the cavitation stage, cells transplanted from the inside to the outside or *vice versa* will develop in accordance with their new location (Ziomek *et al.*, 1982). After cavitation, however, cells from the inner cell mass cannot transform into trophectoderm if exposed on the outer surface, and trophectoderm cells injected into the interior cannot turn into ICM derivatives. However, an alternative explanation suggests that cytoplasmic factors influencing the choice between trophectoderm and ICM fates may become radially arranged in the early embryo. This means that outer cells inheriting the peripheral layer of cytoplasm would normally adopt the former fate, whereas inner cells containing deeper cytoplasm would opt for the latter. Several events during early cleavage in mouse tend to support this view. Thus there is a radial polarisation of blastomeres prior to the asymmetric fourth cleavage, such that ultrastructural distinctions are established between the inner (central) and outer (peripheral) regions of each blastomere during the 8-cell stage. Following the fourth cleavage, these distinctions are passed on to the 16-cell stage, which comprises seven apolar inner cells and nine polar peripheral cells, representing the normal ancestors of the inner ICM and outer trophectoderm respectively (Johnson & Ziomek, 1981a; reviewed by Johnson & Pratt, 1983). The implication is that peripheral cytoplasm in the polarised 8-cell embryo may act as a determinant for trophectoderm formation, while central cytoplasm may lead to ICM development. However, this specification of cell fate is evidently still labile (until at least the 32-cell stage), since blastomeres placed in abnormal locations can repolarise in accordance with their new surroundings (see e.g. Johnson & Ziomek, 1983). The polarisation process is an inductive one requiring contact between cells, an ability which appears as early as the 2-cell stage. Notably, the geometry of polarisation in partial embryos is such that the polarised outer region of each blastomere (which can be labelled with fluorescent-tagged concanavalin-A) is always oriented away from the point(s) of cell contact (Johnson & Ziomek, 1981b). In spherical 8-cell embryos, these contacts occur internally, so it is the apical (outermost) regions of each cell which display concanavalin-A labelling, a feature inherited by the polar peripheral cells at the 16-cell stage (see Johnson & Pratt, 1983).

2.7 Establishment of molecular differences between early blastomeres in sea urchins

When do cells or cell groups first become distinct from each other during early development? In the preceding sections we have looked at this question in terms of restricted developmental capacities; to what extent can these be related to differences at the molecular level? In this section we will focus on early sea urchin development (fig. 2.9). At the 16-cell stage, the three blastomere types already show restricted fates (§ 2.6.2); they can also be isolated in bulk from synchronous 16-cell embryos, taking advantage of their different cell sizes.

2.7.1 Up to the 16-cell stage

Quantitative but no qualitative differences are detectable when protein synthetic profiles are compared between meso-, macro- and micro-meres (Senger & Gross, 1978). Actinomycin treatment suppresses these quantitative differences, so that the protein synthesis patterns become indistinguishable between the three cell types. This suggests that all three contain similar populations of maternal messenger RNAs, since actinomycin blocks transcription of the embryo's own genes. In other words, prevalent maternal mRNAs (those identifiable at the protein level) are distributed homogeneously throughout the 16-cell embryo, and presumably also in the zygote. The fact that quantitative differences in protein synthesis arise only when the embryo genome is transcribed, does not necessarily imply that different amounts of or types of mRNA are synthesised in the nuclei of mesomeres, macromeres and

Table 2.4. *Senger & Gross (1978) model to explain quantitative differences in protein synthesis between meso-, macro- and micromeres*

	Cell type	Cytoplasmic maternal mRNA[a]	Embryonic mRNA[b]	Overall messenger population	
A. In normal embryos	Mesomere	MMMM	NN	4M:2N	Different ratios of M
	Macromere	MMMMMM	NN	6M:2N	and N pro-
	Micromere	MM	NN	2M:2N	teins will be translated from each
B. In actino- mycin-treated embryos	Mesomere	MMMM	—	All M	Only M pro-
	Macromere	MMMMMM	—	All M	teins will be
	Micromere	MM	—	All M	translated from each

[a] Since prevalent mRNAs are assumed to be distributed homogeneously, their contribution in each cell type should reflect cytoplasmic volume (macromere > mesomere > micromere).

[b] This contribution is assumed to be constant for all 16 nuclei, in the simplest case.

micromeres. Even if all sixteen nuclei synthesised identical amounts and types of mRNA, quantitative differences would still be observed because of cell-size differences at the 16-cell stage, resulting in different ratios of maternal to embryonic mRNAs (table 2.4). This shows how quantitative differences could arise between blastomeres without any requirement for differential transcription or for uneven distribution of maternal messengers, but merely as a consequence of unequal cleavage divisions leading to differences in blastomere size. It does not of course explain why cleavage should follow this particular pattern.

In fact, a detailed analysis of the RNA populations in meso-, macro- and micromeres suggests that neither maternal RNAs nor new transcripts are homogenously distributed. Rodgers & Gross reported in 1978 that micromeres lack a proportion of the RNA species present in both meso- and macromeres. This was shown by preparing an egg[+] DNA tracer, comprising all those single-copy sequences complementary to total oocyte RNA (the maternal sequence set; see § 2.3 above). This tracer was hybridised with an excess of total cellular RNA prepared respectively from isolated meso-, macro- and micromeres, as well as from whole 16-cell embryos. Egg$^+$ DNA reacted to the same extent (around 90%) with total RNA from mesomeres, macromeres and 16-cell embryos, but to a markedly lower extent (67–80%) with total RNA from micromeres. This means that the maternal sequence set (egg$^+$ DNA sequences) is less fully represented in the micromeres, i.e. they contain a more limited range of RNA species than do the other cell types in 16-cell embryos. Since almost identical results were obtained with actinomycin-treated embryos, this difference cannot be ascribed to new transcription, but probably reflects an inhomogeneous distribution of maternal RNAs in the zygote.

This prompted Tufaro & Brandhorst (1979) to reexamine in greater detail the spectrum of proteins being synthesised in mesomeres, macromeres and micromeres. Despite a high-resolution two dimensional gel technique above to resolve up to 1000 diferent proteins, they could detect only quantitative differences between the three cell types. Specifically, none of the proteins synthesised in meso- and macromeres were absent from the micromere pattern. These protein data at first seem to contradict the inhomogeneous RNA distribution inferred above.

The apparent paradox here was resolved by Ernst *et al.* in 1980. They found that both meso- and macromeres contain rare-class maternal mRNA species which are non-polysomal (i.e. not translated into protein). By contrast, the micromeres do not contain detectable reserves of non-translated mRNAs. Similarly, both meso- and macromeres contain high-complexity nuclear RNA species (transcribed from the embryo's own genes), whereas micromeres do not express such transcripts until later. Thus the overall complexity of micromere total RNA is indeed less than that of meso- and macromere total RNAs; however, the RNA sequences lacking in micromeres are not translated in meso- and macromeres, hence the qualitative identity in protein synthetic profiles.

The limited RNA complexity in micromeres may underlie their restricted developmental potential, but does not explain the similarly restricted potentials of meso- and macromeres. In a sense this is hardly surprising; micromeres represent a very small fraction of the total volume of 16-cell embryo cytoplasm, so that local differences would be expected to show up clearly. A similar local difference confined, say to the animal pole, would be swamped by the much greater volume of mesomere cytoplasm.

2.7.2 Molecular indices of differentiation in later embryos

Although gut tissue is normally derived from the macromeres of a 16-cell embryo, and primary mesenchyme from the micromeres, these are not the sole fates adopted by these particular precursors. As indicated by the cell lineage of the sea urchin embryo (fig. 2.10; Cameron *et al.*, 1987), the four macromeres not only form the vegetal plate (future gut and secondary mesenchyme) but also contribute to the aboral ectoderm (or supra-anal ectoderm in the case of VOM). Most of the aboral and all of the oral ectoderm cells are derived from the ring of

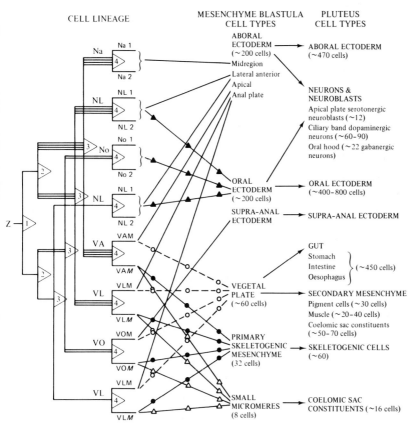

Fig. 2.10 Cell lineage of sea urchin embryo (modified from Cameron *et al.* 1987). Numbers enclosed in triangles indicate cleavage cycle. Z, zygote; N, animal; V, vegetal; L, lateral; M, macromere; *M*, micromere; O/o, oral; A/a, aboral; 1/2, daughters derived from meridional divisions. For further details, see fig. 4.5 in Davidson (1986).

eight mesomeres. The four micromeres again generate two sets of derivatives, namely (i) the primary (skeletogenic) mesenchyme cells, and (ii) the small micromeres (SM cells) which later contribute to the coelomic sacs (see fig. 2.10). As we have seen above, the horizontal third and asymmetric fourth cleavages segregate **tiers** of cells with different development potentials (meso-, macro- and micromeres). Likewise, the fates of both types of macromere derivative (vegetal plate or ectoderm) and both types of micromere derivative (primary mesenchyme or SM cells) become segregated into different tiers of cells by further horizontal cleavages. From the third to sixth cleavage onwards, each cell lineage develops clonally in the sea urchin embryo, with specific tissues deriving from particular groups (often tiers) of early blastomeres (see Chapter IV of Davidson, 1986; fig. 2.10).

The rapid early cleavages slow down in a lineage-specific manner between the sixth and tenth division cycles. In most cells, the cessation of division is marked by the appearance of a cilium and internal polarisation; these cilia are constructed largely from maternal stores of tubulin proteins and mRNAs. In the majority of sea urchin species (though not in other echinoderms), the primary mesenchyme (PM) cells are the first to move inwards into the blastocoel (ingression), becoming rounded and transiently motile. Later, during gastrulation, these cells form bilateral aggregates and begin to lay down triradiate skeletal spicules. This endoskeleton is constructed from calcite and various matrix proteins within syncytial cables formed by the fusion of filopodia extended from several neighbouring PM cell-bodies. The SM cells do not ingress along with the PM cells; rather they remain grouped on the external surface at the vegetal pole (distinguished by their lack of cilia), and are later carried inwards during gastrulation, contributing ultimately to the coelomic sacs. Gastrulation begins when the vegetal plate invaginates to form the archenteron (future gut), a process achieved largely via cell rearrangements and changes of cell shape (Ettensohn, 1985). From the tip of the invaginating archenteron appear secondary mesenchyme cells, whose diverse fates include pigment cells, muscle cells and contributions to the coelomic sacs. Some of these fates may already be specified before invagination begins; thus a monoclonal antibody recognising pluteus-stage pigment cells also picks out the eight or so pigment-cell precursors within the vegetal plate (Gibson & Burke, 1985).

Lining the blastocoel, the inner surfaces of the ectodermal cells secrete a variety of extracellular matrix (ECM) components such as fibronectin, laminin, type IV collagen and heparin sulphate proteoglycan. Drugs which inhibit the synthesis or assembly of the carbohydrate moieties of ECM components (e.g. tunicamycin), severely disrupt the migration of both primary and secondary mesenchyme cells as well as archenteron invagination. This implies that the ECM has several key functions in cell attachment, motility and rearrangement during ingression and gastrulation. Meanwhile, the outer surfaces of the ectodermal cells specifically secrete a coat of hyalin; this protein is synthesised during oogenesis and stored in the cortical granules, from which it is released during the cortical reaction (see § 2.5.1). During embryogenesis, hyalin is also synthesised *de novo* by cells of the ectodermal lineages but not by micromere or vegetal plate derivatives. Notably, when primary mesenchyme cells ingress, they lose affinity for neighbouring cells and for the outer hyalin coat, but gain adhesiveness for fibronectin and ECM components lining the blastocoel (Fink & McClay, 1985). A monoclonal antibody against hyalin disrupts cell-hyalin adhesion and also blocks gastrulation (Adelson & Humphreys, 1988).

From this brief survey, it is apparent that considerable differentiation of cell function has occurred even by the mesenchyme blastula stage, and still more by the end of gastrulation. Molecular correlates of these early differentiative events have been demonstrated in a number of cases, as illustrated in fig. 2.11. Most of these examples involve the transcriptional activation of embryonic gene-sets within specific cell lineages (e.g. primary mesenchyme or aboral ectoderm). Such tissue-specific genes are represented only at low levels, if at all, in the maternal transcript set inherited from oogenesis.

The actin gene family has provided several instructive examples of this type. It comprises six members, one of which encodes a muscle-specific (M) isoform while the other five encode cytoplasmic isoforms (CyI, CyIIa, CyIIb, CyIIIa, CyIIIb; Shott *et al.*, 1984). The coding sequences of the cytoplasmic actin genes are all very similar, but their mRNAs have different 3' untranslated regions, allowing the separate identification of all six types of actin transcript by *in situ* hybridisation (see Angerer & Davidson, 1984; Davidson *et al.*, 1985; Cox *et al.*, 1986). Representative examples are shown for four of these actin transcripts in fig. 2.11, parts *E* to *R* inclusive.

Thus CyI (fig. 2.11*M–P*) and CyIIb are coordinately expressed in vegetal regions of the embryo before and immediately after the ingression of PM cells (though not in the vegetalmost SM cells), then

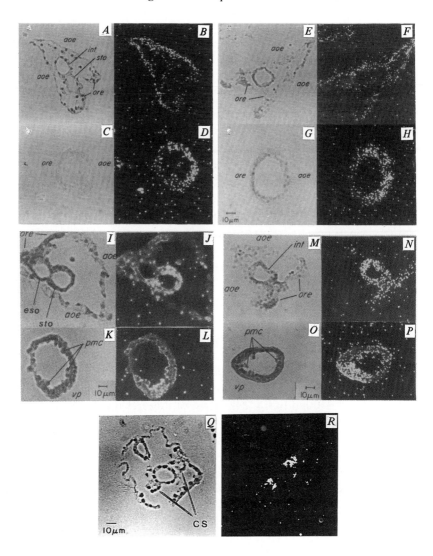

Fig. 2.11 Differential gene expression in sea urchin embryos. The
pairs of fields show sections under light microscopy (*A, C, E, G, I,
K, M, O, Q*) or under dark-field illumination after *in situ*
hybridisation with a labelled gene probe (*B, D, F, H, J, L, N, P, R*).
In the latter fields, bright specks indicate hybridisation with the
corresponding mRNA, and the intensity of speckling indicates the
relative level of expression. All photographs kindly supplied by
Professor E. H. Davidson, California Institute of Technology, and
reproduced by permission of the copyright holders. Parts *A* to *P*,
from R. C. Angerer & E. H. Davidson (1984), *Science*, **226**, 1153–60,
copyright 1984 by the AAAS; parts *Q* and *R*, from E. H. Davidson,

later in the invaginating archenteron and oral ectoderm. By contrast, CyIIa transcripts are expressed transiently in PM cells and again in secondary mesenchyme cells, in both cases appearing when these cells first ingress towards the blastocoel (fig. 2.11*L*); CyIIa is later expressed in the stomach region of the gut (fig. 2.11*J*). The two CyIII genes (a and b) are expressed specifically in aboral ectoderm, where they may assist in constructing a rigid cytoskeletal framework (fig. 2.11*E–H*). The same aboral ectoderm cells also express the Spec1 gene (fig. 2.11*A–D*), which encodes a Ca^{2+}-binding protein. Finally, the muscle-specific M actin gene is expressed only late in embryonic development in a few cells associated with the bilateral coelomic sacs on either side of the oesophagus (fig. 2.11*Q–R*). Note that the constituents of these sacs later give rise to the adult sea urchin when the larval tissues are eventually discarded. The β-tubulin gene family also shows evidence of tissue-specific regulation. Thus one isoform (β1) is confined to pluteus ectoderm, while another (β4) is restricted to meso- and endoderm; however, two further isoforms (β2 and β3) are less tissue-specific (Harlow & Nemer, 1987).

Fig. 2.11 (*cont.*)

C. N. Flytzanis, J. J. Lee, J. J. Robinson, S. J. Rose III & H. J. Sucov (1985), *Cold Spring Harbor Symp. Quant. Biol.* **50**, 321–8, copyright 1985 by Cold Spring Harbor Press.

Parts A to H: expression of Spec1 (*B, D*) and CyIIIb actin (*F, H*) mRNAs in pluteus (*A, B, E, F*) and mesenchyme blastula (*C, D, G, H*) embryos. Both probes detect gene expression mainly in aboral ectoderm.

Parts I to L: expression of CyIIa actin mRNA (*J, L*) in pluteus (*I, J*) and late gastrula (*K, L*) embryos. Note expression of CyIIa in secondary mesenchyme cells beginning to move into blastocel (*L*), and later in stomach region of gut (*J*).

Parts M to P: expression of CyI actin mRNA (*N, P*) in pluteus (*M, N*) and early mesenchyme blastula (*O, P*) embryos. Note expression of CyI in primary mesenchyme cells in the process of ingression (*P*), and later in the intestinal region of the invaginating archenteron as well as in oral ectoderm (*N*).

Parts Q and R: expression of M actin mRNA (*R*) in parts of the coelomic sacs on either side of the oesophagus in a late pluteus stage embryo.

Abbreviations: ore, oral ectoderm; aoe, aboral ectoderm; cs, coelomic sac; int, intestine; sto, stomach; eso, oesophagus; vp, vegetal pole; pmc, primary mesenchyme cells.

For each group of photos, bar shows 10 μm.

A number of different genes are expressed in mesenchymal lineages, many of them specifically in skeletogenic PM cells. One such gene encodes a 208 kd collagen which is localised mainly in the periphery of the PM-derived endoskeleton, though some transcripts from this gene are also detectable in other mesenchymal lineages (Angerer *et al.*, 1988). A major spicule-matrix protein is encoded by the SM50 gene, whose transcripts are rare in total embryonic mRNA but represent as much as 1% of the mRNA in the skeletogenic PM cells, the only lineage in which this gene is expressed (Benson *et al.*, 1987; Sucov *et al.*, 1987). Transcription of the SM50 gene is activated during PM cell ingression (Killian & Wilt, 1989), and this gene is also active in adult skeletogenic tissues in a variety of echinoderms (Drager *et al.*, 1989). There is a major changeover in the pattern of protein synthesis accompanying the differentiation of 16-cell-stage micromeres into PM cells (Harkey & Whiteley, 1983); this change occurs on schedule in the progeny of micromeres cultured *in vitro*, as well as *in vivo*. Harkey *et al.* (1988) have isolated cDNA clones representing five mRNAs expressed abundantly in PM cells but not in larval ectoderm nor in 16-cell-stage micromeres; all five transcripts are induced coordinately just before the ingression of PM cells and are expressed abundantly thereafter in the skeletogenic lineage. Another marker expressed by ingressing PM cells is Meso1, a 380 kd protein which is first detectable by antibody staining when these cells begin to delaminate from the inner surface of the blastula (Wessel & McClay, 1985). However, the Meso1 protein appears to be translated well in advance of this stage, and the immunodetectable form may result from post-translational modification – perhaps involving the addition of a carbohydrate moiety recognised by the antibody. Note that all of these skeletogenic markers first appear at the time of PM cell ingression, some time *before* overt differentiation (spicule production), but several cell divisions *after* the initial specification of the PM lineage. Indeed, the program of skeletogenesis has been determined in precursors as early as the micromeres of the 16-cell stage, since these cells autonomously execute this program *in vitro* without further reference to cells of other lineages. In microcosm, the PM lineage exemplifies the roles of (i) putative maternal determinants and (ii) localised activation of tissue-specific batteries of embryonic genes, in the genesis of a single differentiated cell-type. How far this example can be generalised remains to be seen.

In the vegetal plate lineage, the Endo1 protein (320 kd) first appears on endodermal cells when the archenteron invaginates, and later becomes restricted to mid- and hind-gut regions; it is translated from embryonic mRNAs immediately before gastrulation (Wessel & McClay, 1985). Likewise, Endo 16 gene-products are first expressed just prior to gastrulation, specifically in the vegetal plate cells (Nocente-McGrath *et al.*, 1898). By contrast, both tropomyosin and muscle myosin heavy-chain genes are expressed only in the coelomic sacs (cf. M actin; Rose III *et al.*, 1987).

These observations suggest that distinct gene sets are activated in different groups of precursor cells; e.g. Spec1 + CyIIIa/b in aboral ectoderm, SM50 etc. in PM cells, Endo1 + Endo16 in gut cells, and M actin + tropomyosin + myosin heavy-chain in muscle cells formed in the coelomic sacs. This may imply common regulation of genes in each set via related *cis*-acting DNA sequences and shared *trans*-acting protein factors. One approach to studying these patterns of regulation is to fuse the regulatory sequences of a tissue-specific embryonic gene onto a foreign 'reporter' structural gene (encoding, for example, β-galactosidase or chloramphenicol acetyltransferase [CAT]). Such constructs are injected into unfertilised eggs, which are then fertilised and allowed to develop, so that the level and spatial distribution of reporter gene-product can be monitored. A CyI-promoter/CAT fusion gene is activated on schedule in early embryos (Katula *et al.*, 1987); correct temporal activation can be detected with as little as 254 bp of CyI flanking sequence, but maximal levels of activation require at least 850 bp. Regulatory sequences lying between −440 and +120 relative to the start of the SM50 gene are able to confer a lineage-specific pattern of expression (confined to primary mesenchyme cells) on a CAT reporter gene (Sucov *et al.*, 1988). Similarly, 2.5 kbp from the regulatory regions of the *Strongylocentrotus purpuratus* CyIIIa gene confer the same pattern of developmental activation on a CAT reporter gene as that normally observed for the endogenous CyIIIa gene (Flytzanis *et al.*, 1987). Furthermore, this CyIIIa/CAT fusion construct is expressed only in aboral ectoderm tissue in *S. purpuratus* embryos, showing correct tissue- as well as stage-specificity. However, if an *S. purpuratus* CyIIIa/CAT fusion gene is introduced into *Lytechinus pictus* embryos, then activation occurs on schedule, but the normal tissue specificity is lost; strong expression of the construct is detectable in PM cells, gut, and oral as well as aboral ectoderm (Franks *et al.*, 1988). The developmental appearance of protein factors that bind to 14 sites in the *cis*-acting regulatory sequences of the CyIIIa gene has been elucidated by Calzone *et al.* (1988b). One binding factor is present at similar concentrations in unfertilised eggs, in 7-hour embryos (CyIIIa gene not yet active), and in 24-hour embryos (CyIIIa gene fully active). Factors binding at five other sites are undetectable in unfertilised eggs, but are present in similar amounts in both 7- and 24-hour embryos. Finally, factors binding to five further sites are again absent from eggs, but are present at *c*.10-fold higher levels in 24-hour compared with 7-hour embryos. It is likely that some proteins in this third class are involved in the transcriptional activation of the CyIIIa gene; future studies should determine whether these proteins also help to activate other genes (such as Spec1) expressed specifically in aboral ectoderm. If such transcription factors are also **inducible**, for instance via altered cell:cell contacts, then many of the regulative properties of early sea urchin embryos can be explained. By activating and/or repressing tissue-specific **sets** of structural genes, such inducible factors could switch the cells affected into alternative fates (see review by Davidson, 1989).

Before leaving this topic, it is worth reconsidering a few aspects of regulative development in sea urchins in the light of these lineage-specific patterns of gene activity. As mentioned earlier, whole embryos can become animalised or vegetalised as a result of various chemical treatments. In particular, Zn^{2+} ions cause animalisation (blocking gastrulation and perhaps exaggerating ectodermal development), a process which can be reversed to some extent by treatment with chelating agents. Cell division is relatively unaffected in Zn^{2+} animalised embryos, but the expression of early ectodermal markers (Spec3, Blastj1) is both increased and prolonged. (Conversely, these markers are suppressed in LiCl-vegetalised embryos.) However, zinc treatment delays the appearance of later ectodermal markers such as Spec1 and CyIIIa actin. Thus the overall effect of Zn^{2+} is to suspend the ectodermal differentiation program at an early stage, preventing its progression to later stages (Nemer, 1986). In the starfish *Asterina*, LiCl treatment between 7 and 10 hours of development increases the proportion of cells directed towards meso/endodermal fates by about one third, at the expense of certain ectodermal fates (but by no means *all* ectodermal cells are so redirected: Kominami, 1984). The expression of vegetal-specific meso-/endodermal markers (such as SM50 and alkaline phosphatase) can be evoked by lithium treatement in animal blastomeres isolated from sea urchin embryos (Livingstone & Wilt, 1989). In embryos from which the PM cells have been deleted, late-ingressing secondary mesenchyme cells are able to substitute for them and produce skeletal spicules (Ettensohn & McClay, 1988). These examples suggest that instances of regulation in post-cleavage echinoderm development do not require wholesale respecification of cells across the entire embryo, but probably involve the reallocation of fates among quite limited groups of cells, while other lineages continue along their normal differentiation pathways. This view of regulative development fits well with the suggestion (above) of limited alterations of cell fates via inducible transcription factors (Davidson, 1989).

2.8 Induction

The extensive cell movements which occur during gastrulation (and later) juxtapose groups of cells that were separated at earlier stages. These contacts between different groups of cells often alter the path(s) of development followed by one or both groups; i.e. they undergo changes in morphology and/or differentiation which do not occur if contact is prevented. The interactions required to achieve such changes are termed **inductive**, and the agents responsible, **inducers**. Several features distinguish induction from other cell signalling systems in animals (see reviews by Gurdon, 1987; Jacobson & Sater, 1988):

(i) It is a short-range interaction, acting either through diffusible inducer molecules (perhaps secreted by the inducing cells and binding to receptors on the induced cells), or else mediated via actual cell contact (in which case the inducer might be an extracellular surface component of the inducing tissue). Induction can occur even when the interacting tissues are separated by a semipermeable membrane in explant cultures; however, this does not necessarily prove that diffusible inducers are involved, since fine cytoplasmic processes can still make contact through the pores of the membrane (see e.g. Smith & Thorogood, 1983).

(ii) Both the availability of inducer and the ability to respond to it (competence) are strictly limited in duration, such that a particular inductive event can only take place within a narrow time window during development. In this respect, induction differs radically from hormonal switches (see § 3.3), where the approriate receptors are usually a permanent feature of the responding tissue. Limited periods of competence and/or inducer availability are essential in many instances of organogenesis, where a complex structure such as the vertebrate eye is built up through a hierarchical series of inductive events, each of the earlier inductions being prerequisite for the later ones. The region of tissue able to respond to an induction (the morphogenetic field) is often much larger initially than that part of it which eventually forms the induced structure (see review by Jacobson & Sater, 1988).

(iii) Commonly, the observable response to induction occurs some time (hours or days) after the inductive event has taken place. Indeed, if the interacting tissues are juxtaposed *in vitro* for relatively brief periods and then separated, the induced pathway(s) of differentiation will nevertheless proceed on schedule. When studying molecular correlates of induction, it is important to select markers which are expressed early in the induced pathway. Even so, it may be difficult to disentangle the effects of cell division and other processes which have occurred since the induction event took place (see Gurdon, 1987).

(iv) Many inductive events involve interactions between mesenchymal (mesodermal) and epithelial (ecto- or endodermal) components, frequently resulting in the differentiation of specific glandular structures from the epithelial tissue. In some cases, this can be achieved by fairly non-specific inductive influences, which may be available from several heterologous types of mesenchyme different from that normally used *in vivo* (the homologous mesenchyme). Such inductions are often referred to as permissive, since they enable the completion of a process that is already under way in the responding tissue. In other cases, only homologous mesenchyme can achieve the requisite induction, radically altering the pathway of epithelial differentiation. Such examples are more properly described as instructive, since at least part of the specificity resides in the inducing tissue (see e.g. Wolff, 1968). However, this distinction is not absolute; on occasion the induction process is at first instructive but later becomes permissive. Thus pancreas induction from endoderm in mouse embryos requires pancreatic mesenchyme at the 9-somite stage, but by the 15-somite stage any type of mesenchyme is effective. Hierarchies of inductive events may be involved in such cases, the early ones being instructive and the later ones permissive (see Gurdon, 1987).

Many of the classical examples of induction are more complex than is at first apparent. Thus the formation of an eye lens from competent head ectoderm in vertebrate embryos was long thought to be directed by inductive influences emanating from the underlying optic cup (future retina). It now appears that this alone is insufficient to achieve lens induction, which also requires complex prior inductions from both endoderm and head mesenchyme (see Grainger *et al.*, 1988). The account below will describe two early inductive events which occur in amphibian embryos, namely the induction of mesoderm and the subsequent induction of neural tissue (fig 2.12). Both processes are complicated by regional differences, implying differential responses either to two (or more) distinct inducers, or else to different concentrations of the same inducer.

In *Xenopus* embryos, early signs of muscle differentiation towards the end of gastrulation are provided by the appearance of transcripts from the muscle-specific α-skeletal and α-cardiac actin genes (Mohun *et al.*, 1984). A still earlier muscle marker is the expression of a myogenic switch gene, MyoD (see § 3.1; Hopwood *et al.*, 1989a). A cytoskeletal actin gene (type 8) is also expressed exclusively in embryonic muscle, yet its promoter is very similar to that of another such gene (type 5) which is expressed throughout the embryo (Mohun & Garrett, 1987). It remains to identify the precise sequence elements required for the activation of muscle-specific genes, and the *trans*-acting proteins which interact with them. The muscle differentiation program requires both prelocalised maternal components and induction. Although α-actin expression in future muscle cells does not begin until late gastrulation, all of the components required for this activation are already prelocalised in the subequatorial dorsal region of the fertilised but still uncleaved egg (Gurdon *et al.*, 1985b). Nucleated egg fragments containing this region of cytoplasm can later differentiate muscle cells autonomously in culture. Muscle and other mesodermal cells also differentiate non-autonomously as a consequence of inductive interactions between animal and vegetal blastomeres in early embryos (Gurdon *et al.*, 1985a); this can be studied *in vitro* by culturing conjugates of animal and vegetal cells. When cultured alone, animal blastomeres develop only into ectoderm, but when cocultured with vegetal blastomeres some of the animal cells are induced to form mesoderm. The expression of muscle markers (α-actins) can first be detected some 7 hours after induction; the minimum contact time required for muscle induction is only 1.5–2.5 hours, but the response is greater if this period is prolonged up to about 5 hours. Competence is strictly limited in this system; animal cells lose their ability to respond to mesodermal induction by the early gastrula stage, while vegetal cells lose their inductive ability by the late blastula stage. However, the time interval before expression of muscle-specific genes depends on the developmental stage of the responding tissue (animal cells) and not on the timing or duration of the inductive contact (Gurdon *et al.*, 1985a).

The induction of mesoderm from animal blastomeres by vegetal influences shows regional specificity (Dale *et al.*, 1985). Single identified vegetal blastomeres from 32-cell embryos can induce differ-

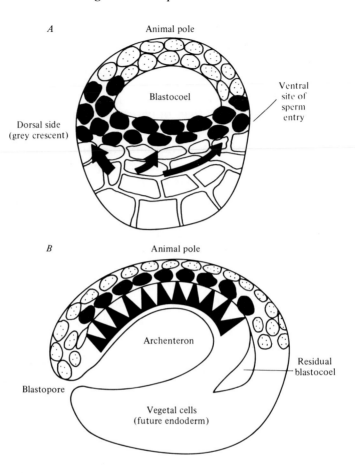

Fig. 2.12 Early inductions in *Xenopus* (modified from Gurdon, 1987).

A Mesoderm induction in midblastula. ⊕ , responsive animal cells (future ectoderm); ⬤ , induced mesodermal tissue (cells responding to induction); ◯ , inducing vegetal cells (future endoderm); ▲ ▲, inductive influence(s).

B Neural induction in gastrula. ⊕ , responsive ectodermal cells (future epidermis); ⬤ , induced neural tissue (inner layer of ectoderm); ▶ , inducing mesodermal tissue (notochord + prechordal plate) producing one or more inductive influences.

In both parts, the cells competent to respond to induction are indicated by dotted ovals, while those which actually respond are shown as black ovals; arrows indicate the direction and source of inductive influences.

ent mesodermal derivatives from a tier of lineage-labelled animal blastomeres (Dale & Slack, 1987). Thus the dorsovegetal D1 blastomere induces dorsal-type mesoderm (notochord and muscle), whereas the laterovegetal and ventrovegetal (D2, D3, D4) blastomeres induce intermediate or ventral-type mesoderm (including some muscle, mesenchyme, blood and mesothelium). These and other results suggest that two different inductive influences may be involved, one producing dorsal and one ventral mesodermal derivatives; intermediate structures might arise where these abut through the dorsalisation of ventral-type mesoderm by dorsal-type mesoderm (Dale & Slack, 1987).

The nature of the inducer(s) involved here (or indeed in other induction systems) remains obscure. Isolation and identification of putative inducing agents is rendered difficult by the fact that many heterologous substances display inducer activity, including such unlikely chemicals as LiCl and the dye methylene blue. This re-emphasises the point made earlier, that much of the specificity of induction resides in the responding tissue, and that non-specific agents can often substitute for the genuine inducer used *in vivo*. Unfortunately this also means that protocols designed to isolate natural inducers may identify the wrong molecule, since so many different substances can exert similar inductive effects (see Gurdon, 1987). However, in the case of mesoderm induction, a potent mesoderm-inducing factor (MIF) has been isolated from a *Xenopus* cell line (Cooke *et al.*, 1987). Exposure to MIF for as little as 15 minutes causes animal pole cells to differentiate into axial mesoderm several hours later. MIF exerts this inductive effect when injected into the blastocoel but not when injected intracellularly into individual blastomeres, implying that this agent acts on its animal-pole target cells via extracellular receptors (Cook *et al.*, 1987). Basic fibroblast growth factor (bFGF) can also induce limited mesodermal differentiation from animal blastomeres, but this effect is synergistically enhanced by transforming growth factor-β (TGF-β), which is ineffective alone. That two such growth factors (acting in concert) might be natural mesodermal inducers is suggested by the fact that *Xenopus* embryos contain mRNAs encoding close relatives of both bFGF and TGF-β (Kimelman & Kirschner, 1987); in the latter case, one major mRNA species (Vg1) is of maternal origin and is localised in the vegetal region of the mature egg (see § 2.4.2c; Weeks & Melton, 1987). Similarly, a 4.2 kb transcript present in the *Xenopus* oocyte encodes a protein related to human bFGF; this protein is stored in the oocyte in amounts sufficient to induce mesoderm formation (Kimelman *et al.*, 1988). Vegetal blastomeres might secrete several such factors, which could interact with appropriate receptors on the surfaces of the animal blastomeres in order to achieve mesodermal induction. However, the overall process is undoubtedly more complex than this. Although MIF is related to the TGF-β family, it is not the species encoded by Vg1 mRNA. Moreover, MIF possesses 'organiser' activity (such that MIF-treated cells can induce extra sets of axial structures), whereas bFGF shows no such ability (see Cooke, 1989).

It must be remembered that cell properties are changing rapidly during cleavage, and that flexibility as to cell fate is lost progressively during early development. If single vegetal blastomeres are taken from successively later cleavage stages, labelled with a lineage marker and transplanted into the blastocoel, an increasing proportion of them express only endodermal fates, in spite of their new cellular environments. In other words, these vegetal cells become increasingly committed to an endodermal fate which cannot then be altered following transplantation (Wylie *et al.*, 1987). In similar experiments with animal blastomeres, there is a progression from early pluripotency (cells able to contribute to any germ layer after transplantation) through labile to firm ectodermal commitment (cells forming ectodermal derivatives only, independent of their new surroundings: Snape *et al.*, 1987). Thus mesodermal induction is possible only during the early pluripotential stage before animal cells become committed to ectodermal fates.

During gastrulation in *Xenopus*, the involuting mesodermal tissues (specifically the notochord and prechordal plate) induce the overlying ectoderm to form neural plate, which later rolls up into the neural tube. This induction shows several of the features noted above. Thus there is regional specificity such that the first material to involute (prechordal plate) induces forebrain (archencephalic) structures while the later-involuting notochord induces hindbrain (deuterencephalic) and then spinal cord structures. This has been variously explained in terms of two (or more) inducers, or of different responses to high and low concentrations of a single inducer. Once again there is strict spatial and temporal specificity; only the inner layer of ectoderm differentiates into neural tissue while the outer layer remains epithelial. There is also a sharply defined lateral boundary between those cells which respond to neural induction (the future neural plate) and those which do not (which remain epidermal). Various markers are expressed in both layers of the ectoderm prior to neural induction, including peanut lectin receptors (Slack, 1985) and embryonic epidermal keratins (Jamrich *et al.*, 1987). In both cases, induction suppresses expression of these markers in the inner layer (future neural tissue; see fig. 2.12), while a variety of neural-specific markers appear in this layer within hours of the inductive event. These markers include the neural cell-adhesion molecule N-CAM (Balak *et al.*, 1987) and transcripts from the *Xenopus* homeobox gene *XlHbox6* (see § 5.6). However, there is evidence for some predisposition of dorsal ectoderm to form neural tissue, and of ventral ectoderm to form epidermis. Expression of the neural marker gene *XlHbox6* can be induced by placing ectodermal and mesodermal tissues in contact in culture (neither tissue expresses this gene if cultured alone); however, *XlHbox6* expression is most readily inducible in dorsal ectoderm – the region normally destined to form neural tissue – implying a predisposition of this region to undergo neural induction (Sharpe *et al.*, 1987). The epidermal marker Epi 1, by contrast, is expressed more actively on ventral micromeres (future epidermis) than on dorsal micromeres (future neural tissue); this difference can be detected well before gastrulation (even as early as the 8-cell stage), implying prelocalisation of maternal factors required to activate Epi 1 expression in the appropriate region (London *et al.*, 1988). It is worth noting that animal cells from *Xenopus* blastulae offer a classic example of instructive induction; left alone, they will differentiate into ectoderm, but vegetal cells can induce them to form mesoderm, while mesodermal tissue can induce them to undergo neural differentiation. In this sense, they are truly pluripotential.

3

Differentiation in vertebrate systems

3.1 Introduction: determination versus differentiation during myogenesis

As discussed previously (§ 2.2), the specialised cell types in an adult animal differ fundamentally in their utilisation of 'luxury' genes but overlap to a great extent in their 'housekeeping' functions. Both quantitative and qualitative changes may alter the pattern of gene expression during the course of differentiation in a given cell type. However, the appearance of tissue-specific **markers** (e.g. enzymes or other proteins) in significant amounts is usually taken as diagnostic, and for practical purposes will serve as a definition of differentiation. Many adult vertebrate tissues have been characterised in great detail in terms of their final differentiated states and the underlying patterns of gene expression. However, the molecular events associated with tissue specification during early development remain little known in most vertebrate embryos, although rapid progress is now being made in both *Xenopus* (see e.g. § 2.8) and mouse (see e.g. § 5.6). This situation contrasts with that in many invertebrate systems, where the molecular correlates of early developmental events are often better characterised than the differentiation of adult tissues (see e.g. § 2.7; also chapters 4 and 5).

Cells often commit themselves irrevocably to a particular fate well before they express any of its differentiated characteristics, a phenomenon known as **determination** or **commitment** (the two terms tend to be used interchangeably). A striking example of determination in advance of overt differentiation is provided by the imaginal discs involved in insect metamorphosis (see § 5.2.2). In molecular terms, determination might involve the **preselection** of tissue-specific gene sets, making them available for expression e.g. in the form of smooth-fibre chromatin regions containing DNaseI-hypersensitive sites (cf. Groudine & Weintraub, 1982). Significant transcription from these genes would only ensue when some further signal is received to initiate differentiation. Such signals may be inductive (see § 2.8) or hormonal (see § 3.3), or may involve polypeptide factors acting on particular precursor lineages (e.g. in the blood system; see § 3.2). Directly or indirectly, such signals must exert their influence at the DNA level by activating the preselected target genes. This could be accomplished by the *de novo* synthesis or activation of *trans*-acting transcription factors which interact in a modular fashion with *cis*-acting promoter and/or enhancer sequences shared by the target genes (cf. the Ig genes in lymphoid cells: § 2.2.2). Note that differentiation signals may down-regulate previously expressed genes (see below) as well as activating new tissue-specific genes.

Myoblast cells do not express differentiated muscle characteristics such as contractile proteins and their mRNAs while actively dividing, although these replicating cells already possess a distinctive phenotype (Kaufman & Foster, 1988). They will only proceed to accumulate muscle-specific mRNAs and proteins after passing through a terminal division prior to fusion into syncytial myotubes, as shown e.g. for myosin heavy-chain mRNAs (John *et al.*, 1977). The dividing myoblasts are determined in that they cannot develop into other cell types, but final differentiation is delayed until the postmitotic stage and subsequent fusion. The terminal myoblast division is followed by the sequential activation of transcription from a set of muscle-specific genes (Lawrence, J.B. *et al.*, 1989), encoding for example myosin heavy and light chains, α-actins, tropomyosin, troponin, etc. At least in the case of the actins, this involves an isoform switch, since cytoplasmic β-actins are expressed in dividing myoblasts, whereas post-mitotic myoblasts and myotubes express cardiac and later skeletal α-actins. Both of these α-actins are markers for the earliest stages of muscle differentiation in mouse embryos (cf. in *Xenopus* mesodermal induction: § 2.8); notably, the cardiac isoform predominates in the somites as well as in the developing heart (Sassoon *et al.*, 1988). A chick α-actin gene transfected into rat myogenic cells is expressed at vastly higher levels during muscle differentiation in culture (Nudel *et al.*, 1985). This activation occurs even if the introduced actin genes are methylated; they undergo the same set of demethylations as that shown by the endogenous α-actin genes in myoblast cells *in vivo* (Yisraeli *et al.*, 1986). By contrast, the cytoplasmic β-actin genes are down-regulated during muscle differentiation (Seiler-Tuyns *et al.*, 1984), a feature mediated through an intragenic 40bp sequence which lies just upstream of the poly-adenylation signal. This sequence can confer similar down-regulation during myogenesis when placed at a corresponding position in a heterologous gene (De Ponti-Zilli *et al.*, 1988). Among the genes activated during muscle differentiation is that encoding the muscle-specific form of creatine kinase (CK). This activation occurs at the level of transcription and is mediated by 5'-flanking sequences, 3.3 kbp of which can confer a similar pattern of transcriptional activation during myogenesis when fused to a bacterial reporter gene. The sequences immediately upstream of the CK cap site share homology with the promoter regions of several other muscle-specific genes such as those coding for α-actin and myosin heavy chain (Jaynes *et al.*, 1986), suggesting that such sequences may mediate in the activation of their transcription.

Although the muscle-specific contractile protein genes are not expressed in dividing myoblasts, these cells are nevertheless firmly committed to a muscle fate. One approach to studying the determination pathway has been opened up by transfecting cDNAs from myogenic cells into a mouse fibroblast cell line (3CH 10T1/2). This particular cell line is able to undergo myogenic conversion when treated with the demethylating agent 5'-azacytidine (Lassar *et al.*, 1986), suggesting some innate predisposition to adopt the myogenic pathway. (Other such cell lines are predisposed to convert e.g. into adipocytes.) Among the cDNAs prepared from myogenic 10T1/2 derivatives or normal myoblasts, the MyoD1 species is able to convert parental 10T1/2 fibroblasts into myogenic cells at high frequency following transfection (Davis *et al.*, 1987). The MyoD1 gene is located on mouse chromosome 7 and on human chromosome 11; it encodes a nuclear phosphoprotein with a limited region of homology to the MYC product of the c-*myc* cellular oncogene. Expression of a protein fragment comprising only 68 amino acids derived from key parts of the MyoD1 gene (the basic and MYC-related domains) suffices for myogenic conversion of 10T1/2 cells (Tapscott *et al.*, 1988). A few other cDNAs such as *myd* are also able to convert 10T1/2 cells into muscle derivatives (Pinney *et al.*, 1988), suggesting a hierarchy of genes involved in myogenic determination (see also Blau, 1988). A third factor regulating myogenesis, termed myogenin, has been characterised using a different strategy (Wright, W.E. *et al.*, 1989). Again, a cDNA clone encoding myogenin can induce myogenic conversion when transfected into 10T1/2 cells. Myogenin appears to be specific for skeletal muscle cells; it is expressed transiently during the early phases of muscle differentiation in culture, and also during myotome formation in the somites of 8.5 day mouse embryos (see § 5.6). Thus MyoD1, myogenin, etc. may have different functions in different types of embryonic muscle tissue (Sassoon *et al.*, 1989). The MyoD1 protein binds via its MYC-related domain to the promoters of many muscle-specific genes and thereby activates their expression (Lassar *et al.*, 1989), although one such gene may be activated independently of MyoD1 (Baldwin & Burden, 1989). Overexpression of MyoD1 fusion constructs can force a variety of cell types to activate muscle-specific genes that are normally kept silent (Weintraub *et al.*, 1989). These include the endogenous myogenin and MyoD1 genes, both of which are positively (auto) regulated by MyoD1 (Thayer *et al.*, 1989). The cDNA sequence of myogenin reveals a MYC-related region similar to that of MyoD1 (Wright, W.E. *et al.*, 1989), suggesting a family of related genes involved in myogenesis. The domain common to the MYC, MyoD1 and myogenin proteins is also shared by two factors which bind to the immunoglobulin κ-light chain enhancer, and by the product of the *Drosophila* gene *daughterless* (involved in sex determination and cell fate; see § 5.2.3). This domain comprises the helix-loop-helix motif involved in dimerisation and binding to DNA (Murre *et al.*, 1989).

The remainder of this chapter will consider two vertebrate differentiation systems in greater detail. Section 3.2 will deal with the blood system, concentrating mainly on erythroid (red blood) cells and the developmental regulation of their major protein products, the globins. Section 3.3 will deal with hormonal regulation of differentiation, as exemplified by egg-white- and yolk-protein production under sex-steroid control in egg-laying female vertebrates.

3.2 Erythroid differentiation

Red blood cells (erythrocytes) are among the most specialised of all vertebrate cell types, devoting most of their protein synthesis to the production of globin polypeptides (α and β chains in adults).

Each globin polypeptide becomes complexed to an iron-containing haem molecule which acts as a prosthetic group for oxygen-binding. Two α-type and two β-type globin chains each with attached haem bind together as subunits of the tetramer **haemoglobin** (Hb), which is characterised by cooperative allosteric oxygen-binding. This means that binding of O_2 to the haem of one subunit causes conformational changes in the other subunits which facilitate O_2-binding at the three remaining haem sites. Oxygen release in the tissues is promoted by CO_2 and H^+ there, which bind to different but allosterically linked sites in the haemoglobin subunits. The converse happens where oxygen is freely available (in the lungs or gills), with CO_2 and H^+ being released and new O_2 bound. Vertebrate erythrocytes are circulating blood cells packed full of haemoglobin; they are thus highly specialised for carrying oxygen from the lungs or gills to the tissues where it is required, and also for CO_2/H^+ transport in the reverse direction (for details, see Stryer, 1981, chapter 4). Although haemoglobin is the major differentiation marker in erythroid (i.e. erythrocyte precursor) cells, a series of haem-synthesising enzymes and other proteins are also essential for proper cell function.

In the following subsections, we shall consider: (i) the origin and differentiation of red blood cells in adults; (ii) the switches in haemoglobin type which occur during development; (iii) the expression and control of globin genes; and (iv) their organisation in the DNA.

3.2.1 Differentiation of erythroid cells in adults

Erythrocytes, and indeed all other varieties of blood cell, are constantly renewed throughout life. Each type of blood cell has a characteristically limited life span, averaging 120 days for human erythrocytes, but considerably longer for some granulocytes and lymphocytes. Worn-out cells are phagocytosed by macrophages in the liver and spleen, while new cells are constantly produced from special blood-forming sites. These are located mainly in the adult bone marrow, but accessory sites include, for example, the spleen, In humans, the average rate of erythrocyte production is around two million cells per second! The source of these new cells is a dividing population of **stem cells** which remains within the bone marrow. In fact, a single type of **pluripotent** stem cell (Hall & Watt, 1989), can give rise to a wide variety of blood-cell types, as indicated in fig. 3.1. At each stem-cell division, one of the daughter cells effectively remains a stem cell (hence this population is indefinitely self-renewing), while the other undergoes a limited series of divisions. This latter type of cell becomes committed during its early divisions to follow one of the several available pathways of blood-cell differentiation (fig. 3.1). Injections of bone marrow permit the survival of animals whose own stem-cells have been destroyed by a lethal dose of radiation. The donor stem cells can recolonise the recipient bone marrow and hence renew all types of blood cell in the rescued animal. Immune selection can be used to purify putative stem cells from the bone-marrow population (much of which consists of committed precursors). In mouse, proliferative stem cells are negative for a variety of

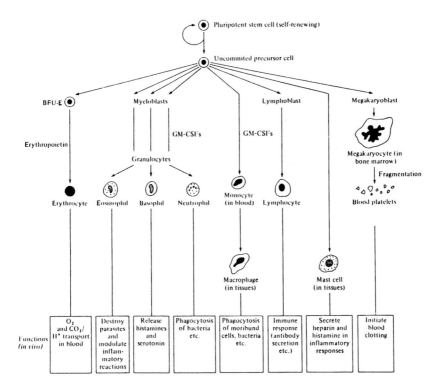

Fig. 3.1 Pathways of blood cell differentiation (adapted from Alberts *et al.*, 1983 and Ede, 1978).

lineage markers, express low levels of the Thy1 antigen, but are positive for the stem-cell antigen Sca1. As few as 30 such precursor cells can rescue a lethally irradiated mouse in 50% of cases, and can reconstitute all blood-cell types in the survivors (Spangrude *et al.*, 1988). Such cells also give rise to monomyelocytic as well as T and B lymphoid derivatives *in vitro*. The initial commitment events remain little understood, although some evidence suggests that different stem-cell clones may be activated sequentially during haematopoiesis (Lemischka *et al.*, 1986).

Lineage-specific glycoprotein factors are required in order to stimulate division and differentiation among particular kinds of committed precursor cell (see review by Sachs, 1987). For instance, the proliferation of myeloid precursors is controlled by the granulocyte/macrophage colony stimulating factors (GM-CSFs), each of which interacts with specific receptors present on the surfaces of these myeloid cells (see reviews by Metcalf, 1985, 1989). The GM-CSF group includes one factor specific for macrophage production (M-CSF or CSF-1), another which stimulates granulocyte production (G-CSF), and two acting earlier on common myeloid precursors (GM-CSF and Multi-CSF); the last of these also acts on precursors in certain other lineages, and is often known as interleukin 3 (IL-3). There is hierarchical cross-regulation between the receptors for these various growth factors, such that occupancy of the IL-3 receptors down-regulates all of the other receptor types, while occupancy of the GM-CSF receptors down-regulates only the G- and M-CSF receptors (see Walker *et al.*, 1985). A further growth factor termed haemopoietin I (HI), functions synergistically with either IL-3 or CSF-1, apparently by allowing these factors to act on earlier precursors than either would normally influence on its own (Stanley *et al.*, 1986). Apart from these humoral effects (mediated by soluble factors), the bone-marrow microenvironment includes both extracellular matrix (ECM) and stromal cells, which may also affect the growth and/or differentiation of particular blood-cell lineages. One identified bone-marrow ECM component is haemonectin, a 60 kd protein which acts as a lineage-specific attachment molecule for granulocyte precursors (Campbell *et al.*, 1987). The later growth-stimulating factors also induce the production of specific differentiation factors within their target cells; in the case of the granulocyte/macrophage precursors, the GM-/G-/M-CSFs induce a DNA-binding protein known as MGI-2 which is required for differentiation (Weisinger & Sachs, 1983; Sachs, 1987).

Similar features characterise the other major blood cell lineages (fig. 3.1). In each of them, a series of **amplification divisions** results in very large numbers of progeny cells from a single committed precursor (fig. 3.2); this proliferative phase is under the control of growth factors such as the GM-CSF group described above. In the erythroid lineage, the major identified growth factor is the glycoprotein erythroprotein (Er); the human Er gene has been cloned and is expressed in a biologically active form when introduced into tissue culture cells (Lin *et al.*, 1985). There is also evidence for differentiation factors involved in terminal erythroid differentiation (e.g. Nomura *et al.*, 1986). High concentrations of erythropoietin stimulate the division of early precursor cells; these are known as BFU-Es (erythroid burst-forming units) because each such precursor gives rise to a large 'burst' containing many thousands of differentiated erythrocytes. This involves eleven or twelve successive cell divisions in each clone founded by a single BFU-E, as shown *in vitro* using low-density cultures of bone marrow cells. Low concentrations of erythropoietin, on the other hand, stimulate division

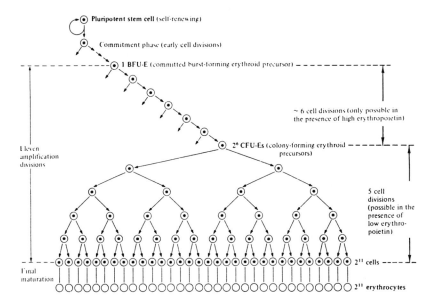

Fig. 3.2 Erythroid cell lineage (modified from Alberts *et al.*, 1983).

only among later erythroid precursors; these are know as CFU-Es (erythroid colony-forming units) because each such cell can undergo only five or six further cell divisions before final differentiation. Hence a small colony of 60 or fewer erythrocytes is produced from each CFU-E at low erythropoietin concentrations, compared with the large burst of many thousand red blood cells formed from each BFU-E at high erythropoietin concentrations (fig. 3.2).

There is a sequence of characteristic changes in cell ultrastructure and staining properties during the last few cell divisions in the erythroid series (fig. 3.3). This leads from the proerythroblast through basophilic, polychromatic and orthochromatic erythroblast stages, the last of which gives rise directly to the reticulocyte without further cell division. Terminal differentiation from reticulocyte to erythrocyte is also accomplished without cell division; indeed in mammals the nucleus is eliminated during the reticulocyte stage (fig. 3.3), though other vertebrates such as birds (fig. 3.4) have nucleated erythrocytes. At least in birds, the erythroid-specific histone H5 is involved in the inactivation of erythrocyte chromatin. Globin mRNA and protein (differentiation markers) first appear in small amounts after the CFU-E stage (e.g. in proerythroblasts), and they are accumulated rapidly during later differentiation. However, precursors from the BFU-E stage onwards are irreversibly committed to an erythroid fate. Haemoglobin

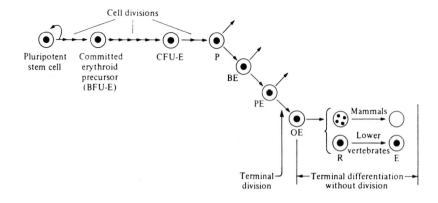

P   Proerythroblast (first signs of haemoglobin synthesis)
BE Basophilic erythroblast
PE Polychromatic erythroblast
OE Orthochromatic erythroblast
R   Reticulocyte (nuclear loss in mammals)
E   Erythrocyte (nucleated in lower vertebrates)

Fig. 3.3 Terminal division and differentiation in the erythroid lineage (modified from Ede, 1978).

synthesis continues during the reticulocyte stage, despite loss or complete inactivation of the nucleus; this implies that globin mRNA is stable and can be translated for some time after transcription of new messengers has ceased. The process of terminal erythroid differentiation can be induced chemically in various lines of mouse erythroleukaemia (MEL) cells cultured *in vitro*. The initial MEL cells are apparently arrested at the committed precursor stage, since they are capable of undergoing only erythroid differentiation, but do not express any of its characteristic features (e.g. globins) in the uninduced state. Induction with agents such as dimethyl sulphoxide commits MEL cells both to globin mRNA production and to passing through a terminal cell division (fig. 3.3), whereas haemin induction commits these cells to the former but not the latter. Commitment to the terminal cell division in this system is associated with a decrease of one specific nuclear protein, p53 (Shen *et al.*, 1983); this remains elevated in haemin-induced cells that synthesise globin mRNA but continue to divide in culture.

3.2.2 Haemoglobin switching

In adult humans the predominant type of haemoglobin is HbA, composed of two alpha and two beta subunits ($\alpha_2\beta_2$). Both α and β globins

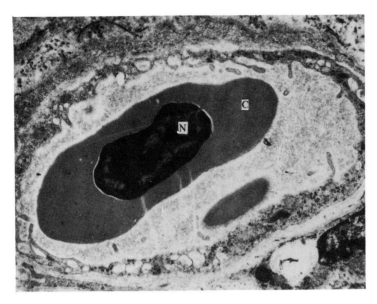

Fig. 3.4 Electron micrograph showing a nucleated avian erythrocyte within a blood capillary supplying the oviduct of a laying hen. Note the homogeneous cytoplasm (C), dark (inactive) nucleus (N) and flattened disc shape (seen in cross section). Magnification ×5500. By courtesy of Dr I. R. Duce (Department of Zoology, University of Nottingham).

are produced by different but related members of the globin gene family. In adults there is also a small percentage of HbA$_2$ (2–3%), composed of two α and two delta subunits ($\alpha_2\delta_2$). The δ chains are derived from a third member of the globin gene family closely related to β. Thus in adult bone-marrow-derived erythroid (E) cells, the β, δ and α globin genes are all actively transcribed. In fact there are two α genes but only one β and one δ gene (see § 3.2.4 below); nevertheless the rates of α and β production are closely coordinated so as to achieve a 1:1 ratio. This is mediated in part by translational control mechanisms, involving for instance the haem precursor haemin (see § 3.2.3 below; Giglioni *et al.*, 1973; Lane *et al.*, 1974). The 40-fold disparity between β and δ levels results in large part from differences in the rate of mRNA production, but δ mRNA is also more rapidly degraded than β mRNA (Wood *et al.*, 1978) and may be intrinsically less stable.

During embryonic and foetal development, however, alternative sites of erythropoiesis (erythrocyte production) and several further globin genes are pressed into service. In human foetuses, definitive E cells are produced by the liver. The α genes are active in these foetal E cells, but the β and δ genes are largely inactive (though at least some of these cells express β at low levels). Instead, two β-related gamma genes are active in the foetal E cells, producing foetal haemoglobin (HbF; $\alpha_2\gamma_2$). The two γ globin genes are not identical, though their protein sequences differ by only one amino acid (glycine in $^G\gamma$, alanine in $^A\gamma$). HbF has a higher oxygen affinity than HbA, hence oxygen can be transferred from the maternal circulation (HbA) to the foetal blood system (HbF) via the placenta. There is also an α-related θ globin gene which is expressed at low levels in foetal but not adult E cells, both in humans and in other primates (Leung *et al.*, 1987; Hsu *et al.*, 1988). In human embryos, two further globin genes are expressed in the primitive E cells originating from the yolk sac (the earliest site of erythropoiesis in higher vertebrates); these are the α-type zeta (ζ) and β-type epsilon (ε) genes. During early human development (5–8 weeks), there is an asynchronous switch from ζ to α production and later from ε to γ (plus some β) production (Peschle *et al.*, 1985). These switches occur in both primitive (yolk-sac-derived) and definitive (liver-derived) erythroid lineages, although ζ expression is largely confined to the primitive lineage and low-level β production to the definitive lineage. This results in a series of embryonic haemoglobins, namely Hb Gower ($\zeta_2\varepsilon_2$), Hb Gower 2 ($\alpha_2\varepsilon_2$) and Hb Portland ($\zeta_2\gamma_2$). The ζ and ε genes are shut down in adult and foetal erythroid cells; likewise, the γ genes are normally shut down in most adult erythroblasts (see § 3.2.4). These changes are summarised in table 3.1.

Similar changes in globin type and erythropoietic site are found in other mammals, and to some extent in other vertebrates. Thus chickens express embryonic types of haemoglobin only in the primitive E cells derived from the extraembryonic blastoderm (blood islands) at early stages of development, but switch to adult types in the definitive E cells produced in later embryos. Similarly, amphibia have distinct tadpole and adult haemoglobin types adapted to the different life styles at these two stages. Mice have no distinct foetal haemoglobin, but rather switch directly from embryonic to adult types during early foetal development. Notably, transgenic mice carrying adult β-globin fusion genes (5′ mouse fused to 3′ human sequences, both flanking and coding) fail to express these constructs in the primitive embryonic E cells but do express them actively in definitive foetal and adult E cells (Magram *et al.*, 1985). A complete human β-globin gene plus flanking sequences shows correct human-type regulation in transgenic mice, i.e. it is inactive in primitive embryonic E cells of yolk-sac origin, slightly active in foetal liver-derived definitive E cells, but significantly

Table 3.1. *Haemoglobin gene switching in man (based on Weatherall & Clegg. 1979)*

Erythropoietic tissue	Stage	β-type genes active	α-type genes active	Globin products
Yolk sac	Embryo	ε changing to $^A\gamma/^G\gamma$	ζ changing to α	Hb Gower 1 $\zeta_2\varepsilon_2$ Hb Gower 2 $\alpha_2\varepsilon_2$ Hb Portland $\zeta_2{}^{A/G}\gamma_2$
Liver	Foetus	$^A\gamma/^G\gamma$	α	HbF $\alpha_2{}^{A/G}\gamma_2$
Bone marrow	Adult	δ and β	α	HbA $\alpha_2\beta_2$ HbA$_2$ $\alpha_2\delta_2$

more active in adult bone-marrow-derived E cells (Behringer *et al.*, 1987; see also Townes *et al.*, 1985).

Simplistically, it might be thought that switches in haemoglobin type result from replacement of one erythroid cell population by another, i.e. yolk-sac-derived primitive E cells would express the embryonic globins, liver-derived definitive E cells the foetal globins, and bone-marrow-derived E cells the adult globins. However, haemoglobin switching can take place within single cells derived from the previous erythropoietic precursor population. This apparently occurs under the influence of circulating 'humoral' factors in the bloodstream, and/or as a result of some internal clock mechanism. For example, human HbA and HbF can be detected within the same red cell at around the time of birth. Indeed, a small proportion of adult BFU-Es give rise to F cells containing some HbF as well as HbA, while most give A cells containing only HbA (see Nienhuis & Stamatoyannopoulos, 1978). Zanjani *et al.* (1979) transplanted liver-derived erythroid precursors from foetal sheep homozygous for the β^B variant, into lethally irradiated adults homozygous for β^A (an alternative allele of the same β gene). Haemoglobin containing β^B chains was found to be expressed in increasing amounts over a 45-day period following transplantation, while production of HbF (containing foetal γ chains) remained low. Since the stem cells transplanted were of foetal liver origin, they would normally give rise to erythrocytes expressing mainly HbF. But in the adult recipient environment, γ chain production is suppressed and β^B expression (donor marker) is activated in the progeny of the transplanted liver cells. These findings could be explained either by an internal clock or by humoral factors circulating in the bloodstream.

Similar evidence is available for the erythroid precursor cells of the embryonic yolk sac. Mouse cells of this type produce only primitive E cells expressing embryonic haemoglobins when cultured in isolation. However, when cultured in combination with hepatic tissue from late mouse embryos (post-28-somite stage), these same yolk-sac precursors give rise to definitive E cells expressing the adult haemoglobins (Cudennec *et al.*, 1981). Chapman & Tobin (1979) used double fluorescence-labelling to show that both early and late haemoglobins are present simultaneously in single chick erythrocytes at around the time of changeover (day 6 of embryonic development). Both primitive (yolk-sac-derived) and definitive (liver-derived) BFU-E precursors from 5–8 week human embryos give rise in culture to progeny expressing all three classes of globin (ϵ, γ and β), suggesting a time-dependent change in the pattern of globin production rather than a change of stem-cell population (Stamatoyannopoulos *et al.*, 1987; see also Peschle *et al.*, 1985). Cell hybrids between human foetal erythroblasts and MEL cells switch from γ to β production, again suggesting a developmental clock mechanism (this switch occurs faster when later erythroid precursors are used: Papayannopoulou *et al.*, 1986). To summarise, an internal clock probably provides the major determinant of which globin genes are expressed in E cells, rather than the tissue source (yolk-sac/liver/bone-marrow) or lineage (primitive/definitive) of the stem-cell population. Indeed, it is likely that a single population of stem cells migrates from one erythropoietic site to another during development.

Haemoglobin switching is regulated primarily at the transcriptional level. Groudine *et al.* (1981) showed that embryonic chick erythroid nuclei synthesise no detectable adult globin transcripts; similarly, no embryonic globin sequences could be detected in the nuclear RNA of adult chick erythroid cells – even under conditions which should block transcript processing and degradation.

The chromatin conformation of non-expressed globin genes has been probed extensively using DNaseI. In chickens the non-expressed embryonic β-globin genes are DNaseI-resistant in adult E cells. By contrast, both adult and embryonic types of β gene are DNaseI-sensitive in embryonic E cells; here the adult β gene is presumably in some kind of 'preactivation' state; since it is not yet expressed (Stalder *et al.*, 1980). Weintraub *et al.* (1981) analysed the switching of α-type genes in chick, from U gene expression in primitive E cells to α_D and α_A expression in definitive E cells. During this transition, the U gene becomes DNaseI-resistant and heavily methylated, as well as losing a DNaseI-hypersensitive site near its 5' end. In the definitive E cells, a DNaseI-sensitive undermethylated region extends from the 5' end of the α_D gene, through a 1.5 kbp spacer and the whole α_A gene, to end 1.5 kbp beyond the 3' end of the latter gene. Sharp boundaries mark both ends of this chromatin region. In humans, Mavilio *et al.* (1983) have shown that undermethylated 5'-flanking sequences occur only at active gene sites among the β-type globin genes in embryonic (ε), foetal (γ) and adult (β) E cells. Similarly, DNaseI-hypersensitive sites are found within 200 bp upstream from the cap sites of the $^G\gamma$, $^A\gamma$, δ and β genes in human foetal E cells: by contrast, in adult E cells such sites are found only near the δ and β genes (Groudine *et al.*, 1983). Note that these findings apply to the majority and not necessarily to all of the E cells in a given population. Thus the minority F cells might retain DNaseI-hypersensitive sites near the active $^{G/A}\gamma$ genes, but this feature would be swamped by the greater number of A cells where these genes are shut down. Four erythroid-specific DNaseI-hypersensitive sites are located 5 to 18 kbp upstream of the ε gene (the most 5' member of the β-type gene cluster; see below and fig. 3.5). These sites are present in all E cells regardless of which β-type gene is active; they act cooperatively to boost transcription from the active globin gene(s) (Talbot *et al.*, 1989; Forrester *et al.*, 1989), and are collectively termed the locus activation regin (LAR). Presumably, successive members of the β-type gene cluster (fig. 3.5) must escape from the influence of this LAR at particular stages in development, first the embryonic ε gene and later the foetal γ genes. Such an effect, invoving the same 5' regulatory region influencing succesively more distal members of a gene cluster, could operate via DNA looping (Ptashne, 1988).

Groudine & Weintraub (1981) examined the chromatin conformation of α-type and β-type globin genes in those regions of the 20–30 hour chick blastoderm which will later (at around 35 hours) initiate primitive erythropoiesis (i.e. the future blood-islands). They showed that even the primitive haemoglobin genes (such as U) are still DNaseI-resistant and heavily methylated, showing no signs of transcriptional activity at this early stage. Thus a change in the chromatin conformation of these genes must occur between 23 and 35 hours of development in those cells destined to form primitive erythroid tissue. Although active globin genes are generally undermethylated in their 5'-flanking regions, it is not clear whether this is a cause, a correlate or a consequence of their transcription. There is some support for the first of these possibilities in the globin system, though elsewhere the balance of evidence favours either of the last two. Busslinger *et al.* (1983)

have shown that demethylated 5'-flanking sequences are permissive for transcription of the human γ-globin gene in heterologous mouse cells, whereas extensive methylation of these sequences prevents transcription. Recent evidence implies that DNA methylation of inactive γ-globin genes in adult E cells may be a secondary event in globin switching (Enver *et al.*, 1988).

3.2.3 Expression and control of globin genes

In erythroid cells a 15S precursor RNA (1500–1900 bases long) is transcribed from the entire split gene coding for β globin. This precursor is processed within the nucleus by removing both intron transcripts to generate mature 9S β-globin messenger RNA (Tilghman *et al.*, 1978a, b). All β-type globin genes possess one long and one short intron located at homologous positions within the gene, and their transcripts are presumably processed in a similar way. The α genes also have two introns at homologous sites, but both are short. As a result, the α primary transcripts are only 11S (about 880 bases long) and are again spliced to yield mature 9S α-globin mRNA for export (see Curtis *et al.*, 1977). Little is known about processing for ζ gene transcripts, though both introns in the human ζ gene are long (Proudfoot *et al.*, 1982).

Although translational controls have an important role in balancing the final quantities of globins and other cell products (see below), the pattern of globin gene expression is controlled primarily at the level of transcription. This applies not only to globin gene switching (§ 3.2.2), but also to the onset of globin synthesis during differentiation. In chick embryos, globin proteins can be detected in low amounts as soon as the corresponding mRNAs begin to accumulate during the proerythroblast stage (Hyer & Chan, 1978). Previous claims that globin messenger appearance might precede the onset of translation (implying storage of the mRNA) are explicable in terms of the lower sensitivity of protein as compared to nucleic-acid detection methods in earlier studies.

Several feedback control mechanisms have been identified in red blood cells, some of which act as links between globin production and the haem biosynthetic pathway. For example, haem is found to inhibit the expression of δ-aminolaevulinate (DAL) synthetase, which is the key enzyme that initiates haem biosynthesis. This is a case of end-product repression; excess haem product will halt further haem synthesis until the free product molecules have been sequestered (as haemoglobin) through binding to new globin chains. Conversely, erythropoietin increases DAL synthetase activity and so stimulates haem production.

This occurs concomitantly with the terminal divisions and differentiation promoted by erythropoietin among CFU-E cells (see § 3.2.1 above). The haem precursor haemin specifically enhances α globin translation (Giglioni *et al.*, 1973), so bringing α and β production into the correct 1:1 balance. These are simple examples from an extensive network of metabolic controls which together ensure that differentiating red blood cells function efficiently and respond appropriately to changes in their environment.

Herpes thymidine kinase (TK) and human β-type globin genes have been introduced by cotransformation into mouse erythroleukaemia (MEL) cells lacking their own TK activity (TK⁻). In stable TK⁺ transformant lines (carrying both introduced genes), human β globin mRNA expression can be induced chemically along with mouse globin mRNAs (from the host cell genes). Human γ and ε globin mRNAs are *not* inducible, and are only transcribed from a viral promoter adjacent to the TK gene. Thus correctly regulated expression of the introduced human β globin gene can be obtained in MEL cells; this applies both to fragments of human DNA containing the β globin gene plus 1.5 kbp of 5'-flanking sequence, and to larger fragments containing the entire β-type gene cluster (see fig. 3.5; Wright *et al.*, 1983). This contrasts with the situation in the K562 cell line, where only ε and γ (but not β) globin genes can be expressed. Chimaeric genes were created by fusing the 5' half of the human ε-, γ- or β-globin gene onto the 3' half of the mouse β-globin gene (both flanking and coding sequences were included in each part). Following the introduction of these constructs into K562 cells, their expression can be induced

A β globin gene cluster (based on Fritsch *et al.* 1980)

B α globin gene cluster (based on Lauer *et al*, 1980, Proudfoot *et al*, 1982, Goodbourn *et al.* 1983 and Leung *et al*, 1987).

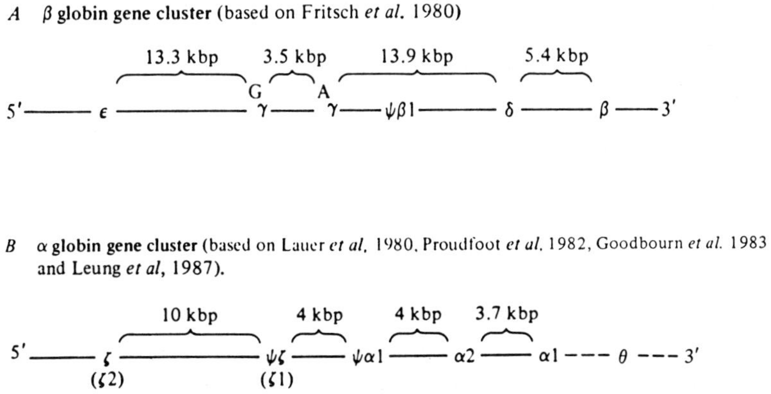

Fig. 3.5 Human globin gene organisation.

through the 5' sequences of the ε or γ gene, whereas the β promoter is non-inducible (Kioussis *et al.*, 1985). This suggests that *trans*-acting factors required to activate ε and γ genes are produced in K562 cells, whereas factors acting on the β gene are produced in the MEL cells studied by Wright *et al.* (1983). These factors presumably interact with promoter and/or enhancer elements in the appropriate target globin genes. Using transient expression assays and various promoter deletions, a series of essential 5' promoter elements has been identified in several globin genes (see e.g. Grosveld *et al.*, 1981; Dierks *et al.*, 1983). For example, the human β-globin promoter contains three key regions: (i) the ATATAA box (-31 to -26); (ii) the CCAATC box (-77 to -72); and (iii) the sequence GCCACACCC (-95 to -87; Cowie & Myers, 1988). The first two of these are common control elements shared by many (CCAAT box) or most (TATA box) genes transcribed by pol II (see chapter 1).

Although human β-globin gene expression is inducible in cotransformed MEL cells (see earlier), this is not true for introduced human α-globin genes; the sequences responsible for this difference lie within the structural genes themselves (Charnay *et al.*, 1984). Similarly, Wright *et al.* (1984) showed that β-globin expression is regulated via sequences both 5' and 3' to the translational start site. More recent work has identified two cooperatively acting tissue-specific enhancer elements, one lying within and one 3' to the human β-globin gene, as well as a 5' promoter element located near position -160 (Antoniou *et al.*, 1988). The human γ-globin gene is similarly regulated via upstream sequences (a distal CACCC element and duplicated CCAAT boxes; Anagnou *et al.*, 1986) as well as a 3' enhancer (Bodine & Key, 1987). Note that DNaseI-hypersensitive sites in the LAR 5' to the ε gene confer high-level erythroid-specific expression on all genes in the β-type cluster (ε, ^{A}γ, ^{G}γ, δ and β; see fig. 3.5; Talbot *et al.*, 1989; van Assendelft *et al.*, 1989; Forrester *et al.*, 1989), but do so at different stages.

In chickens, the erythroid-specific H5 histone gene is regulated through a 3' enhancer (Trainor *et al.*, 1987), and there is another such enhancer located between the adult β^A and embryonic ε globin genes (see § 3.2.4 below) – lying 5' to the latter but 3' to the former (Choi & Engel, 1986, 1988). This enhancer stimulates transcription from both globin genes in a tissue-specific fashion, but its activity is modulated through a stage-specific selector element in the β^A globin promoter. In this way, the β^A-globin gene is strongly activated in definitive erythroid cells, whereas the ε gene is active in primitive erythroid cells (Choi & Engel, 1988). A chicken erythroid-specific protein (Eryf1) binds to the

3' enhancer of the β^A-globin gene, and also to binding sites in the promoter and/or enhancer sequences of all other chicken globin genes (both α- and β-type, embryonic and adult; Evans *et al.*, 1988). The β^A 3' enhancer includes two positively acting domains, one containing an inverted repeat that binds two molecules of Eryf1 (Reitman & Felsenfeld, 1988). A consensus Eryf1-binding sequence is present in all of the chicken globin-gene enhancers/promoters, namely ^T_AGATA^A_G; this same motif is also found in the erythroid-specific 3' enhancer of the H5 histone gene, and also in the 3' enhancer of the human β-globin gene. There are at least four protein-binding sites in the human β-globin 3' enhancer, and these interact with several proteins including an erythroid-specific nuclear factor known as NF-E1 or GH-1 (Wall *et al.*, 1988). NF-E1/GH-1 also binds to sequences in the human β-globin promoter (de Boer *et al.*, 1988). Recent cDNA cloning and sequencing suggests a novel finger-protein structure (see § § 3.3.1 and 5.6) for both GH-1/NF-E1 (Tsai *et al.*, 1989) and Eryf1 (Evans & Felsenfeld, 1989). Taken together, these results suggest the following generalisations: (i) The tissue specificity of transcription for all globin genes is conferred by a *trans*-acting erythroid-specific protein (Eryf1, GII-1/NF-E1), which binds to sequence motifs present in the enhancers and/or promoters of these and other erythroid-specific genes such as H5. (ii) Stage specificity – governing whether embryonic (or foetal) or adult globin gene-sets are active – is regulated via 5' promoter elements which presumably interact with a different or overlapping set of *trans*-acting factors; at least some of these must be developmentally regulated (present only at the appropriate stage).

3.2.4 Organisation of globin genes

Somatic cell fusion has proved a useful technique for analysing the location of globin genes. In brief, when human and mouse cells are fused together in culture, the human chromosomes tend to be eliminated selectively, so that only one or a few of them are retained in particular sublines of hybrid cells. When such cells are analysed for the presence of the human α globin gene, positive hybridisation is found only in those sublines where human chromosome 16 is retained (Diesseroth *et al.*, 1977). Similarly, only hybrid cells retaining human chromosome 11 carry the human β globin gene (Diesseroth *et al.*, 1978). Thus the human α locus lies on chromosome 16, and is unlinked to the β locus which lies on chromosome 11. Both α and β genes are linked to other related globin genes; thus a 40 kb region of chromosome 11 includes one pseudogene together with the active ε, ^{G}γ, ^{A}γ, δ and β genes, while a 27 kb region of chromosome 16 includes both α genes plus one ζ gene and two pseudogenes. A second β-type pseudogene, thought to lie on the 5′ side of the ε gene (Fritsch *et al.*, 1980) now appears to be an artifact (Shen & Smithies, 1982). More recently, the θ-globin gene has been mapped to the 3′ end of the α-globin gene cluster in primates (Leung *et al.*, 1987). The mapping of these human globin genes has involved cloning long stretches of the chromosomal DNA containing them, followed by restriction and sequence analysis. The β globin gene cluster has been studied intensively in many laboratories, and the complete nucleotide sequences of all its active genes are now known (see Lawn *et al.*, 1980; Spritz *et al.*, 1980; Baralle *et al.*, 1980; Slightom *et al.*, 1980; Efstratiadis *et al.*, 1980); an overall map of the β gene region is given in fig. 3.5*A*, based on Fritsch *et al.* (1980). Similar information is available for the human α gene cluster (Lauer *et al.*, 1980; Proudfoot *et al.*, 1982), and is summarised in fig. 3.5*B*.

The sequences and organisation of globin genes have been compared extensively between vertebrate species (see review by Jeffreys, 1982). The arrangement of the β-type genes in the order of their expression during development holds true for other mammals as well as for humans, although the number of genes and the intergenic distances vary considerably. In chickens, however, the two embryonic β globin genes (ε and ρ) are found flanking the two adult β globin genes in the order 5′-ρ-β^{H}-β^{A}-ε-3′ (Dolan *et al.*, 1981). Thus the chick β-type globin genes are not arranged on the chromosome in the order of their expression during development. In *Xenopus* the α- and β-type genes are closely linked together, rather than segregated into separate clusters as in higher vertebrates (including chick).

Despite these differences in arrangement, the globin genes show several constant features. All functional globin genes contain two introns located at homologous positions within the coding sequence. In the case of the β-type genes, a short ~ 120 bp intron occurs close to the 5′ end of the gene, while a longer intron (500–1000 bp in different mammals) occurs towards the 3′ end. In the α globin genes, both introns are short, while in the human ζ gene both introns are long and partially composed of simple repetitive sequences (Proudfoot *et al.*, 1982). Despite these similarities in size and position, the actual sequences of the globin gene introns have diverged rapidly between species (van den Berg *et al.*, 1978). On the basis of their overall sequence relationships one can derive a hypothetical evolutionary tree showing how the present array of human globin genes could have arisen through successive duplication events from a single original globin gene (fig. 3.6) assuming a constant rate of mutational change and divergence. This is, however, an oversimplification, since mechanisms also exist for conserving the sequences of duplicated genes without significant divergence, as in the case of the two α gene copies. Thus the duplication times in fig. 3.6 are at best only provisional.

Several pseudogenes have been identified in human and other vertebrate globin gene clusters (see Little, 1982). They are presumably evolutionary relics, i.e. genes that were once functional but have now fallen into disuse. Several globin pseudogenes have been sequenced; they are mostly found to retain the basic organisation of active globin genes (in terms of intron size, position, etc.) but are mutated in a variety of ways so as to preclude their expression. Examples of this include alterations in the TATA box sequence, nonsense codons occurring early in the 'coding' region, and abnormal intron boundaries which would prevent transcript splicing. One β globin pseudogene (ψβ1) occurs in the human β-type gene cluster, and a rabbit pseudogene (ψβ2) related to the human δ gene has been partially sequenced (Lacy & Maniatis, 1980). In the human α globin gene cluster there is one α-type pseudogene (ψα1; Proudfoot & Maniatis, 1980) and one ζ-type pseudogene (here designated ψζ; Proudfoot *et al.*, 1982); interestingly, the ψζ gene is almost identical in sequence to the active ζ gene, but codon 6 is mutated to a stop codon in ψζ (see Proudfoot *et al.*, 1982).

One exception to this general rule is the mouse ψα3 pseudogene, which represents an α-type gene sequence from which both introns are completely absent (Nishioka *et al.*, 1980; Vanin *et al.*, 1980). Since these

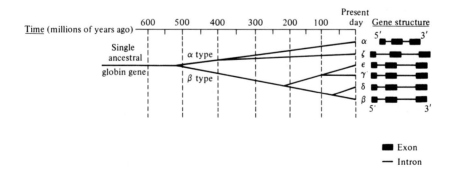

Fig. 3.6 Evolutionary tree for human globin genes (based on Alberts *et al.*, 1983, and Proudfoot *et al.*, 1982).

intron sequences have been deleted precisely, there is an obvious analogy with transcript splicing. This has led to speculation that such 'processed genes' might originate from double-stranded cDNA copies of spliced RNAs, which have become reinserted into the genome in the germ line (Flavell, 1982; Little, 1982). Other examples of 'processed pseudogenes' include members of the tubulin, immunoglobulin and dihydrofolate reductase gene families. In the case of a dispersed human immunoglobulin pseudogene, the DNA sequence ends with a poly(dA) tract and includes spliced J and C regions (§ 2.2.2), both characteristic of processed immunoglobulin messenger RNAs (Hollis *et al.*, 1982).

Nor are 'processed' genes necessarily inactive pseudogenes. In chick, one of the two active calmodulin genes contains introns whereas the other does not (Stein *et al.*, 1983); the latter, though possessing the characteristics of a processed gene, is nevertheless expressed in a tissue-specific manner in muscle cells. However, an alternative evolutionary scenario has been suggested for this particular gene (Gruskin *et al.*, 1987). Another example of a functional 'processed gene' occurs among the human genes encoding 3-phospho-D-glycerate kinase (PGK). The widely expressed PGK1 gene is X-linked and contains 10 introns, whereas the autosomal PGK2 gene contains none and is expressed only in testis. In its lack of introns, the functional PGK2 gene resembles a ψPGK1 pseudogene on the X chromosome, whose own expression is prevented by premature stop codons in all reading frames (McCarrey & Thomas, 1987).

The mouse genome is unusual in that its two α-globin pseudogenes ($\psi\alpha4$ with and $\psi\alpha3$ without introns) are dispersed onto two chromosomes different from that carrying the active cluster of one embryonic and two adult α-globin genes (Leder *et al.*, 1981). If processed pseudogenes are indeed derived from spliced RNA intermediates, then dispersion to different chromosomal locations would be expected; but this does not explain the location of the intron-containing mouse $\psi\alpha4$ pseudogene on a different chromosome.

A variety of blood disorders affecting human globin genes and their expression have been analysed at the molecular level (see review by Weatherall & Clegg, 1979). In homozygous α-thalassaemia I, both α-globin genes are deleted, hence only γ chains are available in the foetus; this is normally a lethal condition (*hydrops fetalis*). In $\delta\beta$-thalassaemia, both the δ and β genes are similarly deleted, again leading to severe anaemia. However, a more extensive deletion involving these same two genes is paradoxically much less severe in its effect; here, the absence of HbA/A$_2$ production can be compensated by a prolongation of HbF synthesis into adult life. Thus the $^G\gamma$ and $^A\gamma$ genes remain active in bone-marrow-derived erythroid cells, a condition known as hereditary persistence of foetal haemoglobin (HPFH). In other cases, HPFH can result from point mutations in the regulatory regions of either γ gene, without deletion of the β or δ loci; in such cases, normal HbA can also be produced, but a significant proportion ($\geqslant 20\%$) of the adult haemoglobin consists of HbF. A point mutation at position -202 relative to the $^G\gamma$ gene results in persistent $^G\gamma$ expression in adult erythroid cells, though the nearby $^A\gamma$ gene is shut down normally after the foetal stage (Collins *et al.*, 1984). Point mutations leading to persistent $^A\gamma$ expression in adults have been mapped to -117 (Greek type) and -196 (Chinese type) relative to the $^A\gamma$ gene. The -117 mutation occurs in the distal CCAAT box (Gelinas *et al.*, 1985); this apparently increases the strength of the $^A\gamma$ promoter but decreases its tissue-specificity, as determined by transient expression assays in a non-erythroid cell type (Rixon & Gelinas, 1988). The -117 mutation affects the interactions of several nuclear factors with the CCAAT box, and specifically decreases the binding of an erythroid-specific protein (NF-E) to sequences flanking the distal CCAAT box (Superti-Furga *et al.*, 1988). This may imply that NF-E normally represses γ globin expression in adult erythroid cells.

In β-thalassaemias, the adult erythrocytes contain little or no β-globin protein. Usually, the amount of β mRNA is reduced or zero, though the β gene itself may remain largely intact, with no apparent deletions. In some of these cases, the defect has been traced to point mutations which alter the splicing pattern of the β globin primary transcripts (see Treisman *et al.*, 1982; Orkin *et al.*, 1982; Mount & Steitz, 1983). Promoter or nonsense mutations could also give the same end result.

The δ and β genes are fused together in Hb Lepore, due to deletion of the sequences that normally separate these genes (see fig. 3.5A). The result is a globin protein comprising part of a δ chain (lacking its C-terminus) joined to part of a β chain (lacking its N-terminus). Thus the 5' portion of the δ gene has become fused to the 3' portion of the β gene. In Hb Kenya, a much larger deletion results in fusion between the 5' portion of the $^A\gamma$ gene and the 3' portion of the β gene. These fusion mutants must have arisen originally by unequal crossing-over between sites within related globin genes (δ and β or $^A\gamma$ and β); for details see Weatherall & Clegg (1979).

Many of these blood disorders could potentially be cured at a somatic level by gene therapy, i.e. by inserting wild-type versions of the mutant gene into the patient's own haematopoietic stem cells and then recolonising his/her bone marrow with the genetically corrected cells. The technical problems to be overcome in this approach are formidable; it would be necessary to select only those cells which show an appropriate number of gene insertions at the correct chromosomal sites, and which regulate these inserts normally in the erythroid lineage. Some progress towards this goal has been achieved, notably by inserting β-globin genes into the chromosomal β-globin locus by homologous recombination (Smithies *et al.*, 1985). Such recombinants can be scored and cloned, and the inserted genes (marked) are often correctly expressed and regulated during induced erythroid differentiation in MEL–human hybrid cells (Nandi *et al.*, 1988). An alternative route uses retroviral vectors to transduce human β-globin genes into mouse bone-marrow cells, which are then transplanted into lethally irradiated mice; significant levels of human β globin are expressed only in the erythroid cells of the survivors (see Dzierzak *et al.*, 1988).

If both globin gene clusters are taken as a whole, it is apparent that only 8% of the available DNA sequences code directly for protein, while a similar percentage is represented by the introns (Jeffreys, 1982). The function of the rest (84%!) remains obscure, although the presence of interspersed repetitive elements has been noted (see Fritsch *et al.*, 1980). It is also possible that the globin gene regulatory sites may include sequences located at considerable distances from the genes whose expression they control, although most such elements identified to date are located within a few hundred bp 5' or 3' to the gene (apart from the LAR). In microcosm, the globin genes highlight a problem that recurs throughout eucaryotic molecular biology – namely, why so much DNA? Many have argued that the apparent excess of DNA (i.e. non-coding and non-regulatory) is either functionless 'evolutionary junk' or else serves merely to space out the genes which it frames.

3.3 Hormonal control of gene expression

3.3.1 General features of hormonal regulation

Hormones can both induce differentiation and change the pattern of gene expression in already differentiated cells. Their characteristic is long-range action, as opposed to the short-range effects of inducing agents (see § 2.8). Polypeptide hormones binding to receptors on the target cell surface usually act via intracellular second messenger systems (e.g. by activating adenyl cyclase and so raising the concentration of cyclic AMP); these in turn act on many metabolic processes including protein phosphorylation within the target cell, with possible consquences for altered gene expression. For example, somatostatin gene expression is stimulated in neuroendocrine cells in response to agents (e.g. various neurotransmitters and neuromodulators) which act through cell-surface receptors to raise the intracellular cAMP concentration. A cAMP-dependent protein kinase responds to increased cAMP by phosphorylating (*inter alia*) a 43 kd nuclear protein, and this phosphorylated form binds specifically to a short cAMP-response element (CRE) located 5′ to the somatostatin gene, thereby activating its transcription (Montminy & Bilezikjian, 1987) The CRE is related in sequence to the AP-1 binding site, except that the former possesses an central C residue not found in the latter (§1.3; Karin, 1989).

A less roundabout route to the same end is taken by many steroid hormones. Their mechanism of action has been worked out in several systems, and the essential steps involved are as follows (fig. 3.7):

(i) Hormone enters the cell; this probably does not require a specific uptake system, since steroids are hydrophobic molecules which pass readily through the cell membrane.

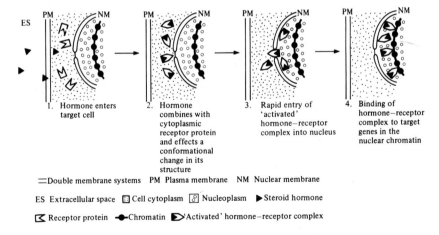

Fig. 3.7 Essential steps in steroid hormone action. Note that unoccupied oestrogen receptor proteins are found in the nuclei and not in the cytoplasm of target cells. This would eliminate stages 2 and 3, allowing the steroid to bind directly to receptor proteins within the nucleus. However, unoccupied glucocorticoid receptors are cytoplasmic in location, and all steps in the above sequence are necessary.

(ii) The hormone binds to a specific *receptor* protein present only in the cytoplasm of target cells. This causes a conformational change (activation) in the receptor protein molecule.

(iii) The activated hormone–receptor complex rapidly enters the nucleus.

(iv) Specific DNA sequences in the 5'-flanking regions of hormone-sensitive genes are recognized by the hormone–receptor complex, either activating or blocking transcription.

There is evidence to suggest that unoccupied oestrogen receptors are located in the nuclei of target cells, i.e. their apparent presence in the cytoplasm prior to hormone binding may be an artefact (see Welshons *et al.*, 1984; King & Greene, 1984). This would simplify the above scheme by combining stages (ii) and (iii). However, unoccupied glucocorticoid receptors are clearly cytoplasmic in location. Hormone binding is required before the glucocorticoid hormone–receptor complex can translocate into the nucleus, a process mediated via two nuclear localisation signals in the receptor protein. One of these (NL2) maps to its hormone-binding domain and the other to its DNA-binding domain (Picard & Yamamoto, 1987). Fusion proteins containing the hormone-binding site and its associated NL2 signal are cytoplasmic in location when hormone is absent, but rapidly move into the nucleus when glucocorticoid hormone is added. The DNA-binding domains of hormone receptor proteins contain clusters of cysteines which probably bind coordinately to a zinc atom, cross-linking a loop of amino acids that can bind to the DNA (Berg, 1989). Somewhat similar 'zinc-finger' loops are found in many DNA-binding proteins, e.g. the *Xenopus* transcription factor TFIIIA (see § 2.4.2b) and *Drosophila* gap-gene poducts (see § 5.4.2 and 5.6; Evans & Hollenberg, 1988). Notably, the N-terminal zinc finger in both the oestrogen and glucocorticoid receptors confers much of their target-gene specificity (Green *et al.*, 1988). The genes encoding receptors for different ligands are all members of a sequence superfamily related by certain common features (including the zinc-finger motif). This family includes genes for the oestrogen, progesterone (Gronemeyer *et al.*, 1987), thyroid hormone (Weinberger *et al.*, 1986a) and retinoic acid (Petkovitch *et al.*, 1987; Giguère *et al.*, 1987) receptors in higher vertebrates, as well as relatives in *Drosophila* (see § § 5.4.2 and 5.6) and other invertebrates. Functional analysis shows that separate protein domains of the progesterone receptor are responsible for DNA binding and for hormone binding (Gronemeyer *et al.*, 1987). The genomic progesterone receptor gene has now been isolated from chicken (Huckaby *et al.*, 1987), and contains eight exons spanning 38 kbp of DNA; notably its promoter lacks TATA or CAAT boxes but shares certain sequence features (e.g. CCGCCC motifs) with the promoters of several housekeeping genes. Similarly, the human oestrogen receptor gene contains 8 exons separated by introns at homologous positions, but spans >140 kbp of DNA (Ponglikitmongkol *et al.*, 1988).

The binding of hormone–receptor complexes to their target genes is mediated through specific DNA sequences termed hormone response elements (HREs). Such HREs usually occur 5′ to the gene whose expression they control, and their locations can be determined by deletion experiments. Typically, 5′-flanking sequences from the gene of interest are linked to a foreign reporter gene and the fusion constructs introduced into cultured cells known to contain appropriate hormone receptors. By using various deletions of the 5′-flanking sequences, one can determine which DNA region is necessary in order to confer hormone-inducible expression on the reporter gene. HREs are properly classed as *cis*-acting enhancer elements, since they can confer hormone inducibility when linked to heterologous promoters (see reviews by Yamamoto, 1985; Berg, 1989).

As well as activating selected target genes in this way, steroid hormones may also repress other target genes. One example of this is provided by the bovine prolactin gene, which is negatively regulated (repressed) by glucocorticoid hormones. There are multiple binding sites for the glucocorticoid–receptor complex within the negative glucocorticoid response element (nGRE) which lies between −51 and −561 relative to the cap site of the prolactin gene (Sakai *et al.*, 1988). These nGRE binding sites are apparently quite different in sequence from the positive GREs identified near many glucocorticoid-inducible genes. The prolactin gene is active in the absence of glucocorticoids, presumably through the interaction of a positive transcription factor with some part of the nGRE sequence. When glucocorticoids are present, hormone–receptor complexes bind to the nGRE and negate the effect of this positive factor (Sakai *et al.*, 1988). The nGRE sequences can confer down-regulation by glucocorticoids when linked to heterologous genes, just as constructs with positive GREs show glucocorticoid inducibility.

The glucocorticoid and progesterone response elements (G/PREs) are identical in sequence (Strahle *et al.*, 1987); specific effects of these two hormones are due to differential expression of their respective receptors (Strahle *et al.*, 1989). The oestrogen response element (ERE) has a different consensus sequence (Klein-Hitpaß *et al.*, 1986), but only two symmetrical base changes are needed to interconvert between the G/PRE and ERE sequences (Klock *et al.*, 1987; Martinez *et al.*, 1987). In some cases, different types of hormone–receptor complex may compete for binding to the same or similar recognition elements. Thus the T3 thyroid receptor binds to a palindromic consensus sequence (the TRE) which is also found in the oestrogen response element (ERE). Binding of thyroid receptors (in an inactive form) to EREs depresses the response of these elements to oestrogen–receptor complexes, decreasing the level of oestrogen-inducible expression; this may reflect competition between the two types of hormone–receptor complex in binding to the same response elements (Glass *et al.*, 1988). Steroid receptors with different recognition elements (e.g. ERE versus PRE) may also interfere with each other. Thus cells expressing both oestrogen and progesterone receptors show diminished progesterone inducibility of reporter genes regulated via PRE sites (see Meyer *et al.*, 1989). This does not involve direct receptor competition for PRE sites, but rather both types of receptor compete for functionally limiting transcription factors which normally mediate in the hormonal induction of target genes. Such indirect interactions between different hormones and their receptors may underlie some of the complexities observed in many hormone-regulated systems (see § 3.3.2 below). Hormone–receptor complexes probably bind to their (often palindromic) recognition elements as dimers (Tsai *et al.*, 1988). Finally, hormone-receptor complexes may act in cooperation with other transcription factors. Upstream from the rat tryptophan oxygenase (TO) gene, a positive GRE occurs close to a CACCC box. In various gene constructs, these two elements can work in either order, but optimal hormone inducibility requires them to be 106 bp apart. This suggests that activation of the TO gene involves protein:protein interactions between the glucocorticoid–receptor complex and a factor binding to the CACCC box (Schule *et al.*, 1988).

In the remainder of this section, we will explore how female sex steroids regulate two groups of genes encoding major egg proteins, which provide important nutrient reserves needed to sustain early development in egg-laying vertebrates (see § 2.4). These are respectively the egg-white protein genes expressed in hen oviduct (§ 3.3.2), and the vitellogenin (yolk-protein precursor) genes expressed in avian and amphibian liver (§ 3.3.3).

3.3.2 Regulation of the egg white protein genes

That familiar foodstuff, the hen's egg, is a remarkable structure. Its two main constituents – the yolk and white – represent an enormous synthetic investment, in that gram quantities of their respective proteins are produced daily by a laying hen. For example, an average rate of egg production involves the synthesis of 3×10^{19} molecules per day of the major egg-white protein ovalbumin (Palmiter, 1975). As mentioned earlier (§ 2.4.1), yolk platelets are assembled in the growing oocytes, but their component proteins are synthesised as longer precursors in the liver under hormonal control (see § 3.3.3 below).

By contrast, the egg-white proteins are produced in the magnum portion of the hen oviduct, and are laid down as layers of albumen (thick, then thin) around the yolk during its passage down the oviduct. Shell membranes enclose each egg as it passes through the isthmus of the oviduct, and finally the shell (mainly calcium carbonate plus some brown porphyrin pigment) is secreted in the shell gland prior to laying. The approximate times involved for each stage are as follows: (i) 3 hours in the magnum, (ii) 1¼ hours in the isthmus, and (iii) 20–21 hours in the shell gland, pigment being added only during the last 5 hours (Sturkie & Mueller, 1976).

The immature oviduct of a young female chicken is a narrow tube weighing only milligrams, composed mainly of a layer of columnar epithelial cells surrounded by connective tissue (fig. 3.8*A*). In a mature laying hen, however, the oviduct is a massive structure weighing 50–60 grams, of which at least half is contributed by the magnum. Although the ovaries and oviducts develop initially as paired structures in birds, those on the right side degenerate before sexual maturity, so that only the left ovary and oviduct are functional in laying hens.

Basically, the mature magnum is a thick-walled tube, whose luminal surface is intricately convoluted into ramifying **tubular glands**. Groups of tubular gland cells synthesise vast quantities of the major egg-white proteins (principally ovalbumin, but also conalbumin, ovomucoid and lysozyme), which are assembled into granules (fig. 3.8*B*) for secretion into the lumen. Ovalbumin is the major storage material in the egg white, while conalbumin acts as an iron-carrier and lysozyme as a bactericidal agent. Although tubular gland cells predominate numerically in the magnum (by about 9 to 1) two other cell types are found in an alternating arrangement in the layer lining the lumen. These are: (i) the goblet cells, which secrete avidin (see below) and also mucopolysaccharides to lubricate the egg's passage; and (ii) ciliated epithelial cells, which assist in moving the secreted material

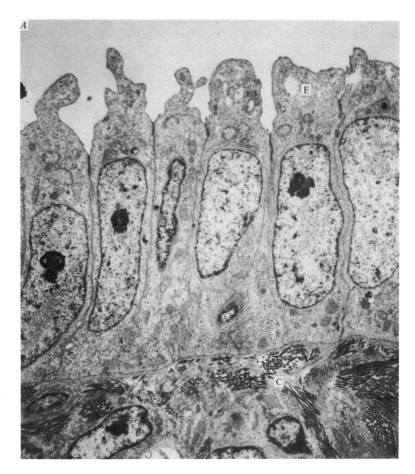

Fig. 3.8 Differentiation of chick oviduct. Electron micrographs showing immature (7 week; part *A*) and mature (50 week; parts *B* & *C*) hen oviduct tissue, by courtesy of Dr I. R. Duce (Department of Zoology, University of Nottingham).

A Immature oviduct. Magnification ×2655. Note single layer of columnar epithelial (E) cells (relatively undifferentiated), with underlying connective (C) tissue (containing collagen fibre bundles).

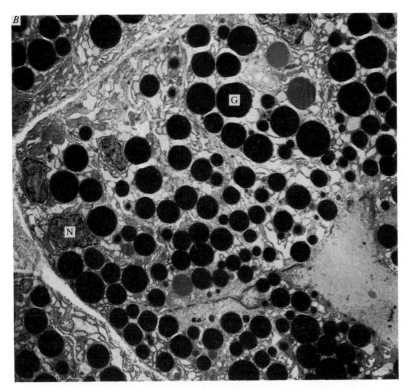

B Mature oviduct magnum – tubular gland cells. Magnification ×1240. Note the electron-dense (dark) granules (G) of egg-white proteins in the cytoplasm, and nucleus (N).

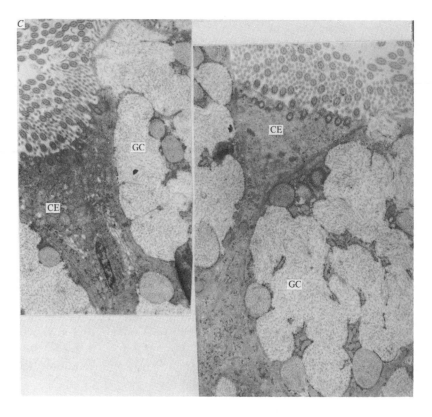

C Mature oviduct – luminal lining layer. Magnification ×1170. Note the regular alternation of ciliated epithelial (CE) cells with goblet (GC) cells packed full of mucopolysaccharides (speckled areas).

(fig. 3.8*C*). The outer wall of the oviduct is composed of connective tissue and smooth muscle, whose contractions are mainly responsible for moving the large egg down the oviduct.

It will be apparent from the contrasts in fig. 3.8 (*A* versus *B* and *C*) that extensive growth and differentiation are required in order to convert a simple immature oviduct into the complex adult organ. Most of this occurs during sexual maturation at around 100 days post-hatching. The whole process is under hormonal control, as is the subsequent synthesis of egg-white proteins. The major hormones involved are the female sex steroids, i.e. oestrogens (mainly 17β-oestradiol) and progestins (mainly progesterone). Both of these steroid classes are produced in the ovary, and are themselves subject to hormonal regulation from the pituitary. The interactions between oestrogen and progesterone are complex, partly reinforcing, partly modulating, and partly antagonising each other's effects in the oviduct (see table 3.3, later). For instance, oestrogen is prerequisite for the differentiation of goblet cells in the luminal lining of the magnum, but only when further stimulated by progesterone can these cells secrete avidin (a minor egg-white protein that binds biotin). Another important distinction is that oestrogen indirectly promotes mitosis among oviduct cells, whereas progesterone does not. The egg-white protein genes in differentiated oviduct cells can also respond to other steroids, namely androgens and glucocorticoids.

When an immature hen is first exposed to oestrogen (either naturally or after precocious injection of hormone), a complex **primary response** is elicited in the oviduct. This involves massive cell proliferation and tissue growth, accompanied by the differentiation of several distinct cell types (at least three in the magnum alone; fig. 3.8), and regionalisation of the oviduct into the infundibulum, magnum, isthmus and shell gland. Although egg-white protein synthesis is induced during this primary response, most workers have preferred to study the simpler secondary response. If oestrogen is withdrawn after the primary response, production of the egg-white proteins ceases, but the differentiated oviduct structure does not regress. When hormone is readministered, egg-white protein synthesis is rapidly induced without the necessity for preparatory differentiation. This **secondary response** is both faster and more extensive (i.e. higher production of egg-white proteins) than the primary response. Successive cycles of hormone administration and withdrawal give a reproducible pattern of induction and repression of egg-white protein expression. Eventually, however, the oviduct becomes post-reproductive and can no longer respond to further hormonal stimulation. In the account below, we shall concentrate specifically on the expression of egg-white proteins in the tubular gland cells of the oviduct magnum during secondary hormonal responses.

(a) *Control of egg-white protein synthesis*

During secondary oestrogen stimulation, ovalbumin production accounts for around 60% of total protein synthesis in oviduct tubular gland cells, while the production of conalbumin, ovomucoid and lysozyme reaches 10%, 8% and 2–3% respectively. In the resting oviduct during hormone withdrawal, these proteins are no longer synthesised, and their corresponding mRNAs are virtually undetectable (less than 0.0002% of the total cellular RNA). Twelve hours after hormone administration, ovalbumin mRNA represents 0.1% of the total cellular RNA, while ovomucoid and lysozyme mRNAs represent 0.0065% and 0.005% respectively (Schutz *et al.*, 1977). At this stage the concentrations of all three messengers are still rising. By contrast, the amount of conalbumin mRNA plateaus after 6 hours at around 0.02% of the total cellular RNA (approximate value based on Palmiter *et al.*, 1977). Since the vast bulk of cellular RNA is ribosomal, these values represent massive new messenger synthesis. In the case of ovalbumin mRNA, for instance, the resting level of 10–30 molecules per tubular gland cell rises to at least 10 000 per cell by 12 hours after hormone administration, and to 70 000 per cell during maximal stimulation (table 3.4 below). The major egg-white protein mRNAs increase in concentration by more than a thousand-fold in response to oestrogen.

Conalbumin mRNA is induced by oestrogen much more rapidly than the other egg-white protein messengers (see table 3.2, fig. 3.9 and discussion below). The time of first appearance for the corresponding proteins suggests that translation begins as soon as the mRNAs reach the cytoplasm; i.e. the kinetics of mRNA and polypeptide appearance are practically indistinguishable. Thus the temporal and (to some extent) quantitative expression of all four egg-white proteins is mainly controlled at the level of mRNA production. Moreover, nuclear precursors to these mRNAs are barely detectable during hormone withdrawal, suggesting that all four genes are regulated transcriptionally. Another important effect of the hormone is to stabilise the egg-white protein messenger RNAs specifically. Many other mRNA sequences are expressed in small amounts in hormone-stimulated oviduct cells and these continue to be expressed during hormone withdrawal. Thus secondary hormone administration elicits high-level expression of a

Table 3.2. *Control of egg-white protein expression (after Schutz et al.; 1977 and Palmiter et al., 1977).*

Protein	% of total cell protein synthesis in laying hen oviduct magnum	mRNA, as % of total cell RNA 12 hours after oestrogen administration	mRNA, as % of total cell RNA in resting oviduct	Time-lag between oestrogen injection and mRNA appearance (hours)
Ovalbumin	50–65%	0.1% rising	0.0002%	2–3
Conalbumin	10%	approx. 0.02%, plateau	approx. 0.0002%	0.5
Ovomucoid	8%	0.0065%, rising	0.0001%	2–3
Lysozyme	2–3%	0.005%, rising	0.0001%	2–3

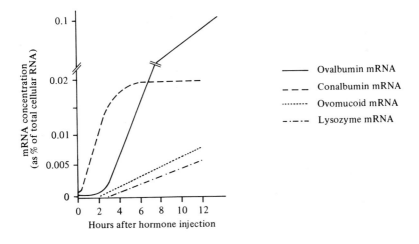

Fig. 3.9 Oestrogen induction of egg white protein mRNAs (after Schutz *et al.*, 1977 and Palmiter *et al.*, 1977).

limited set of egg-white protein genes, this effect being reinforced by a stabilisation of their messengers. These data are summarised in table 3.2 and fig. 3.9.

The 2–3 hour time-lag between oestrogen administration and messenger appearance for ovalbumin, ovomucoid and lysozyme mRNAs, contrasts sharply with the rapid induction of conalbumin mRNA which appears within 30 minutes of oestrogen injection. As mentioned earlier (§ 3.3.1), steroid hormones first bind to specific receptor proteins in the target cells, and the activated hormone–receptor complexes then alter the pattern of gene expression. Several types of steroid receptor protein are found in oviduct tubular gland cells; these include not only oestrogen and progestin receptors, but also glucocorticoid and androgen receptors. All of these hormones can induce egg-white protein synthesis as a secondary response in differentiated oviduct cells, though only oestrogens can induce the primary response. However, the pattern of gene expression is different for each of these hormones. Following administration of radioactively labelled steroid, maximal concentrations of the oestrogen–receptor complex are found inside the nuclei of oviduct cells within 30 minutes of hormone injection, implying that conalbumin gene transcription must be switched on almost immediately by this complex. By contrast, expression of the other egg-white protein genes is minimal during a lag period of some 2–3 hours, but thereafter increases linearly (see fig. 3.9). If progesterone is used in place of oestrogen, a distinctly different pattern of conalbumin induction is observed (see table 3.3); in particular, there is a pronounced 2–3 hour lag before its mRNA first appears. Indeed, if progesterone is administered during the course of oestrogen stimulation, then conalbumin mRNA accumulation is temporarily halted while ovalbumin mRNA synthesis continues (see Palmiter *et al.*, 1977). These complexities apparently reflect direct interactions between the hormones and their target cells, since identical results are obtained by treating oviduct explants (from oestrogen-stimulated birds) with one or both steroids *in vitro* (McKnight, 1978).

In view of the massive amounts of ovalbumin produced by the tubular gland cells in a laying hen, it is worth noting that the major ovalbumin gene is single-copy and is *not* amplified before or during hormone stimulation. Palmiter (1975) has reviewed the transcriptional and translational parameters which permit the production of 3×10^{19} ovalbumin molecules per day in a laying hen's oviduct. There is some uncertainty in the figures quoted in part *C* of table 3.4, due to a very rough estimate of the number of tubular gland cells (probably $2–3 \times 10^{10}$) in each oviduct magnum.

Following hormonal withdrawal, transcription of the egg-white protein genes virtually ceases, and presynthesised cytoplasmic mRNAs are degraded. Indeed, ovalbumin mRNA has a markedly shorter half-life during acute hormone withdrawal than during stimulation, suggesting that this messenger is stabilised in the presence of hormone

Table 3.3. *Induction of conalbumin and ovalbumin mRNAs by oestrogen and progesterone (after Palmiter et al., 1977).*

Hormone	Time between hormone injection and first appearance of mRNA (hours)		Initial rate of mRNA accumulation (molecules/min./cell)	
	Conalbumin mRNA	Ovalbumin mRNA	Conalbumin mRNA	Ovalbumin mRNA
17β-oestradiol	0.5	2–3	20	20
Progesterone	2–3	2–3	8	20
17β-oestradiol + progesterone (administered simultaneously)	2–3	2–3	12	32

(Palmiter & Carey, 1974). Stocks of pre-existing egg-white proteins also disappear from the lumen of the tubular gland in the absence of hormone. However, the chromatin of the ovalbumin gene remains DNaseI-sensitive during hormone withdrawal (Palmiter *et al.*, 1977). Presumably this facilitates the rapid reinitiation of transcription from the ovalbumin gene when a further hormonal stimulus is given. The 5'-flanking sequences of the active ovalbumin gene in oviduct chromatin contain four DNaseI-hypersensitive regions located near the cap site (+1) and at −800, −3300 and −6000 bp (Kaye *et al.*, 1986) Oestrogen induction results in the appearance of all four sites, whereas the third site does not appear during progesterone induction. These sites mostly disappear during hormone withdrawal, although a new site appears between the second and third. Furthermore, four major DNaseI-hypersensitive sites and two minor ones are located towards the 3' end of the active ovalbumin gene and in its 3'-flanking regions. Those sites closest to the 3' end of the gene (one major and both minor) as well as the most distal site are all hormone-dependent, whereas the other two sites are retained during hormone withdrawal (Bellard *et al.*, 1986).

The chicken lysozyme gene is abundantly expressed under sex-steroid control in hen oviduct, but also constitutively at low levels in macrophages (sex steroids have no effect here). Fritton *et al.* (1984) have described alternative sets of DNaseI-hypersensitive sites in the 5'-flanking region of this gene, depending on its expression state. One of

Table 3.4. *Parameters of ovalbumin production in tubular gland cells (after Palmiter, 1975).*

A Transcription rate		*B* Translation rate	
Parameter	Value	Parameter	Value
Initial rate of mRNA$_{ov}$ accumulation (after oestrogen stimulation)	20 mol/min/cell	Polysome size	12 ribosomes/ mRNA$_{ov}$
Steady-state rate of mRNA$_{ov}$ synthesis in laying hens	34 mol/min/cell	Ribosome transit time (i.e. time taken to complete one ovalbumin protein chain)	1.3 min
Half-life of mRNA$_{ov}$	24h	Rate of protein elongation	300 amino acids/min
Mean life-time of each mRNA$_{ov}$ molecule	35h	Rate of ribosome initiation on each mRNA$_{ov}$ molecule	9.2 ribosomes/ min

C Steady state concentration of mRNA$_{ov}$ (a) in oestrogen stimulated
oviduct. 70 000 mol/cell
(b) in laying hen oviduct.
100 000 mol/cell
Rate of ovalbumin synthesis: 6.4×10^5 mol/min/cell
Assuming 2–3×10^{10} tubular gland cells per oviduct, this gives
3×10^{19} mol/day/hen

the three sites associated with oviduct expression appears and disappears according to the presence/absence of hormone (i.e. is hormone-dependent), whereas the other two remain present during hormone withdrawal. The alternative set of sites found in macrophage chromatin shows no alternation in response to hormone. Both sets share a distal site 6.1 kbp 5' to the gene, which apparently represents an enhancer sequence (Thiesen *et al.*, 1986); cell-specific proteins as well as general transcription factors (e.g. the TGCA protein) activate lysozyme gene expression by binding to this enhancer.

(b) *Egg-white protein genes and their organisation*
The egg-white protein genes in chick are each subdivided by several introns, ranging in number from three (lysozyme gene) to sixteen

(conalbumin gene), and each gene is 4–6 fold longer than its mRNA. The structures of these genes are summarised briefly below, followed by a discussion of the 5'-flanking sequences which are required for their regulation by steroid hormones.

(i) *Lysozyme gene.* This gene occupies 3.9 kbp of chromosomal DNA and includes three intron sequences (Lindenmaier *et al.*, 1979). All of these occur in the protein-coding portion of the gene, and the four exons correspond approximately to distinct domains of the protein structure (Jung *et al.*, 1980).

(ii) *Ovomucoid gene.* Seven introns have been identified within a genomic DNA fragment representing the entire chicken ovomucoid gene (Catterall *et al.*, 1979; Stein *et al.*, 1980). Hen oviduct nuclear RNA contains a series of ovomucoid mRNA precursors ranging in size from 1.5- to 5-fold longer than the final messenger (Nordstrom *et al.*, 1979). The largest of these species (5.5 kb) is probably a primary transcript of the entire gene; all of the precursors carry a poly(A) tag, and the removal of intron transcripts follows a preferred order during processing (Tsai *et al.*, 1980). Once again there is evidence that each exon more or less corresponds to a functional domain of the ovomucoid protein; moreover, the present gene appears to have evolved via duplication of a DNA sequence that was already split (see Stein *et al.*, 1980).

(iii) *Conalbumin gene.* This gene is considerably larger and more complex, being subdivided by 16 introns into 17 exons (some as short as 60 bp; Cochet *et al.*, 1979). Gene duplication is again implicated in the evolution of this structure. Repetitive sequences occur both in the 5'-flanking regions and within one of the introns of the conalbumin gene. Oviduct conalbumin is in fact the same protein as liver transferrin (which acts as an iron-carrier in the blood). These two proteins have identical amino-acid sequences, but differ in their carbohydrate moieties attached post-translationally. The regulation of their expression from the single-copy conalbumin/transferrin gene is quite different in the two tissues involved. Thus liver production of transferrin is regulated in part by iron concentrations in the blood, but is much less responsive to the steroid hormones which induce conalbumin synthesis in the oviduct. Nor is this due to a lack of steroid receptors in liver cells, since vitellogenin production in the liver is strongly induced by oestrogen (see § 3.3.3 below).

(iv) *Ovalbumin gene.* The ovalbumin gene in chickens is split by seven introns (300–1400 bp long) into eight exon segments (Dugaiczyk *et al.*, 1979). The ovalbumin messenger RNA contains an unusually long 3' untranslated region, and this part of the mRNA-coding

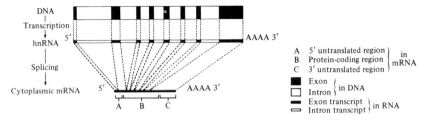

Fig. 3.10 Structure and expression of the ovalbumin gene (not to scale). From Breathnach *et al.* (1978) and Dugaiczyk *et al.* (1979).

sequence is uninterrupted in the genomic DNA. Six of the introns occur within the protein-coding region of the gene, while the seventh (which is long) interrupts the DNA coding for the 5′ untranslated region of the messenger (fig. 3.10; see Breathnach *et al.*, 1978). This delineates a short 5′ exon encoding a signal sequence in the mRNA.

The entire gene of 7.7 kbp is transcribed into a long precursor RNA from which the seven intron transcripts are removed in a preferred order within the nucleus (Tsai *et al.*, 1980). The final messenger is 1859 bases long [excluding the cap and poly(A)] and has been completely sequenced (McReynolds *et al.*, 1978).

(v) *Ovalbumin-related X and Y genes*. Royal *et al.* (1979) have mapped two large cloned segments of chicken DNA which overlap for about 1000 bp within the ovalbumin gene. One of these, 17 kbp long, extends in a 3′ direction beyond the ovalbumin gene; the other, 30 kbp long, extends in a 5′ direction from the ovalbumin gene, so covering some 46 kbp of DNA in all. Within the DNA on the 5′ side of the ovalbumin gene, Royal *et al.* (1979) identified two further genes related both in sequence and in intron/exon pattern to that encoding ovalbumin. These two relatives of the ovalbumin gene are designated X and Y, with Y located between the X and ovalbumin genes; stretches of non-expressed DNA sequences separate these three genes. All three lie in the same orientation, forming a small family of linked genes rather reminiscent of the globin gene-cluster. Both X and Y appear to be functional genes (not pseudogenes), X being transcribed to give a 2400-base messenger after processing, while Y gives a 2000-base mRNA. Neither of these mRNAs is abundant, yet both are synthesised under the same conditions of hormone stimulation as the ovalbumin messenger (see also LeMeur *et al.*, 1981). The function of the X and Y genes remains unclear, but probably they encode minor egg-white proteins.

(vi) *Coordinate hormonal control of the egg-white protein genes*. The above account indicates that at least six genes (X, Y, ovalbumin, ovomucoid, lysozyme and conalbumin) respond to hormonal induction in chick oviduct cells. A variety of different steroids can elicit similar responses, including androgens and glucocorticoids as well as progestins and oestrogens. Induction is coordinate in that the same set of genes is affected by all these hormones, although the resulting patterns of gene expression are somewhat different in each case (table 3.3). This suggests additional controls governing the response of each individual egg-white protein gene. As outlined earlier, (§ 3.3.1), hormone-responsive elements (HREs) can be mapped by fusing various lengths of 5′-flanking sequence from a hormone-inducible gene onto a foreign reporter gene. Following transfection of the fusion constructs into cultured cells containing appropriate receptors, the hormone inducibility of reporter gene expression can be assayed. Constructs retaining the HREs will show elevated expression in the presence of hormone, whereas constructs from which these sequences have been deleted will not. In this way, Dean *et al.* (1983, 1984) have shown that both progesterone- and oestrogen-response elements (PREs and EREs) lie between −95 and −197 bp relative to the cap site of the chicken ovalbumin gene. Likewise, sequences located between −164 and −208 bp are required for progesterone induction of lysozyme gene expression in the oviduct (Renkawitz *et al.*, 1984). In general, the HREs are enhancer-type sequences required for the hormonal activation of a limited set of target genes, with the hormone–receptor (HR) complex acting as a ligand-activated positive transcription factor (see § 3.3.1 and Yamamoto, 1985; Berg, 1989).

The 5′-flanking regions of the lysozyme gene also contain two separate transcriptional silencer elements, one at −250 and one at −1000 (Baniamad *et al.*, 1987). These silencers work largely independently of their position or orientation (like enhancers), but their influence can be overcome by strong activation of enhancers regulating the same gene. Such silencers may help to repress expression of the gene in tissues where its activity is not required. Notably, the −250 silencer is markedly less effective in macrophages, where the lysozyme gene is constitutively active at a low level.

The 5′-flanking sequences of both the conalbumin and ovalbumin genes show cell- and species-specificity. When fused to an SV40 reporter gene, these promoters function only in primary cultures of chicken liver or oviduct cells, but not in other chicken cell types nor in any non-chicken cell. This implies that the transcriptional activation of both genes depends on species- and cell-specific transcription factors, including (but not confined to) the appropriate HR complexes (Dierich *et al.*, 1987). The ovalbumin promoter is repressed in oviduct cells in the absence of hormone, but this is relieved by the binding of HR complex. This repression in the absence of hormone depends on 5'-flanking region from −132 to −425, which overlaps with the HREs that bind the HR complex (Gaub *et al.*, 1987). Other transcription factors may also interact with various DNA sequences essential for ovalbumin gene expression. One such factor required for transcription (at least *in vitro*) binds to the chicken ovalbumin upstream promoter (COUP) region between −70 and − 90 (Pastorcic *et al.*, 1986; Sagami *et al.*, 1986). This COUP-binding factor is not cell-type- or species-specific (it was purified from Hela cells; Sagami *et al.*, 1986; Wang *et al.*, 1987), but appears to be a member of the steroid receptor superfamily (Wang *et al.*, 1989). Interestingly, the binding of this factor to COUP DNA may be stabilised by a second essential factor (S300-11) which does not itself bind to DNA (Tsai *et al.*, 1987).

3.3.3 Vitellogenin genes and their regulation

Within the oocyte, yolk platelets are assembled from fragments of the large precursors known as vitellogenins (170 kd), as previously discussed in § 2.4.2a. Vitellogenins are actually synthesised in the female liver in oviparous vertebrates, and transported to the ovary as dimers in the bloodstream. Like the egg-white proteins in oviduct, vitellogenin expression in liver is under hormonal control by female sex steroids.

(a) *Chicken vitellogenin*
Oestrogen concentrations are normally very low in roosters, and vitellogenin mRNA is almost undetectable in the liver (approx. 0.5 molecules per cell). However, male liver cells do contain oestrogen receptors, just as the egg-white protein genes can respond to androgens in female oviduct. Thus if 17β-oestradiol is injected into roosters, vitellogenin synthesis is induced rapidly in the liver. Note that this is a **primary** response to the hormone, and is much easier to study than the primary response in oviduct since the liver cells are already differentiated. Vitellogenin mRNA levels reach a maximum of around 6000 molecules per cell within 3 days of hormone injection (Burns *et al.*, 1978). Thereafter the amount of this messenger declines exponentially (half-life of 29 hours), falling to less than 10 molecules per cell by the 17th day after injection.

Another yolk protein known as VLDL II (very-low-density lipoprotein) is also synthesised by hen liver cells *in vivo*, and shows a similar pattern of oestrogen response in rooster liver cells. Increased transcription of both VLDL II and vitellogenin genes is detectable within 30 minutes of a single hormone injection (Wiskocil *et al.*, 1980). After these mRNAs have peaked at several thousand molecules per cell, new transcription of their genes ceases, and pre-existing mRNAs of both types disappear over a period of days; this may involve destabilisation of the mRNAs in the absence of hormone (cf. Palmiter & Carey, 1974, discussed above). However, primary cultures of rooster liver cells do not synthesise vitellogenin in direct response to added oestrogen (though they do produce VLDL II); direct vitellogenin induction can be elicited only by serum from roosters treated 4 days previously with oestrogen, or else by growth hormone or prolactin (Boehm *et al.*, 1988). These *in vitro* results suggest that primary induction of avian vitellogenin may be an indirect process, in contrast to *Xenopus* where direct primary induction has been demonstrated in cultures of male liver cells treated with oestrogen [see *(b)* below].

In immature chicken liver cells, but not for example in brain cells, the chromatin of the major vitellogenin gene (VTG II) contains DNaseI-hypersensitive sites within the gene and beyond its 3′ end, although there is no expression at this stage. The tissue-specificity of these sites may reflect the commitment of liver cells to express the VTG II gene once an appropriate hormonal influence becomes available. Primary oestrogen treatment causes three new DNaseI-hypersensitive sites and a single demethylated site to appear in the 5′-flanking region of the VTG II gene (Burch & Weintraub, 1983). The latter and two of the former remain present during hormone withdrawal, so that secondary oestrogen treatment induces only a single DNaseI-hypersensitive site in this region. This may explain why liver cells respond more rapidly to secondary stimulation by oestrogen, and give higher vitellogenin expression than during the primary response (cf. egg-white protein production in oviduct cells; § 3.3.2 above). Following hormonal induction, the single demethylated site 600 bp upstream of the VTG II gene appears in non-expressing oviduct as well as in liver chromatin; even in the latter case, it appears after the onset of transcription and is retained during hormone withdrawal (Burch & Weintraub, 1983). On all three counts, there is no obvious role for this particular demethylation event in the regulation of VTG II expression.

A further demethylation site has been described at position +10, associated with a DNaseI-hypersensitive region just within the VTG II gene; this site is fully methylated in most tissues examined, but is demethylated specifically in liver cells from both laying hens and oestrogen-treated roosters (Saluz *et al.*, 1988). This site is included within a 10 bp protected (protein-associated) region when methylated in untreated rooster liver cells, but within a 20 bp protected region when unmethylated in hen liver cells; this difference presumably reflects the binding of two different proteins to this DNA region. Both the primary (see above) and secondary responses of the VTG II gene to oestrogen induction in rooster liver display unexpected features, suggesting the involvement of factors other than the oestrogen–receptor complex. If nuclei are purified from rooster liver one month after primary oestrogen treatment, then VTG II expression can be reactivated (as a secondary response) by adding a cell extract derived from oestrogen-treated liver (Jost *et al.*, 1986). This response is only partly abolished by anti-oestrogens (tamoxifen) or by antibodies against the oestrogen–receptor complex, implying that additional factors must play some role in this system. Two non-histone proteins have been shown to interact with the ERE located 600 bp upstream of the VTG II gene; neither is tissue-specific, but both increase the binding efficiency of the oestrogen–receptor complex to the ERE (Feavers *et al.*, 1987).

(b) Xenopus *vitellogenins*

Direct primary and secondary responses involving vitellogenin induction by oestrogen have been described for *Xenopus* male liver cells in culture (Searle & Tata, 1981). This system facilitates the analysis of early events in the induction process. *Xenopus* vitellogenin mRNAs are also much more stable in the presence of oestrogen (half-life 3 weeks) than in its absence (half-life 16 hours; Brock & Shapiro, 1983). Thus oestrogen exerts a dual control over vitellogenin levels, acting both on transcription and on mRNA stability.

The existence of a small family of four related vitellogenin genes in *Xenopus* was shown using cloned cDNA sequences derived from the 6.3 kb vitellogenin messengers (Wahli & Dawid, 1979). The mRNA-coding sequences (exons only) of the two B-type genes have diverged in about 20% of their nucleotide residues from those of the two A-type genes. By contrast, the two A-type genes differ from each other in only 5% of their exon sequences, and the same is true for the two B-type coding regions. This implies an evolutionarily ancient duplication of the ancestral vitellogenin gene to give the A and B classes, each of which has subsequently undergone a further duplication to yield the A1 and A2 plus B1 and B2 genes. The A1 and B1 genes are linked in the same orientation, lying some 15.5 kbp apart (Wahli *et al.*, 1982). While one major class of vitellogenin proteins is processed into phosvitin and lipovitellin I + II fragments (see § 2.4.2a), a different class of

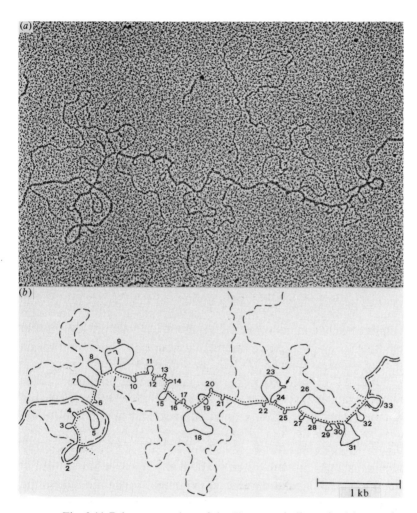

Fig. 3.11 R-loop mapping of the *Xenopus* vitellogenin A1 gene, by hybridisation with the 6.3kb mRNA derived from it. *A* electron micrograph; *B* diagram of loop pattern; bar represents 1kb. For the most part, the displaced DNA strand has not reformed duplex structures with the non-hybridised intron loops, which are numbered from 1 to 33 in part. ———— , coding DNA strand; -------, non-coding (displaced) DNA strand; · · · · · · · ·, mRNA strand. Photographs kindly supplied by Prof. W. Wahli (University of Lausanne, Switzerland) and reprinted with permission from the copyright holders. *Cell*. From W. Wahli *et al.* (1980) *Cell* **20**, 107–17.

vitellogenin polypeptides gives lipovitellin I + II and phosvette I + II fragments (these last being small yolk phosphoproteins of 14 and 19 kd respectively). Thus the evolutionary divergence of the vitellogenin A and B genes is reflected in a functional diversification among their products (see Wahli *et al.*, 1981; Wiley & Wallace, 1981). Note that a similar diversity of vitellogenin genes is found in other egg-laying animals, including the nematode *Caenorhabditis elegans* (see § 4.2).

Wahli *et al.* (1980) have compared the genomic organisation of the two class A vitellogenin genes. Both are highly complex genes split by 33 introns at homologous positions (see fig. 3.11). As mentioned above, the exon sequences (mean length 175 bp) are very similar in both genes, differing in only 5% of their base pairs. Perhaps surprisingly, the two genes are quite different in overall length; 21 kbp for A1, but only 16 kbp for A2. This difference is due entirely to the greater lengths of the introns in the A1 gene (mean intron length 450 bp) compared with the A2 gene (mean intron length 310 bp). Whereas the exons have evolved slowly through point mutations, the introns have evolved much more rapidly and radically – with examples of deletions, insertions and duplications as well as point mutations (Wahli *et al.*, 1980). There are significant regions of homology between the *Xenopus* A1 and chicken VTG II genes, both in their 5' coding exons and within the 5'-flanking regions (six conserved sequence blocks: Walker *et al.*, 1983). These latter suggest common control mechanisms, though oestrogen induction is more direct in *Xenopus* than in chicken.

At least six middle-repetitive sequence-families are represented among the introns of the A1 gene (Ryffel *et al.*, 1981). During oestrogen stimulation, they are of course transcribed as part of the vitellogenin nuclear precursor RNA. They can therefore be classed as nucleus-confined repetitive transcripts (see also § § 1.4 and 2.3). Interestingly, these same six repetitive sequences are also expressed in the nuclear RNA of male liver cells never exposed to oestrogen, i.e. where the vitellogenin genes are silent (Ryffel *et al.*, 1981). Thus members of all six repetitive families must also occur within other transcription units (genes) that are not regulated by oestrogen. Three of these repetitive sequences have been identified within introns of the larger albumin gene, which is expressed in liver cells at all times. However, these shared repetitive elements in both genes are found at scattered locations, and may not fulfil any regulatory role (Ryffel *et al.*, 1983).

Oestrogen is the main (direct) determinant for vitellogenin expression by liver cells in *Xenopus* (in contrast to its indirect effects in chicken; see (a) above). An oestrogen response element (ERE) is located at −331 relative to the *Xenopus* A2 vitellogenin gene; this sequence can confer oestrogen-inducible expression on heterologous promoters, even when introduced into human breast cancer cells (which contain appropriate hormone receptors, albeit of a different species type; Klein-Hitpaß *et al.*, 1986). The minimal effective ERE is 35 bp long and contains a 13 bp perfect palindrome (5'-GGTCACAGTGACC-3'). Similarly, the *Xenopus* vitellogenin B1 gene is regulated via an upstream ERE which is minimally 33 bp long and contains two imperfect 13 bp palindromic repeats; a third such imperfect repeat occurs further upstream. Combinations of any two of these three repeats are able to act independently as an ERE, conferring oestrogen-inducible expression on heterologous promoters (Martinez *et al.*, 1987). By contrast, the perfect palindromic repeat found upstream of the A2 promoter is able to act on its own as part of a functional ERE (further copies of related sequences are not required). Point mutations within the A2 ERE abolish oestrogen inducibility, though a defined combination of two mutations can convert the ERE into a functional GRE (§ 3.3.1; Martinez *et al.*, 1987).

However, the enhancer-like ERE is not the only determinant of vitellogenin expression in *Xenopus*. Dobbeling *et al.* (1988) have described an activator sequence located between −121 and −87 bp upstream of the A2 gene. This activator sequence can stimulate transcription from heterologous promoters, but does so in a cell-specific manner, i.e. it is active in mammalian liver or breast cancer cell-lines, but not in fibroblasts or Hela cells. The activator is composed of at least three sequence elements, two of which can function independently to some extent, whereas all three act cooperatively when combined (Dobbeling *et al.*, 1988). The whole activator sequence interacts *in vitro* with several rat-liver nuclear transcription factors, but these are distinct from the set which recognises the HP1 element common to many liver-specific genes. Thus the hormonal regulation of vitellogenin production is superimposed upon (rather than substituted for) a liver-specific pattern of expression (see end of § 1.3.2). This accords with the fact that oestrogen activates vitellogenin but not egg-white-protein genes in hen liver, but *vice versa* in oviduct.

4

The nematode
Caenorhabditis elegans

4.1 Introduction

Caenorhabditis elegans, unlike *Drosophila*, is a relative newcomer to the ranks of organisms favoured by developmental geneticists. Largely through the foresight of Sydney Brenner in the mid-1960s, this small and free-living soil nematode was singled out to become the 'eucaryotic *E. coli*'. The principal requirements were for a simple organism with a short life cycle, easy to maintain in large numbers under laboratory conditions, yet displaying a variety of cell types and behaviours which would facilitate screening for mutants. Among the specific advantages afforded by *C. elegans* in these respects are the following:

(i)   Its life cycle, which can be completed in three days at 25 °C.

(ii) Its small size and ease of laboratory maintenance (mass cultures can be fed on bacteria grown as suspensions or as lawns on agar plates), with the result that millions of individuals can be screened for rare mutants following mutagenesis.

(iii) Its unusual system of reproduction (see §4.5 below) which allows both self- and cross-fertilisation, so facilitating the isolation and maintenance of mutant strains.

(iv) Its small genome (only half the size of that in *Drosophila*), which simplifies both genetic and molecular analysis (see §4.2).

(v)   Its relative transparency, allowing good resolution of nuclei and other structures in the living animal when viewed by Nomarski interference microscopy.

(vi) The fact that it has a fixed number of somatic cells. This last feature [in combination with (v)] has allowed the complete cell lineage of the animal to be traced from zygote through to adult (see §4.3 below). This gives a unique flavour to most studies of *C. elegans*, since descriptions of developmental events usually involve patterns of cell division, rather than biochemical or morphological features (see e.g. §4.4).

The basic body plan of *C. elegans* is simple, consisting of two concentric tubes, the inner intestine separated by a pseudocoelomic space from the outer layer of cuticle, hypodermis, musculature and nervous system; in the adult, the pseudocoelomic space is occupied by the gonad (see fig. 4.1). The three-layered collagenous cuticle is secreted by the underlying hypodermis, which itself consists of large multinucleate syncytial cells; lateral cords of seam cells running the length of the body produce raised cuticular alae, which act rather like the treads of a tyre during locomotion. Movement is effected by striated body wall muscles which lie in four longitudinal strips (two dorsal, two ventral). Food (normally bacteria) is ingested into the bilobed pharynx anteriorly, then passes through the intestine for digestion and out through the posterior anus. The nerve cells lie mainly (i) around the pharynx, (ii) along the ventral midline and (iii) in the tail; their processes form a ring around the pharynx and also form the dorsal and ventral nerve cords which run the length of the body. The normal mode of reproduction involves self-fertilisation within XX hermaphrodites (see §4.5 below), but there are also morphologically distinctive XO males which can fertilise hermaphrodite eggs to produce outcross progeny; these males arise infrequently in wild populations through X-

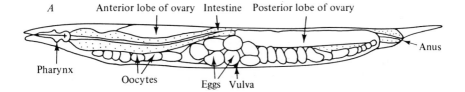

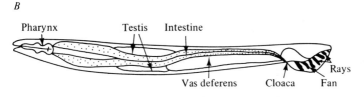

Fig. 4.1 *Caenorhabditis elegans* morphology (redrawn from Sulston & Horvitz, 1977). *A* Adult hermaphrodite. *B* Adult male. ⊡ , Intestinal tissue.

chromosome loss at meiosis. The hermaphrodite gonad is a double U-shaped structure, with two lobes of germ-line cells joined to a central somatic region (uterus and spermathecae), opening externally via the vulva. By contrast, the male gonad is single-lobed and opens posteriorly via the vas deferens and cloaca; there are specialised adaptations of the male tail (rays, fan, etc.) to facilitate clasping and insemination of hermaphrodites.

From this brief summary, it is clear that *C. elegans* has a wide range of specialised cell types carrying out diverse functions. Yet the organism is only about 1 mm long, and comprises precisely 959 somatic nuclei in the adult hermaphrodite (some cells being syncytial) and 1031 in the adult male. As described below in §4.3, these adult totals are attained via largely invariant cell lineages, both during embryogenesis (giving 558 nuclei at hatching in hermaphrodites and 560 in males; Sulston *et al.*, 1983), and subsequently during larval development (Sulston & Horvitz, 1977; Sulston *et al.*, 1980). Certain specific cells undergo programmed cell death in the course of both embryonic and postembryonic development (see §4.4.5). After hatching, somatic cell divisions are mainly confined to the progeny of specific blast cells, whereas most other cells are retained without further division. The postembryonic blast-cell lineages are elaborated during the four larval stages (L1 to L4), which are separated by moults when the cuticle is resynthesised. Some mutations affecting postembryonic cell lineage patterns are described in §4.4. If conditions of crowding or limited food supply are sensed during the L1 stage, the larva can undergo a modified and prolonged L2 stage (L2d) leading to a facultative diapause; this is represented by the non-feeding and stress-resistant dauer larva (see §4.6 below) which can survive for many weeks. If conditions become favourable, the dauer larva can resume normal development via the L3 moult to the L4 and adult stages. *C. elegans* is the only animal in which the complete structure of the nervous system is known, together with the ancestry of all its component cells and the precise pattern of their interconnections (reconstructed from serial EM sections). A number of behavioural mutants have been linked directly to structural and/or functional deficiencies in the nervous system; the example of the mechanosensory neurones is described briefly in §4.7.

4.2 The *C. elegans* genome and its organisation

The *C. elegans* haploid genome comprises only 10^8 base pairs, or about half the size of the *Drosophila* haploid genome. Some 80% of this is single-copy DNA, showing short-period interspersion of repetitive sequences in the *Xenopus* pattern (rather than *Drosophila*-type long-period interspersion). The repetitive sequences are mostly present in 10–100 copies per haploid genome, and include the usual range of transcribed genes as well as mobile DNA elements such as the Tc1 transposon (see below). There are perhaps 3000–4000 different genes in total. The haploid set of genetic material is organised into six chromosomes, one X plus five autosomes. The ratio of X chromosomes to autosome sets (X:A) determines sex in *C. elegans*, as in *Drosophila* (see § 5.2.3). An X:A ratio of 1.0 (in XX diploids) results in hermaphrodite development, whereas a ratio of 0.5 (in XO diploids) leads to male development (see § 4.5). Most nuclei in *C. elegans* are diploid, but many hypodermal nuclei are tetraploid, while the intestinal cells endoreplicate their DNA at the end of each larval stage, becoming 32C in the adult.

One consequence of the self-fertilisation process is that heterozygous individuals (e.g. after mutagenesis) will generate 25% of homozygous mutant offspring, so facilitating the recovery of recessive mutant alleles. Mutations causing premature stop codons (such as UAG in *amber* mutants) can often be rescued by unlinked suppressor mutations. As in procaryotes, such suppressors involve an altered tRNA which can sometimes insert an amino acid at that particular stop codon (Bolten *et al.*, 1984). Phenotypic rescue of *amber* mutants can be effected by microinjecting the suppressor tRNA (Hodgkin, 1985; Kimble *et al.*, 1982). Cloned genes have been introduced into the germ-line by microinjecting them into the syncytial ovary or directly into oocytes; such genes may become integrated into chromosomes or remain as extra-chromosomal tandem arrays (Fire, 1986). In either case, such introduced genes are expressed and hence can rescue the mutant phenotype caused by mutations in the corresponding endogenous genes of the recipient strain (e.g. for myosin genes; Fire & Waterston, 1989). *In situ* hybridisation can be used to establish the chromosomal location even of genes present in a single copy per haploid genome (Albertson, 1985).

Perhaps the most useful feature of the *C. elegans* genome is the Tc1 transposon, which can be used for insertional mutagenesis, for generating genetic mosaics and for cloning genes by 'transposon tagging'. The Tc1 transposon is 1610 bp long, with terminal inverted repeats of 54 bp (cf. *Drosophila* P elements; see § 5.1.3a); it also includes an open reading frame that probably encodes a transposase enzyme of 273 amino acids (Eide & Anderson, 1988). The numbers of Tc1 transposons vary widely according to strain, with only *c.* 30 copies per haploid genome in the widely used Bristol strain but ten times that number in the Bergerac strain. Excision and reinsertion of Tc1 elements occur at relatively high frequencies in the germ line of the latter strain (Eide & Anderson, 1985), causing mutations at new sites of insertion. Such mutations spontaneously revert when the Tc1 element becomes excised precisely; since this occurs frequently in somatic tissues, the result is a genetic mosaic with a clone of wild-type revertant cells on a mutant background. This can be used to investigate which cells need to express a particular gene product in order for development to proceed normally. Tc1 elements rarely if ever transpose in Bristol-strain animals; the strain difference is apparently caused by activation of Tc1 transposition in the Bergerac case. If a gene of interest has become mutated as a result of Tc1 insertion on a Bergerac background, one can proceed to clone that gene by first outcrossing to the Bristol strain (so as to reduce the Tc1 copy number and prevent further transposition), and then looking for a novel restriction fragment hybridising with a Tc1 probe on Southern blots. This fragment should represent the new Tc1 insertion into the gene of interest, hence sequences flanking the insert can be used to identify and clone the wild-type gene (transposon tagging).

The entire *C. elegans* genome is now being cloned as a series of overlapping 20–40 kbp fragments, and over the next few years a great deal of molecular information is likely to accumulate. Most of the studies published to date concern repeated genes or related members of gene families. There are about 70 major rRNA genes (a 7 kbp repeat unit containing the 18S, 5.8S and 28S cistrons) and about 110 5S rRNA genes, both organised into tandem clusters at unique genomic loci. There are 10–12 copies of each of the core histone genes, but these are not organised into tandem clusters like the *Drosophila* or sea urchin α-type histone genes; rather, they are clustered in small groups like those in mammals.

There is a large family of 40–150 related collagen genes in *C. elegans*, several of which have been cloned. They are fairly small genes (*c.* 1.5 kbp), containing two small introns each and encoding proteins of 28–32 kd. These are related to vertebrate non-fibrillar (e.g. type IX) rather than the larger fibrillar collagens. However, collagen gene probes detect some large 4–5 kb RNA species in *C. elegans*, as well as the predominant 1.1–1.4 kb class; it is possible that these larger collagen RNAs encode fibrillar-type collagens. The collagens isolated from *C. elegans* cuticle are mostly of high molecular weight (60–200 kd), apparently as a result of covalent cross-linking between smaller (*c.* 30 kd) monomeric proteins translated from the majority size-class of collagen mRNAs (Politz & Edgar, 1984). The patterns of collagen gene expression are stage-specific, as would be expected from the different structure and molecular composition of the new cuticle synthesised at each moult (Cox & Hirsh, 1985). In general terms, these collagen genes are sequentially activated; once switched on, a given gene tends to remain active during subsequent moults, although its level of expression may vary. Thus the *col-1* gene is expressed abundantly towards the end of embryogensis (during synthesis of the L1-stage cuticle), but its transcripts are also detectable during later moults; by contrast, the *col-2* gene is only expressed during the L2d-to-dauer moult (Kramer *et al.*, 1985). At least two mutants affecting body shape result from mutations in particular *C. elegans* collagen genes, implying a role for their wild-type protein products in normal morphogenesis; these are the *sqt-1* (Kramer *et al.*, 1988) and *dpy-13* (von Mende *et al.*, 1988) genes. The *sqt-1*[+] collagen appears to have several important functions during morphogenesis, since different *sqt-1*[−] mutations cause unusually diverse effects (including shortening, lengthening or helical twisting of the body), act at different stages during development, and display unexpected interactions with several other genetic loci.

Because the strips of body-wall muscle are needed for motility but not for viability, it is relatively easy to isolate mutants affecting the structure and/or function of these muscles (for review, see Waterston, 1988). Such mutants usually show altered or defective mobility and are classed as uncoordinated (*unc*). Among the first muscle-specific genes to be defined in this way was the *unc-54* gene encoding the major myosin heavy-chain (myoB; Macleod *et al.*, 1977). This gene contains introns whose positions do not delineate the major structural domains of the myosin heavy-chain protein (Karn *et al.*, 1983). The conserved regions of the *unc-54* gene have been used to isolate three further myosin genes; one of these encodes the minor body-wall myosin (myoA) while the other two encode pharyngeal myosins (myoC and myoD; Miller *et al.*, 1986). Paramyosin is encoded by the single *unc-15* structural gene. The *unc-22* gene is unusually long (>20 kbp) and encodes a large (500 kd) protein which forms part of the myofilament lattice in the A band (Moerman *et al.*, 1988). There are four actin genes in *C. elegans*, three of which (*act-1, -2, -3*) are clustered together in a 12 kbp region of DNA on linkage group 5 (Files *et al.*, 1983) while the fourth (*act-4*) is unlinked on the X chromosome (Albertson, 1985). The *act-1* and *act-3* genes encode the major muscle actins. One curious feature of three of these actin mRNAs is their possession of a 22-base leader sequence which is not represented in the corresponding gene; the same leader is also found on some other mRNAs in *C. elegans*. This selfsame sequence is found as the first 22 nucleotides of a novel 100-base transcript derived from a gene adjacent, but in the opposite orientation, to the 5S rRNA gene. It therefore seems likely that the 22-base leader is transferred from this transcript to the 5' end of various mRNAs by a process of intermolecular *trans*-splicing (Krause & Hirsh, 1987). Consistent with this is the recent discovery that the leader sequence is present in snRNP particles involved in splicing (Van Doren & Hirsh, 1988).

There is also detailed molecular information concerning genes whose expression is required for gametogenesis. In adult males, some 30 different major sperm proteins (MSPs) are expressed abundantly from 40–60 MSP genes. Note that these genes are not wholly male-specific, since they are transiently expressed during spermatogenesis in late L4 hermaphrodites (see §4.5). Thirteen out of 14 MSP cDNAs have different sequences (87–90% conserved), but they encode proteins that are even more closely related (96–100%). More recently several large genomic clusters of MSP genes have been isolated; each of the six clusters contains 3–13 functional MSP genes plus some pseudogenes (Ward *et al.*, 1988). The 5'-flanking sequences are conserved among the transcribed MSP genes (consistent with common regulation), but the 3' trailers of different MSP transcripts are not (Klass *et al.*, 1988).

During oogenesis, large quantities of yolk proteins (yp's) are accumulated in the growing oocytes. As in vertebrates (see §3.3.3), these yp's are synthesised elsewhere, i.e. in the intestinal cells rather than within the oocytes themselves. There are four abundant yp species produced in hermaphrodites but not in males, two of 170 kd (termed A and B) and one each of 88 and 115 kd. These last two are apparently cleaved and modified *in vivo* from a larger 180 kd precursor; thus *in vitro* translation of hermaphrodite mRNA yields the two 170 kd bands plus the 180 kd precursor, but neither the 88 nor 115 kd species are present (Sharrock, 1984). By analogy with their vertebrate counterparts, the yp primary protein products have been termed vitellogenins; they are encoded by a family of six *vit* genes in the *C. elegans* genome. One of these (*vit-6*) corresponds to the 180 kd precursor and is the least closely related of the six genes (Spieth & Blumenthal, 1985). Since all other cDNA clones so far isolated correspond either to *vit-2* (encoding yp 170A) or to *vit-5* (encoding yp 170B; Spieth *et al.*, 1985a), it is possible that the other three members of the family are pseudogenes. This is clearly the case for *vit-1* which contains an in-frame stop codon. The vitellogenin mRNAs are all about 5 kb in length, are hermaphrodite-specific, and are expressed mainly in intestinal tissue. Conserved sequence elements are found in the 5-flanking (promoter) regions of all six *vit* genes; these include the heptamers TGTCAAT occurring 4–6 times in both orientations in each promoter, and CTGATAA present once or twice in each promoter (Spieth *et al.*, 1985b). The *vit* genes are not clustered, except for *vit-3* and *vit-4* which lie only 4 kbp apart.

4.3 Early development and cell lineage in *C. elegans*

C. elegans embryos and adults contain a fixed number of nuclei (see earlier); moreover, most cell fates are fixed in a mosaic fashion such that the loss of one particular cell (and all its descendants) cannot be made good by regulative growth, except in a very few cases (see later). The combination of these two factors has faciliated the elucidation of the compete somatic cell lineage of *C. elegans* (Sulston *et al.*, 1983), an achievement unique among metazoan animals. One technique of prime importance in this work involves the use of pulsed-dye laser beams to destroy particular cells in the developing worm, followed by careful microscopic observation to determine which cells are absent at later stages. This laser ablation method was pioneered on the postembryonic cell lineages which mainly involve a limited number of blast cells (Sulston & Horvitz, 1977; Sulston *et al.*, 1980); later these studies were extended to include the whole of embryonic development (Sulston *et al.*, 1983). The complete cell lineage is shown in fig. 4.3, while fig. 4.2 shows the sequence of early cleavage divisions giving rise to the six founder cells (see below). In discussing aspects of the cell lineage, the following conventions need to be noted: (i) each cell designation is prefaced by the initial(s) of the embryonic founder cell from which it is derived (see below and fig. 4.2), or, in the case of postembryonic lineages, by letters likewise identifying the appropriate blast cell (e.g. T or V); (ii) the founder/blast cell indication is followed by a series of lower case letters describing the subsequent division pattern, where 'a' means an anterior and 'p' a posterior daughter, while 'l' and 'r' likewise mean left and right daughters, respectively. Thus a designation such as ABprapaapa describes the division history of just one specific cell derived from the AB founder cell. Since the worm is bilaterally symmetrical, there are equivalent left- and right-side sublineages in most cases (the left homologue of the above cell is

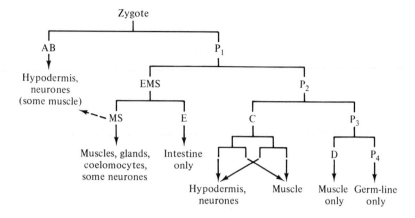

Fig. 4.2 Founder cells generated by early cleavage (modified from Sulston *et al.*, 1983). The six blast cells (AB, MS, E, C, D, and P4) are generated by unequal cleavages. Heavy arrows show the normal range of cell types derived from each blast cell. Bracketed cell types depend on cell:cell interactions indicated by dashed arrows (e.g. AB-derived muscle cells require an influence from MS derivatives).

ABplapaapa). Equivalent or nearly equivalent left and right sublineages are founded by daughters of the MS, C, D and P_4 founder cells and by grand-daughters of AB and E (e.g. ABpl/ABpr and ABal/ABar). In those cases where two daughter cells differ in their developmental fates, the division which gives rise to them is roughly antero-posterior in almost all cases (hence the use of a and p designations in identifying particular cells).

4.3.1 Outline of development

Fertilisation of oocytes occurs internally in an adult XX hermaphrodite, using either its own sperm (produced earlier during the L4 stage and stored in the spermathecae), or else sperm from insemination by a male. During the first 30 minutes post-fertilisation, the zygote produces a tough chitinous shell and a vitelline membrane from presynthesised components stored within the egg. The early embryo is retained within the mother for a variable time, and is then released through the vulva. Gastrulation involves the migration of cells into the interior of the embryo, starting with the intestinal precursors (the two daughters of the E founder cell; see below), followed by the germ-line precursor (the P_4 cell) and then the various muscle precursors (some or all of the progeny of MS, C, D and finally a few AB-derived cells). Nuclear poly(A)$^+$ RNAs representing new embryo transcripts become detectable from about the 100-cell stage (2.5 hours), perhaps corresponding to the midblastula transition in *Xenopus*; prior to this stage most poly(A)$^+$ RNA is cytoplasmic and presumably of maternal origin (Hecht *et al.*, 1981). The process of embryogenesis continues with rapid cell proliferation plus some cell movements and specific cell deaths up to about 7 hours, by which time the embryo consists of a spheroid of about 550 cells. During the next 7 hours, cell division virtually ceases (there are only two divisions and three deaths), while the embryo elongates into a worm writhing actively within the egg shell. During this period of morphogenesis, most tissues differentiate into their mature form, and neuronal processes grow out and form interconnections; later on there is active synthesis of L1 cuticle components (e.g. *col-1* collagen) by the hypodermal cells.

The L1 larva finally hatches at about 14 hours, and can thereafter feed and grow independently. Many gene functions become essential at this stage, presumably reflecting the depletion of maternal gene products. Some 3 hours after hatching, cell division is resumed; however, only about 10% of the cells are involved in such divisions, and these are designated blast cells. In the hypodermis, the postembryonic blast-cell divisions mostly follow a stem-cell pattern such that one daughter remains a stem cell (like its parent) while the other becomes tetraploid and joins the hypodermal syncytium. Similar stem-cell division patterns recur among the hypodermal blast cells at the start of each of the later larval stages. Towards the end of L1 the intestinal nuclei undergo their first round of DNA endoreplication (becoming 4C); this process is also repeated late in the subsequent larval stages so that these cells are 32C in the adult. The L2 stage is mainly one of growth, and there is little further elaboration of body pattern. Most of the sex-specific features in both males and hermaphrodites emerge during L3 (see §4.5), and undergo their final morphogenesis during L4. In hermaphrodites, about 300 sperm are produced from the germ-line precursors during late L4 and are then stored (*c.* 150 in each spermatheca); gametogenesis switches to oocyte production in the hermaphrodite adult (see §4.5).

4.3.2 Early cleavages

The very first cleavage of the zygote is asymmetric, generating a large anterior daughter (AB) and a small posterior daughter (P_1). Character-

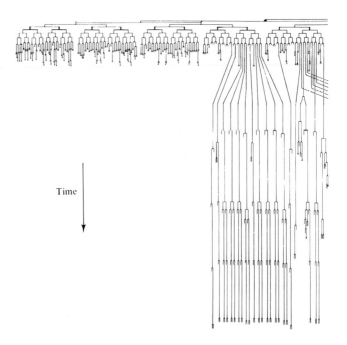

Time

Fig. 4.3 Complete cell lineage of *C. elegans* (photograph kindly supplied by Dr J. Sulston, MRC Laboratory of Molecular Biology, Cambridge).

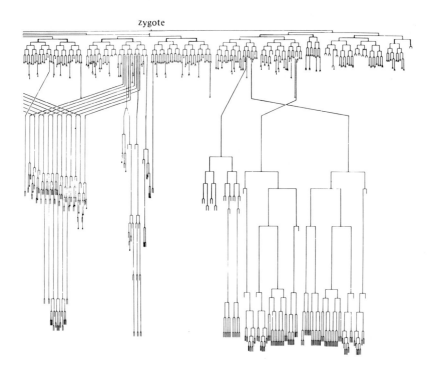

istic cytoplasmic P granules all become segregated into the P_1 cell. This process has been studied in detail by means of monoclonal antibodies which specifically recognise components of the P granules, allowing their location to be determined by immunofluorescence (Strome & Wood, 1982, 1983). P granules are initially distributed throughout the zygote cytoplasm, but as the pronuclei approach each other these granules become concentrated in the posterior periphery of the cell, so that they are all inherited by the P_1 daughter. Colchicine (an inhibitor of microtubule assembly) prevents pronuclear movement and the later assembly of mitotic spindles, but does not block the posterior localisation of P granules. However, this latter process can be blocked by cytochalasin (an inhibitor of actin microfilament assembly), which causes the P granules to coalesce centrally and prevents any of the normal manifestations of antero-posterior (A/P) polarity (Strome & Wood, 1983). This implies that two separate cytoskeletal systems are pressed into service following fertilisation, one based on microtubules (needed for pronuclear movement and later mitosis) and the other on microfilaments (needed for P granule localisation); moreover, this latter system is essential for the establishment of A/P asymmetries. During the next two cleavages of the P_1 cell, a similar sequence of events is observed, such that the P granules all pass to P_2 and then to P_3. But before the P_3 cell divides (into D and P_4; see fig. 4.2), the P granules become associated with the nuclear envelope. They remain associated with this structure in all germ-line cells (derivatives of P_4) throughout the rest of the life cycle (Strome & Wood, 1982); a few P granules are also passed to the D cell, but they are apparently unstable here and disappear before the D cell divides.

It is possible (but as yet unproven) that the P granules contain, or are associated with, germ-cell determinants; similar considerations apply in the case of the polar granules in *Drosophila* (see §5.1.2). However, in the nematode *Ascaris*, which undergoes chromosome diminution in all somatic cells (see §2.1.2), there is evidence from centrifugation experiments that pole plasm specifically prevents diminution from occurring in the germ-line precursor cell, so preserving the full chromosomal complement (Boveri, 1910). The *C. elegans* P granules are maternal in origin; clear indications of this emerge from studies of a fertile mutant in which P-granule staining by one particular monoclonal antibody is abolished. When a homozygous mutant (non-staining) hermaphrodite is outcrossed with a wild-type (staining) male, the first signs of positive antibody staining in the offspring are detected when the gonad starts to proliferate during mid-L1. Thus the (inessential) P-granule component recognised by this antibody is synthesised in developing germ-line cells, and is stored by oocytes as part of the maternal stockpile. Recently, a series of mutants has been isolated (involving genes designated *par-1* to *par-5*) which alter the partitioning of cytoplasmic components such as the P-granules during early cleavage (Kemphues *et al.,* 1988). The general effect of mutations in these genes is to make the early cleavages more equal and/or more synchronous, implying that their wild-type functions are required to ensure the asymmetry and asynchrony which normally characterise these first few divisions of the zygote.

These early cleavage divisions generate a set of six founder cells, designated AB, MS, E, C, D and P_4. Some of these founders give rise to a single cell type; thus all descendants of E become intestinal cells, all progeny of D form muscle, and all of the germ-line cells derive from P_4 (see fig. 4.2). AB mainly forms ectodermal derivatives such as hypodermis and neurons, but also contributes some muscles to the pharynx. These muscle cells appear only among progeny from the posterior daughter of AB, which is itself generated by a tilted cleavage such that one cell (ABp) comes to lie behind and to one side of the other (ABa). This cleavage defines both the dorsal–ventral and left–right axes of the embryo. The two daughters of AB are developmentally equivalent at first, but later their fates diverge as a result of cell interactions. Specifically, some influence from MS derivatives is required to induce the production of pharyngeal muscle cells from among the descendants of ABp, since these muscles do not appear if the P_1 blastomere is deleted at the 2-cell stage or the EMS cell at the 4-cell stage (Priess & Thomson, 1987). A similar lack of AB-derived muscle cells is apparent in the progeny of weak *glp-1⁻* mutants, suggesting a maternal requirement for the *glp-1⁺* product to mediate in this induction process (Priess *et al.*, 1987). As we shall see in §4.5, *glp-1⁺* also has an important function in controlling the choice between mitosis and meiosis in germ-line cells. Conversely, there may be some inductive influence of AB progeny on the MS-derived cells which normally contribute the other pharyngeal muscles, since these are not formed if the AB cell is deleted at the 2-cell stage. The remaining founder cell, C, contributes both hypodermis and neurons from its anterior grand-daughters (Caa and Cpa) but only muscle from its posterior grand-daughters (Cap and Cpp).

One pertinent question which arises in the case of tissues derived from a single founder cell, is whether or not this results from the segregation of specific cytoplasmic determinants. As we have seen, P granules become shunted through several successive cleavages into the germ-line precursor P_4. However, it is not yet clear whether these granules carry information specifying a germ-line fate.

The intestinal cells are exclusively derived from the E founder cell, and in this case there is stronger evidence for gut determinants normally confined to the E cell and its progeny. Two markers of intestinal differentiation are (i) autofluorescent gut granules containing tryptophan catabolites (Laufer *et al.*, 1980), and (ii) a gut-specific carboxyl esterase encoded by the *ges-1* gene (Edgar & McGhee, 1986, 1988). Both markers first become detectable at around 3.5 hours of development, when gastrulation has begun and the two intestinal procursors (Ea and Ep) have migrated into the interior of the embryo. These gut markers still appear on schedule in 2-cell or later embryos where cleavage has been arrested by colchicine or cytochalasin treatment; however, no expression can be detected in cleavage-blocked 1-cell embryos. After cleavage arrest of 2-cell embryos, gut markers appear only in the P_1 cell; likewise, in blocked 4-cell or 8-cell embryos, they appear only in EMS and in E respectively. This suggests that gut determinants are segregated from P_1 to EMS to E during early cleavage (Edgar & McGhee, 1986). If the P_1 nucleus is destroyed, leaving a P_1 cytoplast next to an intact AB cell, then gut granules never develop in the embryo; however, if one of the AB progeny is later made to fuse with the P_1 cytoplast, then in about 50% of cases the embryo goes on to develop gut granules. This implies that some AB-derived nuclei have undergone gut differentiation under the influence of determinants present in P_1 cytoplasm (Wood *et al.*, 1984; Schierenberg, 1985). Note that zygotic gene transcription is required for the expression of both gut markers, since their appearance is blocked by α-amanitin; thus neither is wholly dependent on maternal factors. Recent experiments with the DNA synthesis inhibitor aphidicolin demonstrate that gut marker expression requires a short period of DNA synthesis in the E cell during the first cell cycle after the establishment of the intestinal clone (Edgar & McGhee, 1988).

There is also evidence for the autonomous appearance of body-wall muscle markers (myosin and paramyosin) in the appropriate precursor cells in cleavage-arrested embryos (Gossett & Hecht, 1982). However, this demonstration is perhaps less convincing because muscle markers are expressed only later in embryonic development (from about the 400-cell stage), and also because muscle tissue is derived from several (rather than a single) founder cells. Similarly, cleavage-blocked embryos have been used to study how the potential to form hypodermis (as well as muscle and intestine) becomes segregated during early development (Cowan & McIntosh, 1985).

In view of the late onset of zygotic gene expression in *C. elegans*, it is not surprising to find that many mutations affecting early development act maternally rather than zygotically (see Wood, 1988b). Temperature-sensitive lethal mutations in at least 55 different genes have been shown to cause developmental arrest (i.e. at the restrictive but not at the permissive temperature); these fall into the five categories listed below. Most such mutations belong to classes (i) and (iv), while (v) is represented by only two cases (for details, see Hirsh & Vanderslice, 1976; Wood *et al.*, 1980; Miwa *et al.*, 1980).

(i)   Strict maternals, in which maternal expression of the wild-type gene is both necessary and sufficient for embryo survival (embryos from mutant oocytes *cannot* be rescued by zygotic expression of a wild-type paternal allele after outcrossing).

(ii) Strict zygotic requirement for wild-type gene activity in order to ensure embryo survival (maternal expression *not* needed).

(iii) Requirement for *both* maternal *and* zygotic expression of the wild-type gene (neither is dispensable if embryo is to survive).

(iv) Partial maternals, in which *either* maternal *or* zygotic expression of the wild-type gene is sufficient to ensure survival (in practice, this means that zygotic expression of the paternal wild-type allele can rescue embryos derived from mutant oocytes).

(v)   In the two cases of *zyg-8* and *spe-11*, a paternal effect has been described. Mutant oocytes from a homozygous *zyg-8* mother can all be rescued by sperm from a heterozygous (*zyg-8/+*) male; this is true even for those homozygous *zyg-8/zyg-8* progeny which contain no wild-type copy of the *zyg-8* gene nor any maternal stocks of *zyg-8*[+] product. Rescue in this case depends on paternal supplies of the wild-type *zyg-8*[+] product accumulated during spermatogenesis in the *zyg-8/+* heterozygous father. Similarly, the sperm-supplied *spe-11*[+] product is found to be essential for normal *C. elegans* development (Hill *et al.*, 1989).

Many of the strict maternal mutations cause defects during first cleavage and may thus affect components involved in general processes such as mitosis, rather than in any specific developmental event. However, the *par* genes mentioned earlier seem to function in the partitioning of cytoplasmic components during the early cleavage divisions. Even the cell:cell interactions involved in muscle induction from ABp derivatives (see earlier) seem to require presynthesised maternal components such as the *glp-1*[+] product; thus the failure of this process in weak *glp-1*[-] mutants shows strictly maternal inheritance (Priess *et al.*, 1987). However, the roles of most of the genes identified by ts mutations causing developmental arrest remain as yet obscure.

4.3.3 General features of the cell lineage

Before dealing with several mutations which alter the lineage pattern in various ways (§4.4), it is appropriate to comment upon some general characteristics and peculiarities of this lineage (for further details, see Sulston *et al.*, 1983; Sulston, 1988). As described earlier, laser ablation of a given cell usually has little or no effect upon the fates adopted by its neighbours (in this context, 'fates' may be taken to include specific cell lineage patterns as well as biochemical or morphological features). Particular exceptions to this general rule imply that cell:cell interactions are important for determining certain cell fates; these exceptions fall into several distinct categories:

(i) Interactions involving cells of different ancestry and/or developmental potential; these seem broadly similar to induction between different tissues in vertebrate embryology (§2.8). In hermaphrodites, a graded influence emanating from the anchor cell of the somatic gonad is required to induce both primary and secondary cell fates among the vulval precursor cells, as are interactions between these cells (see §4.4.4). If the anchor cell is ablated, all of these precursors adopt the tertiary fate (see §4.4.4). Similarly, continued mitotic divisions among the germ-line cells require the presence of the somatic distal tip cells; if these are ablated, all germ-line mitosis ceases and meiosis is initiated (see §4.5 below). Yet another example has already been mentioned, whereby MS-derived cells induce some descendents of the ABp cell to develop into pharyngeal muscles. It is particularly interesting that the *glp-1*[+] gene-product should be implicated in both the second and third of these examples (Austin & Kimble, 1987; Priess *et al.*, 1987).

(ii) Interactions involving cells belonging to the same equivalence group. Operationally, an equivalence group may be defined as a group of precursor cells whose fates are in some way interchangeable. Frequently such groups share common ancestry (e.g. they may be generated by homologous sublineages), but they are not exclusive clones. Two or more cell fates are available to the members of an equivalence group; the order of priority among these can be determined by laser ablation experiments. Thus in the male preanal equivalence group (shown in fig. 4.5B, below) there are three precursor cells and three such fates; normally P11.p adopts the primary fate, P10.p the secondary fate (producing a hook plus sensillum), and P9.p the tertiary fate. If P9.p is ablated, neither the primary nor secondary cell fates are affected. However, if P10.p is ablated then P9.p adopts the secondary instead of the tertiary fate, while the primary fate remains unaltered. If P11.p is ablated, P10.p adopts the primary and P9.p the secondary fate. Finally, if both P11.p and P10.p are ablated, the remaining P9.p cell adopts the primary fate. Clearly these three fates are interchangeable among the three precursor cells, but in a strict order of priority (which may involve inhibitory cell:cell interactions; see §4.4.4 below).

(iii) Finally, there are cases where the loss of one or more cells contributing to a particular tissue can cause adjacent cells to grow or proliferate in order to make good the deficit. These examples of regulation have been observed only among groups of cells with very similar developmental potentials.

C. elegans is bilaterally symmetrical, a feature achieved in large measure through the deployment of similar (analogous) sublineages on the left and right sides of the animal. In most cases these involve precursor cells with similar ancestries (homologues), going back to the early cleavage divisions that generate left and right daughters. However, in the embryonic head region some structures arise from analogous left and right sublineages which are founded by non-homologous precursors with dissimilar ancestries. There are also a number of unique cells, which usually arise from the confrontation of homologous left/right sublineages across the midline. In such cases, the choice between primary and secondary fates (for two homologous cells) depends on cell interactions, often acting via the *lin-12*[+] gene product (see §4.4.2 below). Finally, there are two examples of structures with 3-fold rotational symmetry. Whereas the male vas deferens is formed from three precursor cells via identical sublineages, the pharynx is formed partly in a bilaterally symmetrical fashion and partly piecemeal. Although cell ancestry is clearly of great importance in *C. elegans*, one cannot infer that lineage actually determines cell fate; this distinction is shown clearly by several examples where different sublineages and even descendants of different precursor cells can generate similar cell fates.

4.4 Genetics of cell lineage

Mutations in a number of genes cause particular alterations in embryonic and/or postembryonic cell-lineage patterns (for reviews, see

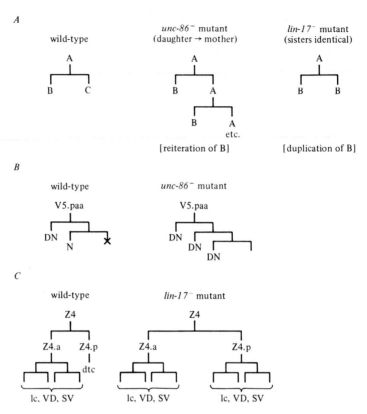

Fig. 4.4. Lineage mutants in *C. elegans* (modified from Horvitz, 1988).

A Schematic summary of cell fates.

B unc-86⁻ mutant effect on V5.paa lineage. In the *unc-86⁻* mutant, the second daughter behaves like its mother, so generating a chain of daughter dopaminergic neurons (DN); this contrasts with a single DN cell, plus a different type of neuron (N) and a programmed cell death (X), generated by V5.paa in wild-type animals.

X, programmed cell death.

C lin-17⁻ mutant effect on the male Z4 lineage. In the wild-type male somatic gonad, Z4.p becomes the distal tip cell (dtc), while Z4.a generates four cells – two precursors of the seminal vesicles (SV), one vas deferens precursor (VD) and one linker cell (lc). In *lin-17⁻* mutants, this Z4.a pattern is reduplicated by Z4.p (giving 8 lc/VD/SV cells), but no dtc is formed.

Sternberg & Horvitz, 1984; Hedgecock, 1985; Horvitz, 1988). Most such genes are given the designation *lin* (for lineage), although some mutants show other phenotypic characteristics leading to a different designation; thus lineage mutations which disrupt the formation of body-wall muscles or of the neurones innervating them will have an uncoordinated (*unc*) phenotype. In §4.4.1 we shall look at two postembryonic lineage mutants which alter division patterns such that normally different daughters either resemble each other (*lin-17*) or else one resembles its parent (*unc-86*). Subsection 4.4.2 will examine the homoeotic *lin-12* gene, which in several instances controls the choice of cell fate among members of an equivalence group. This will be followed in §4.4.3 by an account of the heterochronic gene *lin-14*, which controls the timing of specific developmental events, such as when a particular pattern of divisions (lineage motif) should appear. Evidently, lineage genes do not act in isolation, but rather in conjunction with a variety of other genes controlling morphogenesis and differentiation. The genetic pathway involved in vulval development has been elucidated in some detail, as described in §4.4.4. (A similarly detailed genetic pathway for the development and differentiation of the mechanosensory neurones will be treated separately in §4.7.) Finally, §4.4.5 will describe a set of genes involved in programmed cell death, a fate which recurs in many cell lineages.

4.4.1 Lineage mutants

Both *lin-17⁻* and *unc-86⁻* mutations cause characteristic abnormalities in a number of postembryonic cell lineages (Chalfie *et al.*, 1981). Two examples are shown in detail in fig. 4.4. In several instances where two daughters normally adopt different fates (B and C), the *lin-17⁻* mutant causes both daughters to adopt the same fate (B). This is shown clearly by the effect of *lin-17⁻* mutations on the lineage of the Z4 precursor in the male somatic gonad (fig. 4.4*C*). Normally the posterior daughter of this cell (Z4.p) becomes a non-dividing distal tip cell which regulates mitosis versus meiosis among the germ-line cells (see §4.5), while the anterior daughter (Z4.a) divides to give a set of four cells – two seminal vesicle precursors plus one vas deferens precursor and a linker cell. In *lin-17⁻* mutants, the distal tip cell is not formed (hence there is no germ-line mitosis), but rather Z4.p undergoes the same sequence of divisions as that normally adopted by Z4.a, so duplicating the set of Z4.a derivatives. The wild-type *lin-17⁺* gene product functions as part of a mechanism for generating asymmetry in parent cells, in preparation for the segregation of daughter cells which differ in size (usually) as well as in developmental fate; thus *lin-17⁻* mutations lead to the formation of sister cells which are equivalent in both respects (see Sternberg & Horvitz, 1988). Mutations in certain other genes give rather similar results, but affecting different lineages. Normally the sublineages generated by the V1–V4 and V6 pairs of lateral ectoblasts differ from that of the V5 pair; however, in *lin-22* mutants all six pairs of ectoblasts adopt the pattern normally confined to the V5 pair (Horvitz *et al.*, 1983). The implication is that the (differential?) activity of such genes normally assigns different fates to the daughters of a single precursor (*lin-17⁺*) or among the members of an equivalence group (*lin-22⁺*).

The case of *unc-86* is slightly different, in that mutations of this gene cause one out of two dissimilar daughters to repeat the division pattern characteristic of its parent (daughter-to-mother transformation). This results in multiple reiterations of that fate which normally characterises the other (unaffected) daughter, as shown in fig. 4.4*B*. The effects of *unc-86⁻* mutations are seen predominantly in neuroblast lineages; in the example given, the V5.paa cell normally gives rise to a single dopaminergic neurone (identifiable by catecholamine fluorescence) as one daughter, plus a different type of neurone and a progammed cell death derived from a further division of the other daughter. But in *unc-86⁻* mutants the cell fate characteristic of this second daughter is abolished, being replaced by a repetition of the mother-cell pattern of division, producing another dopaminergic neuron; the same pattern is reiterated further, generating a chain of dopaminergic neurones (fig. 4.4*B*). Thus wild-type *unc-86⁺* activity is required to distinguish one daughter (usually destined for further division) from its mother cell. The sequence of the cloned *unc-86* gene reveals a conserved region encoding a 158 amino-acid (aa) POU domain as part of the 467 aa protein (Finney *et al.*, 1988). Such POU domains have been described in several mammalian transcription factors, including the rat pituitary Pit-1 factor (Ingraham *et al.*, 1988), the human ubiquitous octamer-binding protein NF-A1/Oct-1, and the human lymphoid-specific octamer-binding factor NF-A2/Oct-2 (see also § §2.2.2 and 5.6; Levine & Hoey, 1988). The POU domain is a bipartite structure, including a variant 60-aa homeodomain located towards the C-terminus, separated by a variable spacer from an N-terminal POU-specific sequence of 75 aa; the former element is involved in DNA-binding (Sturm & Herr, 1988). Since both elements of the POU domain are closely conserved between the nematode *unc-86⁺* product and several mammalian transcription factors, it is plausible to suggest that *unc-86* may also encode a cell-specific transcription factor (Finney *et al.*, 1988).

4.4.2 lin-12: *a homoeotic gene*

Broadly speaking, a homoeotic mutant abolishes one particular cell fate (A) and replaces it by another (B). Such cell fates can involve structural features (as in *Drosophila*, see § 5.5) or patterns of cell divisions, as in the cases cited above (§ 4.4.1). Such transformations of cell fate imply that wild-type activity of the homoeotic gene is needed to establish one pathway rather than the other. This is best illustrated by a hypothetical example in which a homoeotic gene *H* is required for pathway A but not for pathway B among a given group of cells. Recessive H^- loss-of-function mutants will abolish fate A and replace it with a reduplication of B. Conversely, there may be dominant gain-of-function H^D mutants in which H^+ is active inappropriately (ectopically) among the B precursors, so converting them to the A fate; these will give a converse phenotype, eliminating B but reduplicating A. This is most simply interpreted in terms of a binary switch mechanism, whereby cells normally choose between the A and B fates by activating the H^+ function only in the A precursors (H^+ must remain inactive in the B precursors). The existence of one or both mutant classes for a given locus is indicative of a homoeotic gene function, although molecular information on the nature and mode of action of its product is needed to establish its precise role. Such information is at present lacking for many *C. elegans* genes of this type, although there are several well-studied precedents in *Drosophila* (§ 5.5).

The *lin-12* gene encodes one such homoeotic function in *C. elegans* (see Greenwald *et al.*, 1983; Lawrence, 1983); mutants involving this gene cause transformations of cell fate among at least 10 sets of cells. Moreover, there are two classes of *lin-12* mutant with opposite effects on certain pairs of cells which normally adopt different fates. If we designate one such pair *a* and *b*, and their normal fates as A and B respectively, then dominant *lin-12*^{D} mutations cause both *a* and *b* to follow the A fate while recessive or null *lin-12*$^-$ mutations cause both to assume the B fate. This implies that high *lin-12*$^+$ activity is necessary for the A fate (hence dominant *lin-12*^{D} mutations causing expression in *b* as well as *a* will reduplicate A), whereas low or zero *lin-12*$^+$ activity is necessary for the B fate (hence null *lin-12*$^-$ mutants abolish A but reduplicate B).

The pairs of cells affected by such *lin-12* mutants are often members of equivalence groups where one cell fate is primary and the other secondary (see §4.3.3ii). One such example involves the left/right homologues ABplapaapa and ABprapaapa, which normally become a G2 ectoblast and a W neuroblast respectively. In dominant *lin-12*[D] (overproducer?) mutants both cells express the G2 fate, whereas in null *lin-12*[-] mutants both adopt the W fate (see fig. 4.5*A*). If a precursor of ABplapaapa is deleted, then its ABprapaapa homologue can change from the W (secondary) to the G2 (primary) fate. It follows that both cells have the potential to follow the G2 pathway, and thus the normal fate of the W neuroblast may depend upon some interaction with the G2 ectoblast (hence in the absence of the G2 cell, W can switch to the G2 fate). Thus the wild-type *lin-12*[+] product is required to establish the primary G2 fate (abolished and replaced by W in null mutants) but *not* for the secondary W fate (abolished and replaced by G2 in dominant mutants).

Rather similar transformations are seen in the male preanal equivalence group (§4.3.3i), where the wild-type *lin-12*[+] product is required for the secondary cell fate normally adopted by P10.p (producing a hook plus sensillum); this pathway is eliminated in *lin-12*[-] null mutants but reiterated in *lin-12*[D] dominant mutants (Horvitz *et al.*, 1983). Moreover, the extent of reiteration depends upon the dosage of the *lin-12*[D] allele (fig. 4.5*B*); a single copy results in P9.p (normally tertiary) duplicating the secondary fate of P10.p, whereas two copies result in triplication of the secondary fate by all three cells of the equivalence group (both primary and tertiary fates are abandoned). These findings suggest a key role for the *lin-12*[+] product in choosing one particular cell fate from among the various options available to members of an equivalence group. Since these options are normally adopted in a strict order of priority (as revealed by laser ablation experiments), it is likely that *lin-12*[+] functions in the cell:cell interactions required to establish such priorities, perhaps as the source or receptor for some intercellular signal.

Such a role is entirely consistent with the sequence of the cloned *lin-12* gene (Greenwald, 1985; Yochem *et al.*, 1988), which reveals extensive homology to the *Notch* gene involved in choosing between epidermal and neuroblast cell fates in *Drosophila* (see §5.3.2). Both sequences predict typical membrane proteins with transmembrane and extracellular domains, the latter including multiple EGF-like repeats. These were first characterised in a family of vertebrate proteins including epidermal growth factor (EGF, which is itself cleaved from a larger precursor protein) and the low-density lipoprotein receptor (LDLr).

A Role of *lin-12* in determining the G2 versus W fate

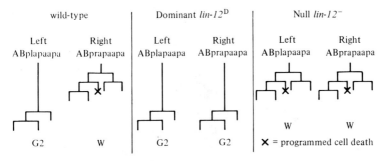

B Role of *lin-12* in the male preanal equivalence group

(i) Wild-type lineage

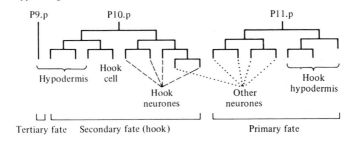

Tertiary fate   Secondary fate (hook)   Primary fate

(ii) Effects of cell ablation

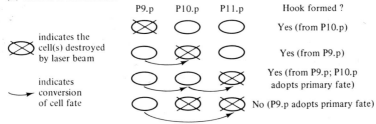

(iii) Effects of *lin-12* alleles on preanal cell fates

	Wild-type	*lin-12⁻*	One copy *lin-12ᴰ*	Two copies *lin-12ᴰ*
P9.p	Tertiary	Tertiary	Secondary (hook)	Secondary (hook)
P10.p	Secondary (hook)	(Tertiary)	Secondary (hook)	Secondary (hook)
P11.p	Primary	Primary	Primary	Secondary (hook)

Fig. 4.5 The *lin-12* homoeotic gene (modified from Horvitz *et al.,* 1983).

Either a signalling (cf. growth factor) or receptor function would be consistent with the phenotypes of both *Notch*⁻ and *lin-12*⁻/ᴰ mutants (see Akam, 1986; Bender, 1985). In the former case, it seems that *Notch*⁻ mutations are non-autonomous in genetic mosaics (Technau & Campos-Ortega, 1987), suggesting a local signalling rather than receptor function. However, there are some disparities between these two proteins which may be functionally important; whereas the *Notch*⁺ product contains 36 tandem EGF-like repeats in its extracellular domain, the *lin-12*⁺ product has only 13, and these are interrupted by other protein sequences (i.e. they are not tandem; Yochem et al., 1988). The *lin-12* and *glp-*1 genes are also related in sequence and are linked 20 kbp apart; both encode similar transmembrane proteins with EGF-like repeats (Yochem & Grennwald, 1989). The distributions of both *lin-12*⁺ (4.6 kb) and *glp-1*⁺ (4.4 kb) transcripts are wider than would be predicted from their respective mutant phenotypes, suggesting broader functions for both genes (Austin & Kimble, 1989; see also § 4.5).

On balance, a receptor function is perhaps more likely for the *lin-12*⁺ product (see Robertson, 1988a). In most cases (though *not* in the G2/W choice outlined above), the *lin-12*⁺ product is required for the appearance of secondary cell fates (i.e. dominant mutants reduplicate secondary fates while null mutants abolish them). This would suggest that *lin-12*⁺ encodes the receptor for an inhibitory signal derived from whichever cell is committed to the corresponding primary fate; this prevents secondary cells from adopting the primary fate unless the normal primary precursor is deleted. It follows that null *lin-12*⁻ mutants cannot respond to the inhibitory signal and hence reduplicate another fate in place of the secondary. However, if this is the case, then dominant *lin-12*ᴰ alleles would have to encode a modified receptor protein that could be activated in the absence of the appropriate signal, i.e. constitutive activation rather than constitutive over-expression. Only in some such way could *lin-12*ᴰ mutations cause a switch from primary to secondary fates. According to this model, activation of the *lin-12*⁺ receptor rather than presence/absence of the protein itself would confer the secondary fate. The contrasting pattern observed in the choice between G2 (primary) and W (secondary) fates, where *lin-12*⁺ function is essential for the former but not the latter, clearly requires some further elaboration of the model. (Thus if *lin-12*⁺ receptors were activated on the G2 cell by some influence from W, consistent with the above model and with the mutant phenotypes, then one would expect the W fate to be primary and G2 secondary; in fact the reverse is true.)

4.4.3 lin-14 *heterochronic mutants*

Whereas homoeotic mutants cause spatial transformations of cell fates (§4.4.2), heterochronic mutants result in temporal transformations such that particular developmental landmarks appear either earlier or later than normal. Some heterochronic mutants cause certain stage-specific events to occur precociously (e.g. in *lin-28* or recessive *lin-14⁻* mutants), whereas others cause such events to be retarded (e.g. in *lin-4*, *lin-29* or dominant *lin-14ᴰ* mutants). Because the two different classes of *lin-14* mutations cause opposite heterochronic transformations in particular lineages, this example will be considered in more detail below; there are clear formal analogies with homoeotic functions such as *lin-12* (reviewed in Ambros & Horvitz, 1984; Horvitz, 1988).

Dominant *lin-14ᴰ* mutants display retarded development of certain features, in that earlier developmental events are reiterated (e.g. super-numerary larval moults) while later events are delayed or even elim-inated (an adult cuticle is never secreted). Conversely, recessive *lin-14⁻* mutants give the opposite phenotype, in which late events occur abnormally early (e.g. an adult cuticle is secreted at the end of L3 rather than L4). One clearcut example is shown in fig. 4.6, involving the postembryonic hypodermal lineages generated by the T blast cells. During the normal L1 stage, each T precursor gives rise to a pair of cells (T.aa and T.ap) from its anterior daughter T.a, whereas its pos-terior daughter T.p produces a group of five cells plus a programmed cell death (these T.p derivatives constitute motif B). Subsequently T.ap is the only cell to divide during L2, thereby generating a further group of four cells (motif A). In recessive *lin-14⁻* mutants, the T precursors adopt motif A precociously (during L1 rather than L2) and motif B is suppressed. Conversely, in dominant *lin-14ᴰ* mutants the T.ap cell fol-lows a lineage pattern during L2 which is the same as that normally followed by the T cell during L1; this results in an extra copy of motif B appearing during L2 (in addition to the one generated during L1), whereas the appearance of motif A is delayed till L3. Note that the A (late) and B (early) motifs represent subprograms executed by other genes, presumably in response to the ambient levels of *lin-14⁺* products (see below).

Broadly speaking, dominant *lin-14ᴰ* mutations cause the retarded adoption of cell fates that would normally appear earlier; these mutants probably overexpress *lin-14⁺* products, hence high expression of this gene seems to specify early cell fates. By contrast, diminished amounts of *lin-14⁺* products (in recessive or null mutants) precipitate

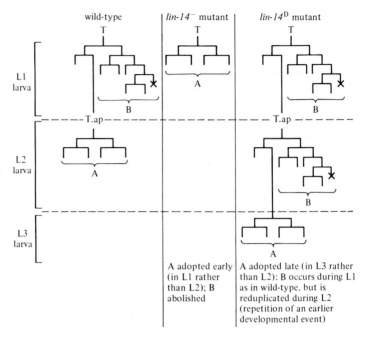

Fig. 4.6 The *lin-14* heterochronic gene (modified from Ambros & Horvitz, 1984).

the early adoption of fates that would normally appear later; this further implies that high *lin-14+* expression delays the appearance of later fates. If late events can take place only when *lin-14+* product levels are low, then one would expect wild-type *lin-14+* expression to decline as normal development proceeds. This is indeed the case; an antibody recognising the *lin-14+* protein stains certain nuclei strongly in *C. elegans* embryos and early larvae, but not in late larvae or adults *except* in the case of *lin-14D* mutants (Ruvkun & Giusto, 1989). This model suggests that postembryonic cells must somehow be aware of the ambient levels of *lin-14+* product in order to decide which subprogram (e.g. early or late) they ought to execute. In the T lineage given in fig. 4.6, two such alternatives are available, suggesting a simple binary switch. However, in other lineages the situation is more complex, with *lin-14+* activity controlling three discrete choices during the L1, L2 and L3 stages (Ambros & Horvitz, 1987). Either cells can respond to three different concentrations of a single *lin-14+* product, or else there may be two distinct *lin-14+* products (active at different stages?) such that cells respond to two concentrations of each. Support for this latter hypothesis is provided by the existence of three further mutant classes, which suggest that there may be two separately mutable *lin-14* gene functions (Horvitz, 1988).

4.4.4 Genetic specification of vulval development

The vulva is a structure formed postembryonically in the hermaphrodite only (the corresponding precursor cells are directed to other fates in males). As shown in fig. 4.7, vulval development involves an equivalence group of six cells (P3.p to P8.p) plus an influence emanating from the underlying anchor cell, which is a somatic component of

A Wild-type (inferred strength of anchor-cell signal shown by thickness of arrow)

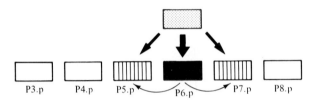

B Wild-type after anchor-cell ablation

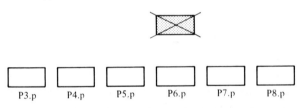

C lin-15 mutant (with or without anchor cell present, i.e. vulval cell fates independent of anchor cell influence)

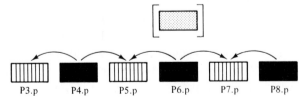

Fig. 4.7 Vulval development (modified from Robertson, 1988a and Sternberg, 1988). P3.p ... P8.p, vulval equivalence group; ▨, anchor cell; ■, primary cell fate (1°); ▥, secondary cell fate (2°); ☐, tertiary cell fate (3°); ➡, anchor cell influence; ⌒, lateral inhibition of neighbouring cells.

the hermaphrodite gonad (see §4.5). The anchor cell is derived from one member of an equivalence group comprising two of the somatic gonad cells (descendants of Z1 and Z4 respectively); the other forms the VU cell, and *lin-12*[+] function is again required for the anchor-cell versus VU-cell decision. The six vulval precursors generate three distinct types of lineage; one cell (normally P6.p) adopts the primary fate and produces a set of eight cells, while the two cells flanking it (P5.p and P7.p) adopt the secondary fate and produce a set of seven cells each. These primary and secondary derivatives contribute towards the vulva proper, whereas the remaining three precursor cells (P3.p, P4.p and P8.p) adopt a tertiary fate in which their descendants contribute to the syncytial hypodermis. The six vulval precursors obey the normal rules governing priority among members of an equivalence group (see §4.3.3i above); if the primary and secondary precursors (P5.p, P6.p, P7.p) are all ablated, the remaining three cells (normally tertiary) can switch to the primary and secondary fates, so forming a vulva from precursors that do not normally contribute to it. The vulval precursor that lies closest to the anchor cell (P6.p) adopts the primary fate while its neighbours become secondary during normal development. Thus wild-type animals show a strict antero-posterior (A/P) order of cell fates among the vulval precursors, namely (A) 3^0, 3^0, 2^0, 1^0, 2^0, 3^0, (P).

Vulval development depends upon an influence emanating from the anchor cell; if this is ablated, both primary and secondary fates are abandoned and all six precursors adopt the tertiary fate by default (fig. 4.7). This implies that high concentrations of anchor-cell influence may induce the primary fate and lower concentrations the secondary fate, a view for which there is some evidence from laser ablation experiments (Sternberg & Horvitz, 1986). However, the primary cell also laterally inhibits its neighbours from following the same path; thus the primary fate cannot be adopted by the immediate neighbours of a primary cell. This is shown clearly in the *lin-15* mutant, where the adoption of primary or secondary vulval fates is almost independent of anchor cell influence. Typically all six precursors adopt the primary or secondary fates alternately, so generating a multivulva phenotype (see fig. 4.7); the tertiary fate is abandoned (Sternberg, 1988). This arrangement is little altered if the anchor cell is ablated, implying that the primary and secondary fates are not dependent on anchor-cell influence in *lin-15* mutants (Sternberg, 1988). If five of the six vulval precursors are ablated, then the single remaining cell always adopts the primary fate; moreover, primary cells are never found adjacent to one another, but are always separated by secondary cells. Taken together, these data suggest a lateral inhibitory influence emanating from the primary cells, causing their immediate neighbours to adopt the secondary fate. Adoption of the secondary fate requires the *lin-12+* function, since *lin-12*^D mutants develop an excess of secondary derivatives while *lin-12−* mutants develop none (see below and §4.4.2).

Mutations in some 28 genes are known to disrupt vulval development. There are two contrasting phenotypic classes, the vulvaless (Vul) mutants in which no vulva appears, and the multivulva (Muv) mutants where several develop (e.g. *lin-15*). Both cause gross morphological abnormalities easily discerned under the dissecting microscope. The Vul group belong to a large class of egg-laying-defective (egl) mutants which are unable to lay their eggs, such that the embryos from self-fertilisation develop internally (their parents become literally bags of worms!). Most Vul or Muv mutants involve single gene mutations, although some Muv phenotypes result from mutations in two unlinked genes (one of which is a *lin-8* and/or *lin-9* allele; Ferguson *et al.*, 1987). Eight of the 28 genes have Muv or Vul null phenotypes, suggesting that these genes normally function only in vulval development and are not required for fertility or viability. Another nine genes have lethal or sterile null phenotypes, their Vul or Muv phenotypes resulting from a partial loss of gene function; most of these genes are involved in a variety of other cell lineages apart from that which generates the vulva (e.g. *lin-12, lin-14, lin-15*; see §§4.4.2 and 4.4.3 above). The remaining genes affecting vulval development are known only from dominant or partial loss-of-function mutations, and their null phenotypes remain unknown. The probable pathway of genetic interactions in vulval development has been elucidated by studying the phenotypes of single and double mutants involving 23 of these genes (Ferguson *et al.*, 1987; results summarised in fig. 4.8). These can be grouped as follows:

(i) Mutations in genes required for the **formation** of a functional anchor cell or functional vulval precursors will cause defects prior to the appearance of the vulval cell lineages.

(ii) Mutations in genes affecting the **determination** of particular vulval lineages will transform the fates of the appropriate precursor cells.

(iii) Mutations in genes needed for the **expression** of particular vulval cell fates will only affect the lineage generated by a single type of precursor cell.

These assignments have been confirmed by studying epistatic interactions in double-mutant combinations; for instance, if two mutations produce distinctive phenotypes in which vulval development is

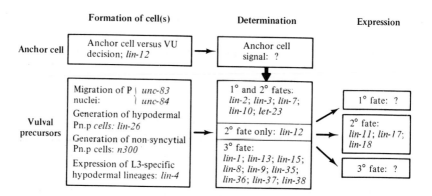

Fig. 4.8 Genes involved in vulval development (after Ferguson *et al.*, 1987). Gene abbreviations are enclosed in boxes together with the processes which they are inferred to control. Unidentified components are indicated by question marks. ➡, anchor cell influence acting on determination of 1° and 2° vulval cell fates.

blocked at different stages, then the phenotype given by both combined should resemble that of the earlier-acting single mutant. Another approach is to use temperature-sensitive (ts) mutant alleles (where available); these result in a gene-product which is active at the permissive temperature (usually 16°C) but inactive at the restrictive temperature (usually 25°C). One can define a ts period during which wild-type gene activity is essential, since exposure to the restrictive temperature during this period will give rise to the mutant phenotype. Hence ts alleles of genes acting early in the pathway would be expected to show earlier ts periods compared with later-acting genes. However, the involvement of the anchor cell influence in vulval development introduces some ambiguities into these assignments. For instance, a Muv phenotype could be caused either by elevated concentrations of anchor-cell signal, or by some alteration of the hypodermal precursors causing them to express primary or secondary vulval fates independently of anchor-cell influence. These possibilities can be distinguished experimentally by ablating the anchor cell; in the former case this will abolish the Muv phenotype, whereas in the latter case the Muv phenotype will not be affected (e.g. *lin-15*; see above). Similarly, Vul phenotypes could be due to a failure of the anchor-cell signal (cf. anchor-cell ablation) or to a lack of response by the vulval precursors. This ambiguity can be resolved by combining a signal-independent Muv mutation (such as *lin-15*) with the Vul mutation in question. If the Vul mutation affects the anchor-cell signal, the Muv phenotype should be unaltered. However, if the Vul mutation affects the vulval precursors, then the Muv phenotype may be modified in some way. As yet, no mutations have been shown to abolish the anchor-cell signal.

One example of the interactions involved is given by the *lin-12* and *lin-11* genes, whose wild-type functions are both required for the secondary vulval fate (though *lin-12+* also plays an earlier role in the choice between anchor and VU cell fates). In dominant *lin-12D* mutants, all six vulval precursors are converted to the secondary cell fate, whereas *lin-11* mutations cause the two secondary precursors (P5.p and P7.p) to adopt an abnormal variant of the secondary fate. In *lin-12D/lin-11* double mutants, all six vulval precursors adopt this abnormal secondary fate. Conversely, null *lin-12−* mutations prevent any of the six precursors from following the secondary fate, hence in *lin-12−/lin-11* double mutants the abnormal secondary fate is not expressed at all. This shows clearly that *lin-12+* functions in the determination of the secondary cell fate, i.e. upstream of *lin-11+* which functions in the expression of that secondary fate. In several such double-mutant combinations, all six precursors adopt the same fate independent of anchor-cell influence. Thus in *lin-12−*/Muv double-mutants all six cells follow the primary fate; *lin-12D*/Vul combinations cause all six to adopt the secondary fate; finally, all six form tertiary derivatives in *lin-12−*/Vul double mutants. All of these outcomes are unaffected by laser ablation of the anchor-cell. This suggests that the normal function of the anchor-cell influence is to set the activity state of certain key genes (e.g. *lin-12*) within particular sets of precursor cells. Once this is accomplished the lineages appropriate to each fate are executed without further reference to the anchor-cell signal.

The genetic pathway shown in fig. 4.8 is incomplete in that there are no candidate genes for certain essential functions such as the anchor-cell signal. Five genes are required for the generation of the vulval precursors (their functions are annotated in fig. 4.8), as well as one (*lin-12*) involved in the formation of the anchor cell. Fifteen further genes act within the precursor cells to determine which of the three available fates should be adopted; among these, *lin-12* is required (again) to determine the secondary cell fate. The products of several of these genes may mediate in the intracellular responses to the anchor-cell and lateral inhibitory signals. Finally, three genes are specifically required for the expression of the secondary cell fate (one being *lin-11*), though no expression-specific genes have yet been identified for the primary or tertiary fates. The three vulval cell fates are thus specified by two intercellular signalling pathways, viz. lateral inhibition and graded induction (Sternberg & Horvitz, 1989). The 3^0 fate is a default option, whereas the 2^0 fate requires *lin-12+* function in both pathways. Wild-type Muv and Vul genes function in the inductive pathway, acting via *lin-12+* to specify the 2^0 fate but independently of it to specify the 1^0 fate.

4.4.5 Genes specifying programmed cell death

Of the 1091 somatic nuclei produced by the somatic cell lineage during hermaphrodite development, 131 undergo programmed death. Since the same cells die reproducibly at specific times in particular lineages, such deaths must result from the execution of a precise genetic sub-program; in other words, programmed death is one option among the available range of cell fates. The determination decisions that cause particular cells to opt for this 'death program' are altered in some lineages (but not in others) by mutations in the *ces-1* gene, resulting in the survival of certain cells that would normally die. Conversely, mutations in *egl-1* or *lin-39* cause additional deaths involving certain cells that would normally survive and differentiate.

Five genes have been identified which actually participate in the cell-death program. In *ced-3* and *ced-4* mutants, all of the cells that would normally die instead survive and differentiate. Perhaps surprisingly, this causes no gross phenotypic abnormalities, implying that programmed cell deaths are not essential in order to ensure normal function and morphology. On the basis of their mutant phenotypes, the wild-type *ced-3+* and *ced-4+* functions are probably involved in initiating the cell-death program (Ellis & Horvitz, 1986). By contrast, the *ced-1+* and *ced-2+* functions are required for the engulfment of dead cells by their neighbours. In *ced-1* or *ced-2* mutants, appropriate cells undergo the early morphological changes characteristic of programmed cell-death, but they are not phagocytosed by their neighbours (Hedgecock *et al.*, 1983). Finally, in *nuc-1* mutants the dead cells are engulfed but their DNA is not degraded, such that extra pycnotic nuclei can be seen in the phagocytosing cells (Sulston, 1976); the wild-type *nuc-1+* gene apparently encodes the principal endodeoxyribonuclease.

The order of gene action inferred from double mutant combinations is *ced-3+*/*ced-4+* > *ced-1+*/*ced-2+* > *nuc-1+*; this pathway is shown diagrammatically in fig. 4.9. The fact that dying cells are normally engulfed by their neighbours might suggest murder rather than suicide as the cause of death. However, in *ced-1* or *ced-2* mutants, the fact that

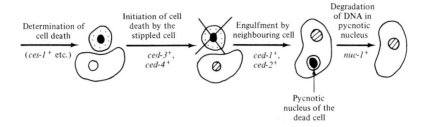

Fig. 4.9 Genetic control of programmed cell death (adapted from Ellis & Horvitz, 1986).

phagocytosis is blocked does not prevent the vast majority of cell-deaths. In fact, only two male-specific cell-deaths are prevented by *ced-1* or *ced-2* mutations, suggesting that these two alone are cases of murder by a neighbour, whereas all other instances of cell-death are suicides (internally programmed). In this context, it is worth noting that the *ced-3*[+] and *ced-4*[+] functions (which initiate the cell-death program) are required autonomously within the cells that are destined to die, as revealed by mosaic analysis. As yet, similar information is lacking for the other three cell-death genes, which could act either in the dying cell itself or else in the cell that engulfs it.

4.5 Gonadal development and sex determination

4.5.1 Development of the gonads and fertilisation

The gonadal primordia are established during L1 and L2, but their proliferation and differentiation occur mainly during L3, with final morphogenesis during L4 (when there is also limited spermatogenesis in hermaphrodites). The germ-line cells are all derived from Z2 and Z3 (daughters of the P_4 founder cell), whilst the somatic components of the gonad are derived from the Z1 and Z4 precursor cells. The final gonad comprises a single reflexed (U-shaped) arm opening posteriorly via the cloaca in males, and two such arms opening centrally via the vulva in hermaphrodites. Distal and proximal refer to positions along these arms relative to the opening of the gonad in each sex.

In distal regions of the gonad, the germ-line tissue is syncytial – each nucleus occupying an alcove partially surrounded by membranes but still open to a central anucleate core of cytoplasm. In the most distal regions the germ-line nuclei divide mitotically, but in more proximal regions they enter meiotic prophase. At around the loop region (where the gonad arm reflexes back on itself) the germ-line nuclei become completely enclosed by membranes and begin to differentiate into germ cells. Moving proximally, successively later stages of germ-cell maturation can be observed. Mature sperm cells are produced by the gonad in males and transiently in late L4 hermaphrodites, while oocytes are formed only in adult hermaphrodites. As we shall see below, mitosis in the distalmost regions of the gonad is regulated by the distal tip cells (dtc's), which are somatic cells located at the extreme distal tip(s) of the gonad arm(s).

In hermaphrodites, twelve somatic descendants of Z1 and Z4 are generated during L1 and L2; ten of these form the central somatic region of the gonad during L2, so separating the germ-line tissue into two blocks. These central somatic derivatives include the anchor cell (see also §4.4.4), and precursors for the uterus and flanking spermathecae (in which sperm produced transiently during late L4 are stored). The other two somatic gonad cells produced during L1 (Z1.aa and Z4.pp) occupy distal tip positions in the two blocks of germ-line tissue, as described above. These dtc's do not divide themselves, but influence nearby germ-line nuclei to undergo mitosis rather than entering meiosis (see below); they also act as leader cells in the morphogenesis of the two arms of the gonad, guiding their outward growth away from the central somatic region, and then later their turning back (during early L4) towards the centre. In this way the double U-bend structure of the adult hermaphrodite gonad is established (fig. 4.10*A*), with the anterior and posterior lobes opening centrally through a vulva induced by the anchor cell.

In males, ten somatic cells are generated from Z1 and Z4 during L1 and L2, but the symmetrical development of the gonad is soon abandoned. Two of these somatic cells become distal tip cells (Z1.a and Z4.p), in this case located together at the distal end of the single-armed gonad; once again these non-dividing dtc's promote mitosis among nearby germ-cell nuclei, deferring their entry into meiosis until more proximal regions are reached. The remaining eight somatic precursors are located together at the elongating (proximal) tip of the gonad; these include the linker cell and the precursors of the seminal vesicle and vas deferens (these structures are formed during L3 and L4). The male gonad grows anteriorly at first but then reflexes posteriorly to produce the single U-bend structure shown in fig. 4.10*B*; note that this pattern of growth is led by the linker cell at the proximal tip, not by the dtc's as in the hermaphrodite gonad. Thus one of the two functions ascribed to the dtc's in hermaphrodites (namely its leader activity) is taken over by

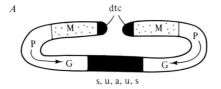

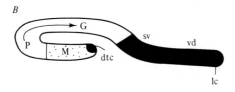

Fig. 4.10 Anatomy of adult male and hermaphrodite gonads (after Kimble & White, 1981). *A* Hermaphrodite (XX), *B* Male (XO). ■■■ , somatic tissue, ☐⋮ , germline tissue; ⋮⋮ , mitotic zone; ☐ , meiotic zone (in sequence from pachytene to gamete-forming regions). In *A*: dtc, distal tip cell; s, spermathecae (2); u, uterus; a, anchor cell. In *B*: dtc, 2 distal tip cells; vd, vas deferens; lc, linker cell; sv, seminal vesicle.

the linker cell in males, leaving male dtc's with a single role in regulating germ-line proliferation.

The role of the dtc's was established by means of laser ablation (Kimble & White, 1981; Kimble, 1981a,b). In either sex, if both dtc's are deleted at any stage after their formation, all germ-line nuclei cease mitosis and enter meiosis. Both dtc's must be destroyed in males in order to obtain this effect; in hermaphrodites, elimination of a single dtc abolishes germ-line mitosis only in the arm led by the ablated dtc. In males, it is possible to alter the position of the dtc's (by destroying their sister cells) such that both lie at the anterior rather than posterior end of the early testis; the net result is to reverse the normal polarity of the gonad, such that anterior germ-line nuclei remain mitotic while those in more posterior locations enter meiosis. This shows that it is proximity to the dtc rather than position within the developing gonad which determines mitosis rather than meiosis. Thus the dtc's directly regulate gonad polarity by controlling the proliferation of germ-line cells.

The proliferation control is mediated in both sexes via the wild-type *glp-1*+ function; *glp-1*⁻ mutants have essentially the same phenotype as animals in which both dtc's have been ablated, except that both cells remain present. In such mutants, the hermaphrodite Z1.aa/Z4.pp cells, and male Z1.a/Z4.p cells, assume their normal positions and morphologies, but are apparently non-functional as regards proliferation control, since all of the germ-line nuclei are meiotic (see Austin & Kimble, 1987). Notably the hermaphrodite dtc's retain their morphogenetic leader function, since a normal double U-shaped gonad is still formed in *glp-1⁻* mutants, even though the structure is filled only by a few sperm cells (the early germ-line derivatives). The *glp-1⁺* product is expressed and functional in germ-line cells, perhaps acting as a receptor for some signal emitted by the dtc's (see end of §4.4.2). Using ts mutations of *glp-1*, several extragenic suppressor mutations have been isolated, all of which allow germ-line mitosis in the absence of *glp-1⁺* function. These suppressors are mutations of known *dumpy (dpy)* genes, including the *sqt-1* gene (§4.2) which encodes a collagen involved in morphogenesis. The suppressor gene-products are probably components of the extracellular matrix, suggesting that this may interact with the *glp-1⁺* product to impose spatial constraints on mitosis versus meiosis in the germline (Maine & Kimble, 1989). As we have seen earlier (§4.3.2), maternal *glp-1⁺* function is required for certain cell:cell interactions in the early embryo, for instance when MS derivatives induce descendants of the ABp cell to form pharyngeal muscles (Priess *et al.*, 1987).

In hermaphrodites, sperm cells mature as they pass from the L4 gonad arms into the spermathecae, where they are stored. As oocytes mature in the later gonad, each one is pushed into the spermatheca on that side, where it becomes surrounded by sperm. Only one of these fertilises the oocyte, while the rest are retained in (or return to) the spermatheca. The self-fertilised egg passes on into the uterus, from which it is laid during early development. This process continues during adult life until all of the *c.* 300 available sperm have been used up (only about 150 are produced per gonad arm during L4). In older hermaphrodites, oocytes still pass through the empty spermathecae; such oocytes remain unfertilised and undergo endoreduplication, becoming highly polyploid. If hermaphrodites are mated with males, the male sperm cells take precedence so that outcross progeny are produced preferentially; however, the hermaphrodite's own sperm are retained and used when the male sperm have been consumed.

Several mutants are known in which the process of spermatogenesis is defective. Mutant males mate normally with hermaphrodites but fail to yield outcross progeny; such mutants are described as either sperm or fertilisation-defective (*spe* or *fer*). In many cases, these mutants produce normal numbers of sperm cells which are either non-motile, or else normal-looking but infertile. In other cases, spermatogenesis may be blocked, e.g. at the spermatocyte stage in *spe-6* mutants or at the spermatid stage in *fer-15* mutants. Most mutants in this class affect spermatogenesis in both males and in late L4 hermaphrodites, but *spe-12* mutants affect only the latter and not the former. The *spe-11* paternal-effect lethal mutation has been mentioned earlier (§4.3.2).

4.5.2 Sex determination in C. elegans

Over 650 of the somatic cells in *C. elegans* are sexually indifferent; that is, their phenotypes are unaffected by the sex of the animal. The remaining 30–40% of the somatic cells, plus all of the germ-line cells, are sex-specific in their phenotype. In *C. elegans*, as in *Drosophila*, the sexual phenotype is dependent on the ratio between X chromosomes and autosome sets. Where this ratio is equal to 1.0, as in XX diploids, hermaphrodite development ensues. A ratio of 0.5, as in XO diploids, results in male development. In wild populations, males arise at a low frequency as a consequence of X-chromosome loss during meiosis (XX⟶XO). The hermaphrodite can be regarded as a female modified so as to produce sperm during the late L4 stage. Embryonic development differs little between males and hermaphrodites, but they diverge increasingly during the postembryonic stages (see e.g. Kimble & Hirsh, 1979). Most sex-specific differentiation is autonomous (i.e. set internally by the X:A ratio); thus hermaphrodite intestine cells continue to secrete vitellogenins (§4.2) even if the entire gonad is ablated. However, development of the hermaphrodite vulva requires a short-range inductive influence from the anchor cell of the somatic gonad (§4.4.4).

A number of genes affecting the sexual phenotype of *C. elegans* have been identified and their probable order of action established by studying epistatic interactions in double-mutant combinations (see reviews by Hodgkin, 1987a, 1988a; Meyer, 1988). The model which has been proposed on the basis of this work is shown in fig. 4.11; this pathway is similar in outline but not mechanism to that controlling sex determination in *Drosophila* (see §5.2.3 and fig. 5.11, later). Mutations in most of the sex determination genes cause sexual transformations in one sex but not in the other. Thus the three *tra*[+] functions are required in hermaphrodites but not in males; XX individuals carrying homozygous *tra-1*[−], *tra-2*[−] or *tra-3*[−] mutations are transformed into phenotypic males, whereas XO individuals show no gross phenotypic

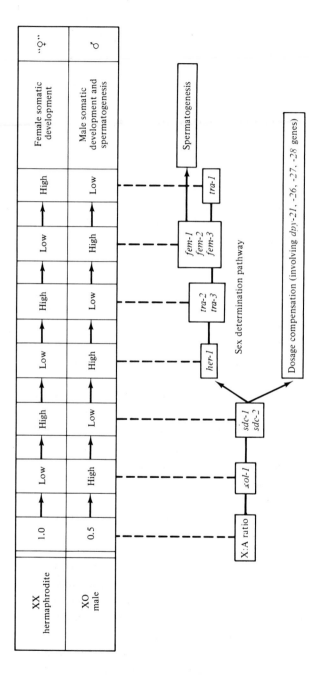

Fig. 4.11 Sex determination in *C. elegans* (modified from Hodgkin, 1988a and from Miller *et al.*, 1988).

Upper part: inferred level of gene activity in XX hermaphrodite ('female') somatic tissues, and in XO male somatic and germline tissues. For pattern of gene activity in hermaphrodite germline, see note on right.

Lower part: inferred hierarchy of genetic control. ——, repression or activation of next gene (see upper part).

Note: in the XX germline, the initial pattern of gene expression during L4 is similar to that in males (except that *her-1* and presumably *xol-1* are not active), allowing transient spermatogenesis to occur. The definitive adult pattern is the same as that in XX somatic tissue, so that gametogenesis switches to oocyte production. The delay in activating *tra-2/-3* in the XX germline may be mediated by the *fog-2* gene product (not shown).

abnormalities. The *tra-3⁺* product is required maternally, but apparently only in very small amounts, perhaps as a cofactor for the *tra-2⁺* product. By contrast, the *her-1⁺* function is required in males but not in hermaphrodites; thus XO individuals carrying *her-1⁻* mutations are phenotypically hermaphrodite, whereas XX individuals appear normal. The three *fem⁺* genes are needed for somatic sexual development in males and for spermatogenesis in both sexes, since *fem⁻* mutants abolish spermatogenesis in both sexes but also cause feminisation of XO individuals. The *fem-3* gene has recently been cloned by transposon tagging and its transcript distribution analysed (Rosenquist & Kimble, 1988). Three transcripts are produced (one embryo-specific, one L4/adult-specific, one present throughout), but they are expressed at 6-fold higher levels in XO compared with XX embryos; moreover, they are only detected in germ-line tissue in XX hermaphrodites, which is the site of spermatogenesis during L4.

Epistatic relationships between these mutants suggest the following hierarchies of gene action:

(i)   in somatic tissues: *her-1 > tra-2/-3 > fem-1/-2/-3 > tra-1;*
(ii) in the germ line: *her-1 > tra-2/-3 > tra-1 > fem-1/-2/-3.*

In hermaphrodite somatic tissues, the high X:A ratio represses *her-1⁺*, so permitting *tra-2⁺/-3⁺* activity. These gene products in turn repress the three *fem* genes, so allowing *tra-1⁺* to become active; this both promotes female and inhibits male somatic development (see Hodgkin, 1987b). The pattern in the hermaphrodite germ line is similar, except that *tra-2⁺* expression is initially repressed by a germ-line-specific product (probably that encoded by the *fog-2* gene). This allows the transient expression of the three *fem* genes in the germ line (see above), so inhibiting *tra-1⁺* activity and permitting the L4 burst of spermatogenesis. Later, *tra-2⁺* escapes from inhibition, thereby repressing the *fem⁺* functions and activating *tra-1⁺*, which switches the germ line over to oogenesis in the adult. In the male XO soma and germ line, the low X:A ratio activates *her-1⁺*; this represses *tra-2⁺/-3⁺* and so allows the three *fem* genes to become fully active. These *fem⁺* functions promote spermatogenesis and repress *tra-1⁺*; the lack of *tra-1⁺* activity ensures male as opposed to hermaphrodite somatic development (see fig. 4.11). Note that repression and activation in the above account do not necessarily imply 'off' and 'on' states, respectively; the data are equally consistent with low and high levels of gene expression. Overall, sexual phenotype is controlled via differential activity of the same set of regulatory genes.

The *tra-1*[+] gene product controls the activities of batteries of sex-specific target genes, repressing those required for male development but activating 'female'-specific functions (note that *tra-1*[+] is transiently repressed in hermaphrodites to allow spermatogenesis during L4). Plausible targets for *tra-1*[+] regulation are genes with specific (usually postembryonic) functions in only one sex. Several genes involved in vulval development (e.g. *lin-2* and *lin-7*) appear to be hermaphrodite-specific, since mutations of these genes cause a vulvaless phenotype in hermaphrodites but have no discernable effect in males. However, several mutants displaying a multivulva phenotype in hermaphrodites likewise generate extra hooks in males (from the same set of pre-cursors), suggesting that these genes act in both sexes. Some of the genes affecting egg-laying are also hermaphrodite-specific (no function in males), and of course the vitellogenin genes are only expressed in hermaphrodite intestine. These genes may be activated (directly or indirectly) by the *tra-1*[+] product.

As for genes expressed only in males, there are several male-abnormal (*mab*) mutations which have little or no effect in hermaphrodites; in males these mostly cause defects in the morphogenesis of male-specific tail structures. Two of these genes have been studied in some detail recently. The *mab-3* gene has two distinct male-specific functions; firstly in repressing vitellogenin synthesis (this takes place in *mab-3*[−] mutant males, but never in normal males), and secondly in the male V ray-cell lineage (Shen & Hodgkin, 1988). Moreover, analysis of epistatic relationships suggests that *mab-3*[+] function is regulated by *tra-1*[+] activity, i.e. high levels of the latter repress the former. The second gene in this class has a wider function in both sexes, since only some of the defects caused by *mab-5*[−] mutations are male-specific. However, the cells affected by such mutations are all posterior in location, although belonging to diverse lineages. In wild-type animals of both sexes, *mab-5*[+] RNA is localised specifically in posterior regions of the body (Costa, M. *et al.*, 1988). Moreover, the cloned *mab-5* sequence contains a homeobox related to that in the *Drosophila Antennapedia* gene (see § § 5.5 and 5.6), suggesting that *mab-5*[+] may play a role in antero-posterior regionalisation in *C. elegans* just as many homeobox-containing genes do in *Drosophila* (see § § 5.4.2, 5.5 and 5.6). Note that genes involved in spermatogenesis (including the MSP genes; see § 4.2) are not strictly male-specific, since limited numbers of sperm are formed during later L4 in hermaphrodites.

The *her-1*[+] gene is at the apex of a regulatory cascade governing sex determination specifically (fig. 4.11). However, it does not respond directly to the X:A ratio, nor is sex determination the only process reg-ulated by this ratio. Dosage compensation describes the process whereby the activities of genes on the X chromosome are regulated differentially between the sexes in such a way that similar amounts of product are obtained from a single copy in males (here XO) as from two copies in XX females (here hermaphrodites). In *Drosophila*, both dosage compensation and sex determination are regulated by the mas-ter gene *Sxl*, which acts upstream of the genes solely converned with sex determination (see §5.2.3 and fig. 5.11, later). A similar situation pertains in *C. elegans*, where the *sdc-1* gene regulates both processes in hermaphrodites (Villeneuve & Meyer, 1987); in XX individuals, re-cessive *sdc-1⁻* mutations cause both masculinisation and hyper-expres-sion of X-linked genes (as occurs normally in XO males). A second gene, *sdc-2*, also participates in this dual control of sex determination and dosage compensation; the mutant phenotypes are similar, except that strong *sdc-2⁻* mutations cause the death of XX animals (quoted in Miller *et al.*, 1988).

A converse pattern is found in XO males carrying recessive *xol-1⁻* mutations, in which both the sex determination and dosage com-pensation pathways are shifted towards their hermaphrodite mode. As well as 'feminising' such XO animals, the *xol-1⁻* mutations are also lethal, apparently due to underexpression of X-linked gene functions. The mutant *xol-1⁻* phentotype can be fully suppressed by *sdc-1⁻/-2⁻* mutations. These and other interactions suggest that the wild-type *xol-1⁺* function promotes male development in XO animals by repressing the genes required for hermaphrodite sex determination and dosage compensation (Miller *et al.*, 1988). It acts upstream of the other genes involved in these pathways (see fig. 4.11), and might even respond directly to the X:A ratio. DNA sequences involved in recognition of the X:A ratio appear to be widespread on the X chromosome, since microinjection of cloned DNA from many parts of this chromosome can 'feminise' XO males, whereas autosomal, phage or plasmid DNAs have no such effect. One such 'feminising region' of X-chromosomal DNA is a sequence of at most 131 bp found within an intron of the X-linked *act-4* gene (McCoubrey *et al.*, 1988). Probably this represents one member of a family of discrete sequence elements associated with many X-linked genes, acting as signals for sex determination.

4.6 The dauer larva

The dauer (= enduring) larva is an example of facultative diapause; if

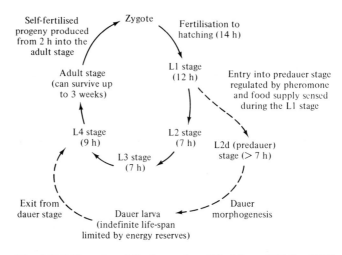

Fig. 4.12 Life cycle of *C. elegans* (modified from Riddle, 1988). Times refer to growth at 25°C. ———▶ , normal life cycle; – –▶ , dauer pathway (facultative disapause).

adverse conditions (crowding and/or limited food supply) are experienced during the L1 stage, larvae can proceed through a modified L2 stage (L2d) and subsequently moult into the dauer state (fig. 4.12). This is characterised by: (i) the absence of feeding; (ii) altered energy metabolism; (iii) resistance to environmental stress; (iv) morphological changes including a thickened cuticle; and especially (v) survival well beyond the normal *C. elegans* lifespan (3–6 months rather than 3 weeks). Dauer larvae seem to be arrested both in development and ageing, and their survival period is limited mainly by the exhaustion of internal energy reserves. If favourable conditions are encountered by dauer larvae (e.g. renewed food supply), then they can exit from the dauer state by feeding, completing L3 development, and re-entering the normal developmental sequence from the L3 moult through the L4 stage to adulthood; the adult lifespan of such individuals does not seem to be affected by the length of time spent in the dauer state. The resistance of dauer larvae to detergents such as SDS provides a means of screening for mutants affecting dauer formation.

Dauer larvae have unique cuticle characteristics, including radial striations not found at other stages, and a dauer-specific form of collagen not expressed during normal development (Cox & Hirsh, 1985). There are also morphological changes in several neurones, including those which control pharyngeal pumping. The excretory gland cell is apparently inactive during the dauer state, in contrast to its obvious activity when filled with secretory granules during normal larval development. The pattern of energy metabolism (production of high-energy phosphates) is also markedly different. Normal larvae switch from the glyoxalate pathway in L1 to the tricarboxylic acid (TCA) cycle, and then utilise this increasingly during later larval stages; by contrast, in L2d predauer larvae the activities of TCA cycle enzymes decline, until in dauer larvae high-energy phosphates are almost undetectable (see review by Riddle, 1988).

There are several environmental cues which, if sensed during L1, lead on through the L2d stage to dauer larva formation. The most important of these are food availability and a specific pheromone produced by other individuals, although temperature also plays a modulating influence. The *C. elegans* pheromone is a fatty-acid-like compound produced throughout the life cycle. A high concentration of this signal (indicative of crowding) promotes dauer formation from young larvae, provided it is sensed from mid-L1 onwards. The food signal is a carbohydrate-like substance produced by *E. coli* and also present in yeast extract. Its effects are opposite to those of the pheromone, i.e. high concentrations inhibit dauer formation while low concentrations promote this process. The relative concentrations of both factors influence dauer development, while food supply becomes increasingly important for initiating recovery from the dauer state (older dauer larvae apparently change in their sensitivity to these two factors). Temperature also influences the choice between normal and dauer pathways; if the concentrations of pheromone and food signals are held constant, a greater proportion of young larvae will opt for dauer formation at high compared with low temperatures.

Two major classes of *daf* mutants show altered dauer development:

(i)   Recessive temperature-sensitive mutants which always form dauer larvae (independent of environmental cues) at restrictive temperatures, but which can develop normally at permissive temperatures; these are known as dauer constitutives.
(ii) Dauer-defective mutants which are unable to form dauer larvae at all. Some of these mutants show a variety of other chemosensory defects (affecting the amphid cells), implying that they may be unable to sense the relevant environmental signals.

The interrelationships between these two mutant classes are complex, since some dauer-defectives can act as epistatic suppressors of dauer-constitutives while others cannot. Overall, it seems likely that the dauer-defective (dd) mutants are blocked at some point in the genetic program leading to dauer formation, whereas dauer-constitutive (dc) mutants generate a false internal signal that activates this program even in the absence of appropriate environmental signals. If the dauer pathway is blocked by a dd mutation at some point *after* the false signal generated by a dc mutation, then the double-mutant combination will be dauer defective. However, if a dd mutant blocks the dauer program at some point *before* the false signal generated by a dc mutant, then the double combination will be dauer constitutive. By analysing all possible combinations of the two mutant classes, the genetic pathway shown in fig. 4.13 can be derived (see reviews by Riddle, 1987, 1988).

The functions of many of these genes remain unknown. However, *daf-22⁻* mutants are apparently unable to produce the dauer-inducing pheromone (Golden & Riddle, 1985), which is the main factor required to initiate the dauer program, and thus acts upstream of all other *daf* functions. The majority of *daf* genes seem to act at various levels in the sensory processing of appropriate environmental signals (upper line in fig. 4.13). Such genes might affect the generation or functioning of particular neurones which participate in the dauer signal-processing pathway; these cells would probably have other roles to perform during non-dauer development. The *daf-2* gene appears to define a

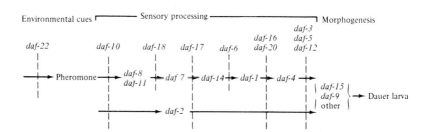

Fig. 4.13 Genetic pathway for dauer larva formation (modified from Riddle, 1988). Mutations in genes connected by horizontal arrows cause false signals at that point, leading to dauer-constitutive phenotypes. Mutations in other genes block the pathway at the point indicated by vertical dashed lines, leading to dauer-defective phenotypes. The probable order of gene action has been inferred from double-mutant combinations of defective and constitutive mutations (see text). Only some of the defective mutants affect the alternative pathway via *daf-2*, as indicated by the lengths of the vertical dashed lines.

separate branch of the pathway, as suggested by the unusual pattern of interactions between *daf-2* (dc) and various dd mutants (lower line in fig. 4.13); this branch could represent the response to a different environmental signal from that which cues in the main pathway (see fig. 4.13). Finally, there are two candiate genes which probably participate in dauer morphogenesis, namely *daf-9* and *daf-15* (Albert & Riddle, 1988). These act downstream of the other *daf* genes, and dc mutations in either gene result in dauer larvae of abnormal morphology. Those produced by *daf-9* mutants are the more dauer-like, since they are non-feeding and show more-or-less normal cuticle changes; those produced by *daf-15* mutants are non-dauer on both counts, but do display some other dauer-like features. These two genes seem to act in parallel rather than sequentially in dauer morphogenesis.

No mutants are yet known which specifically block exit from the dauer state. However, there is evidence for differential regulation of the cell lineages which are executed on leaving the dauer state in order to complete L3 development (prior to re-entering the normal developmental sequence at the L3 moult). Thus heterochronic *lin-14* or *lin-28* mutations, which severely disrupt non-dauer larval lineages, have little effect on post-dauer larvae. This suggests that post-dauer development may be regulated differently from the normal L3 stage.

4.7 Genes specifying the mechanosensory neurones

The *C. elegans* nervous system is the most completely characterised of any animal, comprising 302 neurones plus 56 glial and supporting cells in the adult hermaphrodite (381 plus 92, respectively, in the male). The complete pattern of neuronal interconnections has been established by serial reconstruction from electron micrographs, and there is a wealth of information on the neurotransmitters and other differentiated features of many neurones. The ancestry of each cell in the nervous system is known, and large numbers of genetic mutations affect the generation and/or functioning of particular groups of neurones. In many cases, such mutations result in an uncoordinated (*unc*) phenotype because muscles are wrongly innervated, e.g. by insufficient or excessive numbers of neurones (as in *unc-86*$^-$ mutants; see §4.4.1), or by inappropriate or non-functional nerve cells. The major task now is to integrate this knowledge into a functional description of how the *C. elegans* nervous system actually works (see review by Chalfie & White, 1988). There is an obvious need for some way of simplifying this vast undertaking; one such approach is to ask how groups of neurones (together with the muscle cells which they innervate) cooperate to produce simple behavioural patterns.

One such model system concerns the response of the animal to touch (e.g. by a fine hair). Normal individuals writhe away from the stimulus (avoidance behaviour), and it is relatively easy to screen for mutants which fail to do so. These touch-insensitive mutants define a series of genes required for the generation, differentiation and functioning of the set of touch-sensitive (mechanosensory) neurones which mediate in the animal's avoidance response. There are five functional touch receptor cells of this type, designated ALML, ALMR, PLML, PLMR and AVM; a sixth cell, PVM, can also differentiate into a mechanosensory neurone if displaced to a more anterior position, as occurs in *mab-5⁻* mutants. The sensory processes of these cells extend along the length of the lateral lines and dorsal midline; they are characterised ultrastructurally by the presence of prominent microtubules with 15 (rather than the usual 11) protofilaments, hence their alternative name of microtubule cells. If all of these cells are deleted by means of laser ablation, the animal is unable to respond to touch. By deleting subsets of these cells, one can show that they are organised into two functional groupings; ALML, ALMR and AVM together confer anterior touch sensitivity, while PLML and PLMR are responsible for posterior touch sensitivity.

The neural connections underlying touch sensitivity appear to constitute a simple reflex circuit, namely receptor cell ⟶ interneurone ⟶ motor neurone, this last innervating the body muscles that execute avoidance behaviour. The mechanosensory receptor cells are connected via gap junctions to the interneurones (PVC cells posteriorly, AVD cells anteriorly), which also receive chemical synapses from the opposing set of touch receptors (Chalfie *et al.*, 1985). Thus each set of touch receptors activates one set of interneurones but inactivates the other, so that the animal moves away from the touch stimulus (e.g. an anterior stimulus will activate the AVD interneurones via one or more of the anterior receptors, but these will also inactivate the PVC interneurones posteriorly).

Mutations affecting any step in this reflex will display touch insensitivity; some also show more generalised defects, such as complete immobility if there is a major deficiency in the body muscles or motor neurones innervating them. In other mutants only the touch avoidance reflex is absent. Numerous mutants of this latter type have been isolated, and they fall into 18 complementation groups (Chalfie

& Sulston, 1981). These are: *mec-1* to *-10, mec-12, mec-14* to *-17, egl-15, unc-86* and *lin-32*. Both *unc-86+* and *lin-32+* functions are required for the generation of touch receptors; these cells fail to appear in mutants involving either gene, although there are multiple (pleiotropic) effects on other cell lineages as well. The homoeotic *mec-3+* function is uniquely required for the normal differentiation of mechanosensory cells; in *mec-3−* mutants these cells are formed in their normal positions (i.e. there is no lineage defect), but they lack all of their normal ultrastructural features such as microtubules. Recent analysis of the *mec-3* gene shows it to contain a variant homeobox sequence (see § 5.6, later) which encodes a DNA-binding protein domain (Way & Chalfie, 1988). These features suggest that *mec-3+* is a master switch gene which governs mechanosensory cell identity; presumably this control is exerted through a series of subordinate (downstream) genes which specify particular aspects of the structure and function of these cells (Chalfie & Au, 1989). Precedents from *Drosophila* imply that homeo-domain-containing proteins (in this case the *mec-3+* protein) may bind directly to their target genes.

Four genes (e.g. *mec-7*) affect specific differentiated features of the mechanosensory cells, such as the extracellular matrix overlying the touch receptors and the structure of the 15–protofilament micro-tubules. These are obvious candidates as target genes for the *mec-3+* protein. Also in this category may be several further genes whose mutant phenotypes are characterised by normal-looking but appar-ently non-functional mechanosensory cells (e.g. null mutants of *mec-4*). However, gene-functions in this last group need not be specific to the touch receptor cells; (for instance, mutations affecting the produc-tion of neurotransmitters by these cells would probably have effects on several other types of neurone). This would appear to be the case for both *mec-1−* and *mec-8−* mutants, where defects can be observed in some cells other than the touch receptors. However, the *mec-4+* func-tion seems to be required autonomously within the mechanosensory cells. This emerges from an unusual dominant *mec-4^D* mutation which causes all six cells to degenerate and die soon after their formation; this defect does not affect any other cells, nor can it be rescued by any of the *ced* mutations that normally prevent programmed cell deaths (see § 4.4.5 above). This further suggests that the deaths of these cells are a direct consequence of the *mec-4^D* mutation and are not due to the inappropriate activation of the cell-death program mediated by the *ced+* functions. Mosaic analysis confirms that autonomous expression of *mec-4+*, and probably also of *mec-3+* and *mec-7+*, is required within the mechanosensory cells themselves in order for their development to proceed normally (see e.g. Herman, 1987). Recent cloning and sequencing of the *mec-7* gene show that it encodes a specific β-tubulin protein, namely that from which the 15-protofilament microtubules are assembled (Savage *et al.*, 1989).

Thus overall there are three groups of genes involved in this pathway: *lin-32+* and *unc-86+* functions are required in the cell lineages which generate the mechanosensory cells; *mec-3+* function is essential for specifying their unique identity; finally, a larger set of differentiation functions is needed in order for these cells to express appropriate ultrastructural and biochemical specialisations (see review by Chalfie & Au, 1989).

A similar analysis indicates that some 35 genes are required to define the developmental pathway giving rise to the hermaphrodite-specific neurones (HSNs; Desai *et al.*, 1988). Mutations in some of these genes affect single HSN traits such as cell migration, axon outgrowth or serotonin expression. In other cases multiple traits are affected, and the wild-type products of certain genes in this category may execute regulatory functions (e.g. *egl-5*; cf. *mec-3* in the mechanosensory neurone pathway described above). Nearly all of the genes influencing HSN development are pleiotropic in effect, since mutants also show defects in other cell types. This indicates that these genes are involved in several developmental pathways, not just in that specifying the HSN phenotype.

5

Insect development

5.1 Overview of development in *Drosophila*

5.1.1 Introduction

It is perhaps appropriate that *Drosophila* should now occupy centre stage in any account of developmental genetics; over sixty years ago, Thomas Hunt Morgan left embryology to study genetics in the fruit-fly *Drosophila*. Though many of the mutations discovered then were found to affect developmental processes (see e.g. Morgan, 1934), it is only in the last decade that a coherent molecular picture of early *Drosophila* development has begun to emerge. Many details remain obscure, but it is now clear that the segmentation and homoeotic genes (governing the number/polarity and identity of metameric units respectively) represent 'master switches' controlling key aspects of early development.

The advantages which initially drew Morgan and others to *Drosophila* are not dissimilar to those afforded by *C. elegans* (see §4.1), namely its short life cycle, small genome size and ease of maintenance under laboratory conditions. Indeed, a formidable array of mutants – affecting every aspect of *Drosophila* development – is available from specialist stocks. However, *Drosophila* embryogenesis does not result in a fixed number of cells whose fates are largely dependent on their lineage (as in *C. elegans*); rather, the *Drosophila* embryo comprises an indefinite (but very large) number of cells whose fates depend more upon position than ancestry. Because the developmental patterns of insects are so different from those of other metazoan animals (especially sea urchins and vertebrates; chapter 2), the next subsection of this chapter (§5.1.2) will outline *Drosophila* development from oogenesis through to metamorphosis. This is followed by a brief consideration of two techniques fundamental to our present understanding of *Drosophila* development, namely P-element-mediated transformation (creating transgenic flies) and the genetic marking of clones of cells by X-ray-induced somatic recombination (§5.1.3). Section 5.2 will consider three aspects of later *Drosophila* development; the puffing of polytene chromosomes (§5.2.1) and the study of imaginal discs (§5.2.2) have provided classical insights into gene activity and cell determination, respectively, whereas the elucidation of the sex determination pathway (§5.2.3) exemplifies the mutual reinforcement between developmental genetics and molecular biology in this organism. The next three sections will concentrate largely on early development; dorso-ventral polarity and neurogenesis in §5.3, antero-posterior polarity and metamerisation in §5.4, and homoeotic gene expression in §5.5. Finally, §5.6 will consider the significance of the 'homeobox' and other sequence homologies which have become apparent between several vertebrate genes and many of the segmentation and homoeotic genes in *Drosophila*. Whether these vertebrate genes truly fulfil the criteria expected of developmental 'master switches' remains to be seen, though their patterns of expression during early development (often showing regional specificity) afford tantalising indications that this may be the case.

5.1.2 An outline of Drosophila development

Oogenesis

As described earlier in § 2.4.1, *Drosophila* undergoes meroistic oogenesis, with 15 polyploid nurse cells pouring their synthetic products into the oocyte via the ring canal system. Normally the nurse cell cluster is joined to the future anterior pole of the oocyte, which occupies a terminal position within the egg chamber (see fig. 2.4). In the maternal mutant *dicephalic*, however, two smaller clusters of nurse cells are joined to both anterior and posterior poles of the central oocyte; frequently this results in embryos with duplicated head/thoracic structures and no abdomen (Frey *et al.*, 1984). One possible explanation would invoke 'anterior determinants' of nurse cell origin (perhaps mRNA from the *bicoid* gene; see §5.4.2 below) which become sequestered close to their site of entry into the oocyte.

The nurse-cell/oocyte complex is enclosed by a layer of follicle cells, within which polytenisation of the chromosomes and selective amplification of the chorion protein (CP) genes take place during the later stages of oogenesis (§2.2.1). *In situ* hybridisation shows that some CP genes are spatially regulated; in particular, three minor CP mRNAs accumulate in patterns related to eggshell specialisations (micropyle, dorsal appendages, etc.), while transcripts from the X-linked cluster appear initially in a small group of dorsally located follicle cells. Temporal regulation is also apparent, the X-linked cluster being expressed during stages 10–13 of oogenesis while the 66D cluster is expressed from stage 13 on (Parks & Spradling, 1987). Mutations at the X-linked *dec-1* locus result in a failure to synthesise a 130 kd protein during stage 10 of oogenesis; normally, this polypeptide is cleaved to an 85 kd product which becomes incorporated into the chorion, and is later processed to a 67 kd component during stages 13–14. It is at this later stage that gross eggshell deformities begin to appear in *dec-1⁻* mutants (Hawley & Waring, 1988). The wild-type *dec-1⁺* gene generates a 4 kb transcript during stage 9–10 (encoding the 130 kd primary protein) and a 5.8 kb RNA by alternative splicing during stages 11–12.

During oogenesis the nurse cells synthesise a wide range of maternal RNAs and proteins which are subsequently passed to the oocyte (maternal RNA complexity approx. 12×10^6 nucleotides; Davidson, 1986). Among these products are determinants for dorso-ventral and for antero-posterior polarity, to be reviewed in § § 5.3 and 5.4 below. Other maternal products include mRNAs coding for actin, tubulin, and enzymes such as 6-phosphogluconate dehydrogenase (a long-lived mRNA which will be translated during embryogenesis: Gerasimova & Smirnova, 1979). The EH8 transcription unit is expressed both maternally during oogenesis (2 mRNAs) and zygotically during the blastoderm stage (3 mRNAs), using different combinations of exons (Vincent *et al.*, 1984). Prior to stage 10 of oogenesis, histone mRNAs are present in relatively low amounts and are utilised mainly to provide histones during nurse-cell polyploidisation. After stage 10, histone mRNAs accumulate rapidly in the oocyte, coinciding with the onset of nurse-cell degeneration and cessation of DNA synthesis. Much of the histone mRNA is associated with polysomes throughout, suggesting active translation and storage of newly synthesised maternal histones in the egg (Ruddell & Jacobs-Lorena, 1985).

Cleavage

The *Drosophila* egg is described as centrolecithal, comprising a central nucleus surrounded concentrically by cytoplasm, a thick layer of yolk, a thin shell of periplasm, and the outer chorion sheath. Fertilisation is effected by a single sperm which penetrates the chorion via the micropyle. Within the yolk mass, the zygote nucleus undergoes eight synchronous divisions (at *c.* 9-minute intervals at 25°C) without cytokinesis. Most of these nuclei migrate outwards to the periphery of the egg along with their associated cytoplasm (fig. 5.1*A*), until by the ninth division only a few scattered nuclei remain in the yolk mass. These latter cease division and become polyploid yolk nuclei. The peripheral nuclei form a single syncytial layer (the **syncytial blastoderm** stage; fig. 5.1*B*) and undergo four further synchronous divisions at a slower rate. It is only after the 13th division cycle that cell walls begin to separate each nucleus from its neighbours, and even then the base of each cell remains open to the underlying yolk mass (the **cellular blastoderm** stage; fig. 5.1*C*). Nuclear divisions prior to the interphase of cycle 14 are dependent on maternal factors; thereafter, the zygotic *string (stg⁺)* function is required for further mitoses. Thus null *stg⁻* mutants show cell-cycle arrest after the G2 phase of cycle 14 (Edgar & O'Farrell, 1989). Notably, the *stg⁺* product is related to the *cdc25* protein (required for cell division in fission yeast), and is expressed in dynamic patterns preceding mitosis in *Drosophila* embryos. After the globally synchronous divisions have ceased (cycle 14), 25 locally synchronous mitotic domains are established, correlating with commitment to specific developmental fates (Foe, 1989).

Fig. 5.1 Early *Drosophila* embryogenesis (based on Akam, 1987).

A Nuclear migration (1.25 h). 7th to 8th cleavage. Nuclei migrate outwards towards periphery of egg.

B Syncytial blastoderm (2.0 h). 12th cleavage. Most nuclei reach periphery but remain syncytial (no cell walls separating them). Pole cells form at posterior pole (containing polar granules).

C Cellular blastoderm (2.5 h). 14th cleavage. Cell walls begin to form between nuclei but do not close off the basal ends of cells, which remain open to the central yolk mass.

D Gastrulation (3 h). Mesoderm invaginates ventrally, while anterior and posterior midgut invaginations form endoderm. Pole cells lie within the pmg. Cell walls (not shown) enclose most nuclei.

E Germ band extension (4.5 h). The pmg moves dorsally and forwards, so extending the germ band. This becomes subdivided into metameric units (see § 5.4.2), the most posterior of which are apposed to the head.

•   nucleus; ⌐-⌐ yolk; ⌐·⌐ polar granules; yn, yolk nucleus; pc, pole cells; pmg, posterior midgut; amg, anterior midgut; ms, mesoderm; st, stomodeum; gb, germ band (metameric); ec, ectoderm.

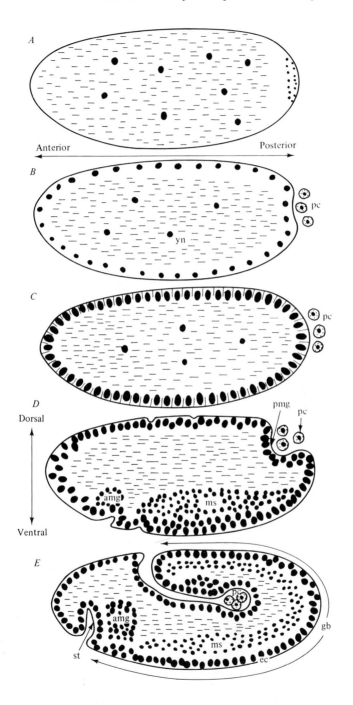

The late syncytial blastoderm stage marks the onset of active transcription for many zygotic genes, with the synthesis of r-, t-, 5S-, sn- and poly(A^+) RNA species beginning during the 11th and 12th division cycles and reaching a maximum by the 14th (Edgar & Schubiger, 1986). These RNAs are all transcribed during the (lengthening) G2 phase of the cell cycle, whereas histone genes are transcribed during S phase (i.e. histone synthesis is now coupled to DNA synthesis, in contrast to the situation during late oogenesis). However, maximal transcriptional activation can be induced precociously as early as the 10th division cycle (but not before) by cycloheximide treatment, which slows down the cell cycle. Thus many genes may become competent for transcription during cycle 10, but their subsequent activation is linked to the lengthening of the mitotic cycle. Some genes are transcribed even earlier than this, in preblastoderm embryos (despite the rapid rate of nuclear division); labelled RNAs from such embryos have been used to screen a *Drosophila* genomic library so as to identify genes expressed during the initial stages of embryogenesis (Sina & Pellegrini, 1982). There are also developmental changes in the pattern of protein synthesis in early *Drosophila* embryos, as revealed by 2D gel analysis (Summers *et al.*, 1986). Ribosomal proteins are prominently synthesised in preblastoderm embryos (Santon & Pellegrini, 1980), though few become incorporated into ribosomes until after cellularisation (which marks the onset of new rRNA synthesis).

Polar granules and germ-cell determination

A few blastoderm nuclei (5–10) do not remain syncytial, namely those entering the posterior pole region containing the characteristic **polar granules**. Such nuclei become segregated into **pole cells** (along with the polar granules) at the time of the 10th nuclear division cycle in the rest of the blastoderm (fig. 5.1*B*). These pole cells divide asynchronously once or twice, and about half of them eventually reach the gonads to become germ cells. UV irradiation of the posterior pole of a *Drosophila* zygote (prior to outward migration of the nuclei) results in a failure of pole-cell formation and hence sterility in the adult (no germ cells). This defect can be rescued by microinjecting a UV-irradiated zygote with cytoplasm from the posterior pole region of an unirradiated egg; pole cells then develop normally from any nuclei migrating into the injected region, and these give rise to functional germ cells. This implies that germ-cell determinants must be localised in the posterior pole cytoplasm, possibly associated with the polar granules. A more rigorous proof that these determinants are indeed cytoplasmic was provided by Illmensee & Mahowald in 1974; their experiment, involving transplants between three genetically marked strains of *Drosophila*, is summarised in fig. 5.2. Note that pole cells are formed even when posterior pole cytoplasm has been injected at an unusual (anterior) site.

Polar granules have been purified biochemically from early *Drosophila* embryos (Waring *et al.*, 1978), but neither these nor their major protein constituent (a basic polypeptide of 95 kd) have been shown to restore pole or germ-cell formation in UV-irradiated embryos. More recent evidence suggests that several interacting factors are required. A subcellular fraction prepared from *Drosophila* eggs can induce the formation of pole cells (but not germ cells) when injected into the posterior (but not anterior) pole of UV-irradiated eggs (Ueda & Okada, 1982). The UV-sensitive component present in this fraction is probably maternal poly(A)$^+$RNA (Okada & Togashi, 1985; Togashi *et al.*, 1986), but evidently this must interact with one or more UV-resistant factors present only in the posterior pole region in order to induce pole-cell formation. Additional factors would be required to ensure that these pole cells go on to develop into germ cells.

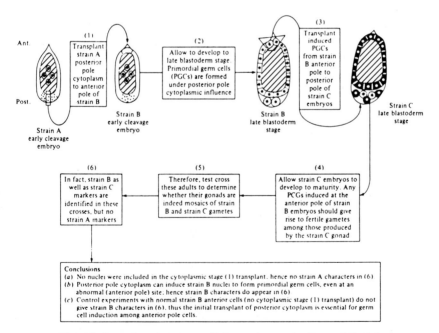

Fig. 5.2 Cytoplasmic germ cell determinants in *Drosophila* (after Illmensee & Mahowald, 1974). Ant, anterior pole; Post, posterior pole; • nucleus; ▨ yolk; ▦ polar granules.

A genetic approach to this problem utilises the so-called grand-childless group of maternal mutations: all offspring of mutant mothers are sterile. Strong mutant alleles of the *tudor* gene lead to an absence of polar granules in all eggs, with the result that no pole cells are formed during embryogenesis (Boswell & Mahowald, 1985); however, these embryos also show abdominal segmentation defects, suggesting that the maternal *tud*+ product functions in the determination of A/P pattern (see § 5.4.2) as well as in pole-cell formation. Another gene in this class is *vasa*, whose wild-type product is required specifically in the female germ line; mutant *vasa*− males, however, are phenotypically normal. The wild-type *vasa* gene encodes a protein related to the eucaryotic translational initiation factor 4A (eIF4A) and other ATP-dependent helicases (Lasko & Ashburner, 1988; Hay *et al.*, 1988), and may function in binding to/unwinding RNA. How this in turn participates in the development of the female germ line remains to be seen. However, it is noteworthy that the *vasa*+ transcript is abundant only in the female germ line and in early embryos (Lasko & Ashburner, 1988). Although *vasa*+ transcripts are present throughout the egg, a monoclonal antibody localises the corresponding protein primarily at the posterior pole of the oocyte, to be sequestered later into the pole cells during early embryogenesis (Hay *et al.*, 1988). Another product which becomes localised in the pole cells is the mRNA encoding cyclin B, whereas the cyclin A messenger is distributed throughout the early developing embryo (Whitfield *et al.*, 1989).

Blastoderm fate map

A fate map of the cellular blastoderm stage (fig. 5.3) indicates which tissues and structures derive from which regions of the two-dimensional sheet of blastoderm nuclei (note that fig. 5.3 is a side view, representing one half of the blastoderm). The central region (enclosed by a thick line) is destined to give rise to the metameric germ band, from which derive the segmented structures characteristic of the larva and adult (see § 5.4). Hatched areas are those which will invaginate during gastrulation. Mesoderm invaginates ventrally, while anterior and posterior midgut invaginations form the endodermal primordia (fig. 5.1*D*). Subsequently, about one quarter of the cells within the ventral neurogenic region begin to migrate inwards and develop into neuroblasts, while the remainder form ventral epidermis; this choice will be considered in more detail in § 5.3.2 below. It will be noted that position along the D/V axis (§ 5.3.1) is a primary determinant governing whether cells within the germ band become mesoderm, ventral

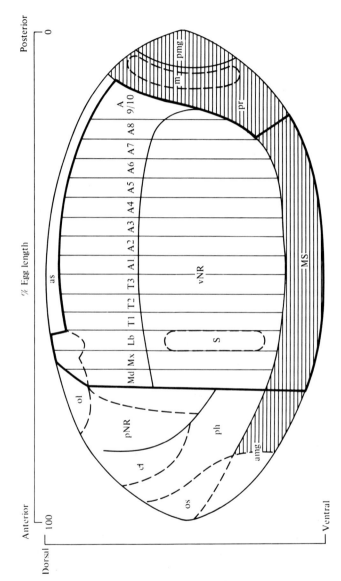

Fig. 5.3 Fate map of the *Drosophila* blastoderm (modified from Campos-Ortega & Hartenstein, 1985, and from Akam, 1987). Hatched region will invaginate at gastrulation. Region enclosed by heavy line will form the metameric germ band, comprising: Md, mandibular segment; Mx, maxillary segment; Lb, labial segment; T1–T3, thoracic segments; A1–A10, abdominal segments. as, amnioserosa; ol, optic lobe; pNR, procephalic neurogenic region; cl, clypeolabrum; ph, pharynx; os, oesophagus; amg, anterior midgut; dE, dorsal epidermis; vNR, ventral neurogenic region; S, salivary gland; MS, mesoderm; pr, proctodeum; m, Malpighian tubules; pmg, posterior midgut.

neuroblasts/epidermis, or dorsal epidermis. After gastrulation, the germ band becomes extended as the posterior midgut invagination moves dorsally and forwards (carrying the pole cells with it), until the the posterior abdomen comes to lie adjacent to the dorsal head structures (fig. 5.1*E*). At this stage the metameric subdivision of the germ band is definitively established, as discussed below in §5.4. The germ band later shortens again, and the embryo hatches into a first instar larva.

Larval development and metamorphosis
During larval life the majority of nuclei become **polytene** through repeated rounds of DNA replication without separation of the daughter chromatids, such that numerous DNA duplexes (chromatids) come to lie side by side in each chromosome. In some larval tissues (particularly salivary gland) the high degree of polytenisation and perfect alignment of chromatids result in giant chromosomes with characteristic banding patterns clearly visible under the light microscope. Highly active gene sites within such chromosomes can be distinguished as expanded regions called puffs; before the advent of recombinant DNA technology, these provided one of the few systems in which differential gene activity could be investigated. A few classic studies in this area are reviewed briefly in §5.2.1 below.

One consequence of the polytenisation process is that many larval cells cannot divide, but merely grow larger as development continues. In *Drosophila* there are three larval instars separated by moults when the exoskeleton is shed and regrown to accommodate extra growth. Each moult is triggered by the steroid hormone ecdysone (β-ecdysterone), whose action is modulated during larval life by high concentrations of the juvenile hormone. The level of ecdysone rises and then falls during moulting, but remains low through the intermoult periods. At the end of the final larval instar, holometabolous insects such as *Drosophila* undergo **metamorphosis**, which results in a radical change of body form. This process is triggered by two massive surges in ecdysone concentration at a time when the juvenile hormone titre has fallen to low levels. The first surge stimulates the secretion of the pupal case (puparium), within which the transformation from larva to adult (imago) takes place under the influence of the second surge. In some insects (e.g. the cecropia moth) there is a prolonged diapause between these two ecdysone surges, such that the insect overwinters as a pupa.

The polytene larval tissues are not simply remodelled during metamorphosis; rather, many of them are discarded wholesale and broken down. The new structures of the adult develop instead from groups of diploid **imaginal cells** which have been present throughout larval life, though playing no active part in larval development. These cells are set aside in small groups during embryogenesis; they remain diploid and divide repeatedly without overt differentiation during the larval instars. Imaginal cells are present in the larva both as **imaginal discs** (which will give rise to the adult cuticular structures of the head, thorax and genitalia) and as **imaginal histoblast nests** (which will give rise to adult abdominal structures). Groups of imaginal cells become **determined** in a stepwise fashion during larval life, such that each group forms one particular structural element of the adult cuticle during metamorphosis; this topic will be reviewed briefly in § 5.2.2 below.

5.1.3 Two basic techniques in Drosophila developmental genetics

(a) P-element-mediated transformation of the germ line

P elements are transposable genetic elements approximately 3 kbp in length, which are present in multiple copies at dispersed locations in the genomes of most *Drosophila melanogaster* individuals from wild populations (P strains). However, most established laboratory stocks of this species lack P elements and are designated M strains; possible reasons for this difference are discussed in a recent review by Engels (1988), as are P-element insertion/excision mechanisms and other topics beyond the scope of this brief account.

Crosses between P strain males and M strain females result in **hybrid dysgenesis**, which is manifested as a failure to produce many viable germ cells in the offspring. This is due to the widespread mobilisation of paternal P elements, such that they transpose to many new genomic positions, frequently causing mutations in essential genes at their sites of insertion. This transposition is specifically restricted to germ-line cells, and does not occur at appreciable rates in somatic cells. By contrast, crosses within P or M strains, or between M males and P females, do not undergo P-element transposition in the germ line.

The P element DNA is largely occupied by a single four-exon gene encoding a transposase enzyme (see fig. 5.4*A*), together with terminal and subterminal inverted repeats (of 31 and 11 bp): there are also 8 bp direct repeats of recipient site DNA on either side of the P element, presumably formed during insertion (O'Hare & Rubin, 1983). P-

A

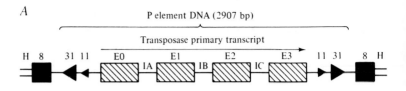

Structure of a P element (after Engels, 1988). H, host DNA; 8, 8 bp direct repeats of host DNA; 31 and 11, 31 bp terminal and 11 bp subterminal inverted repeats; E0–E3, four exons encoding (mostly) the transposase enzyme; IA–IC, 3 introns in transposase gene (IC is not spliced out in somatic cells).

B Use of P elements for germline transformation.

(i) Insert gene X plus controlling sequences (or gene X regulatory regions fused to bacterial reporter gene) into P element.

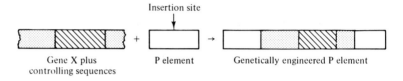

(ii) Introduce both normal and modified P elements into early *Drosophila* embryos (homozygous X⁻ strain if using gene X, rather than a reporter gene). Transposase enzyme encoded by the normal (helper) P element will allow P elements of both types to transpose to many sites in the genomes of germ cells only.

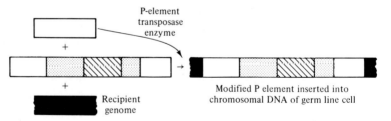

(iii) Screen surviving progeny from transformed germ cells for presence and expression of gene X (or of bacterial reporter gene subject to spatial/temporal/quantitative regulation typical for wild-type gene X).

Key (part *B* only) ▢ . P element DNA; ▨ , gene X or bacterial reporter gene; ■ , recipient genomic DNA; ▦ , regulatory 5' and/or 3' sequences controlling expression of gene X.

Fig. 5.4 P elements.

element mobilisation (excision and insertion) absolutely requires the functional P-element transposase enzyme, a protein of 87 kd which can be produced in germ-line but not somatic cells. This has been inferred from the discovery that P-element transcripts are present but incorrectly spliced in somatic cells (Laski *et al.*, 1986), in that the intron between exons 2 and 3 is not removed. This intron is correctly spliced out in germ-line cells, a feature dependent on key sequences close to and within this 190 bp intron (Laski & Rubin, 1989).

While this explains the mobilisation of P elements in the germ-line cells of progeny from crosses between P males and M females, it does not account for the stability of such P elements within P strains. One model suggests the presence of a transposase repressor protein in P strains, perhaps encoded by part of the transposase gene itself (by analogy with bacterial transposons). This repressor would be present in the oocytes of all P-strain flies (P cytotype, hence no P-element mobilisation), but not in the oocytes of M-strain flies which lack P elements (M cytotype, allowing widespread mobilisation of paternal P elements from P-strain sperm).

To create transgenic flies, P elements are first genetically engineered by inserting copies of a cloned *Drosophila* gene plus all of its putative regulatory regions (at least 12 kbp of extra DNA can be accommodated). Often this will disrupt the transposase gene, hence a mixture of normal (helper) and engineered P elements is introduced into early M-strain *Drosophila* embryos – the former supplying the essential transposase enzyme. Among the (few) viable germ cells which develop, some will contain the modified P element inserted into a non-essential DNA site. Progeny derived from such germ cells can then be screened for the presence and expression of the introduced gene (fig. 5.4*B*). In this way, functional genes encoding alcohol dehydrogenase (Goldberg *et al.*, 1983), xanthine dehydrogenase (Spradling & Rubin, 1983) or dopa decarboxylase (*Ddc*; Scholnick *et al.*, 1983) have been introduced via P elements into recipient embryos which were homozygous mutants for that function; in all three cases the introduced gene is expressed in the progeny of such recipients with largely normal stage- and tissue-specificity.

Of course, viable homozygous mutants are not available if the gene under study is essential for normal development. In these cases, the regulatory regions from such a gene can be linked to a bacterial reporter gene (e.g. that encoding chloramphenicol acetyltransferase or β-galactosidase) and introduced via P elements into wild-type embryos. Some progeny of these embryos will express the foreign bacterial gene with the same spatial and/or temporal specificity as the essential gene under study (whose normal product is still available, hence allowing full development). This approach can be combined with promoter deletion studies (see §1.3.1) to identify DNA regions essential for particular aspects of a gene's regulation (e.g. Hiromi *et al.*, 1985).

Although the positions and numbers of P-element insertions are unpredictable, genes contained within such elements mostly function autonomously (however, position effects are detectable in some cases; see e.g. Spradling & Rubin, 1983). Overall, the expression of transgenes carried on *Drosophila* P elements mirrors the specificity of their normal counterparts more faithfully than is often the case in transgenic mice (see §1.3). This may reflect control mechanisms involving interactions between widely spaced DNA elements in vertebrates, whereas most *Drosophila* genes are regulated by *cis* elements acting over distances of no more than a few kilobase pairs. However, some of the *Drosophila* genes involved in metamerisation have extraordinarily long regulatory regions (see §§5.4, 5.5).

(b) *Genetic marking of cells and their clonal descendents*

When *Drosophila* embryos or larvae are exposed to moderate doses of X-rays, somatic recombination is induced in a small proportion of cells. (Normally, recombination is confined to meiosis in the germ cells, but very rarely it also occurs during mitosis in somatic cells; both processes involve crossing-over between homologous chromosomes.) One can take advantage of this if the animals to be irradiated are heterozygous for a suitable recessive mutation, i.e. one which when homozygous alters the adult cuticle characteristics in a cell-autonomous manner. In such a situation, somatic crossing-over during mitosis will sometimes result in one double-dominant and one double-recessive daughter cell (fig. 5.5). Provided this occurs among the ancestors of the cuticle-secreting epithelial cells, the phenotypes of these daughter cells will be respectively wild-type (like the heterozygous background) and mutant in character. Descendants of the mutant cell will form a **marked clone** in the adult structure(s) derived from that group of cells; that is, a region of cells distinguished by the mutant character of the overlying cuticle (e.g. altered colour or presence of abnormal hairs). The principle is illustrated in fig. 5.5 for a recessive mutation (*mwh*) which when homozygous causes multiple rather than single hairs to appear on the wing blade (see Garcia-Bellido *et al.*, 1979). The rarity of X-ray-induced somatic recombination events

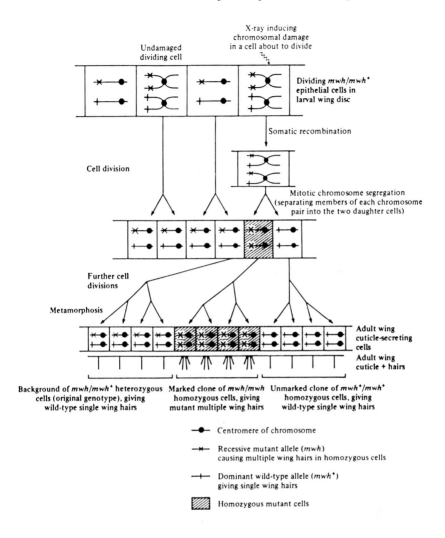

Fig. 5.5 Marking cell clones by means of X-ray-induced somatic crossing-over (modified from Garcia-Bellido *et al.*, 1979).

ensures that only a single cell (and its clonal progeny) will become marked; the probability of two adjacent cells undergoing similar recombination events at the same time is exceedingly remote. A modification of this technique involving the dominant *Minute* mutation will be discussed later (§5.4.1) in the context of compartment boundaries in *Drosophila*.

5.2 Aspects of later development in *Drosophila*

5.2.1 Puffing

The giant polytene chromosomes of the salivary glands in dipteran larvae have provided a unique opportunity to visualise patterns of gene expression under the light microscope. Though largely superseded by the advent of recombinant DNA technology, this approach has elucidated several classic examples of gene switching, some of which are dealt with below. The denser **bands** visible on polytene chromosomes broadly correlate in number and position with known genetic loci as identified by linkage studies (e.g. Judd & Young, 1973; Hochmann, 1973), although each band contains far more DNA in linear terms than would be required for a typical gene. Indeed, some bands have been shown by molecular analysis to contain small clusters of genes with related functions; examples include the 66D group of chorion protein genes (§2.2.1) and the 68C locus described below. The less-condensed interbands seem to represent non-coding DNA which may have a regulatory or spacer function. Recently, this has been demonstrated directly for the large *Notch* gene (§5.3.2), which itself occupies a single cytological band while its 5′ regulatory regions lie in the adjacent interband (Rykowski *et al.*, 1988).

Bands which are transcriptionally active show several distinctive features compared with inactive bands.

(i)   Radioactive uridine is incorporated into RNA at active but not inactive band sites, as revealed by *in vivo* labelling and autoradiography.

(ii) There may be gross changes in the state of DNA condensation at active sites, giving rise to expanded regions known as **puffs**. These are formed from the condensed DNA of a band by a process similar to untwisting the strands of a rope (fig. 5.6*A*; see Beerman, 1964). Such puffs are sites of active RNA synthesis as defined by criterion (i) above. However, RNA synthesis also occurs at some non-puffed band sites (Bonner & Pardue, 1977a). Probably puffing represents a modification of the DNA organisation required for very high levels of transcriptional activity (cf. lampbrush chromosome loops; §2.4.2).

(iii) *In situ* hybridisation of the polytene chromosomes with a specific labelled probe (whether RNA, cDNA or cloned DNA) allows identification of the active puff site which is engaged in synthesising the corresponding primary transcript (fig. 5.6*B*i; Lambert, 1972). In tissues *not* expressing the probe sequence as RNA, this same approach can be used to map the gene to one particular band on the polytene chromosomes (fig. 5.6*B*ii).

(iv) RNA polymerase II is present on polytene chromosomes at active puff sites. However, fluroescent-tagged antibodies against this polymerase label many inactive interband regions as well as the puffs, whereas similar antibodies against RNA polymerase I label only the nucleoli (fig. 5.6*C*i and ii; Jamrich *et al.*, 1977a, b).

Puffing patterns are both stage- and tissue-specific, as would be expected if the puffs represent gene sites engaged in particularly active transcription; some examples of this specificity are outlined below.

(a) *Balbiani ring 2 in* Chironomus

In the larval salivary glands of the midge *Chironomus*, several puffs attain enormous sizes and are termed Balbiani rings (BRs). These are sites of intense transcription for RNAs encoding a range of large proteins which are secreted in massive quantities by the glands (Grossbach, 1973). Probably BRs 1 and 2 encode polypeptide fractions 3 and 2, respectively (Pankow *et al.*, 1976). All of the BR genes are expressed coordinately and maximally during the second larval instar, but subsequently show different patterns of expression in prepupae (Lendahl & Wieslander, 1987).

BR2 is located on chromosome IV; in salivary glands this puff is so large that it can be microdissected out in order to study its structure and RNA products. In other polytene larval tissues such as Malpighian tubules this chromosomal site is not puffed, and *in situ* hybridisation with BR2 RNA identifies a single band in region 3B10 on chromosome IV (fig. 5.6*B*ii; Derksen *et al.*, 1980). This band contains some 470 kbp (linear length) of DNA, far greater than the known length of the major BR2 transcription unit (37 kbp; Lamb & Daneholt, 1979). However, at least two different transcripts are derived from the BR2 locus, designated BR2α and BR2β RNAs; both are expressed in second instar larvae, but whereas the latter accumulates transiently in prepupae, the former does not (Lendahl & Wieslander, 1987), implying differential control.

After spreading for EM, each BR2 transcription unit can be visualised as a Christmas-tree-like structure (fig. 5.7) reminiscent of active rRNA genes in *Xenopus* oocytes (fig. 2.1*B*). Very active transcription is implied by the close packing of polymerases and product

Fig. 5.6 Transcription in polytene chromosomes.

A

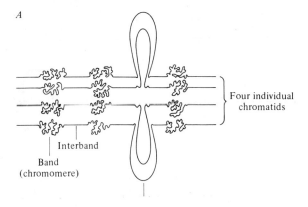

A Puffing (diagrammatic, after Beerman, 1964).

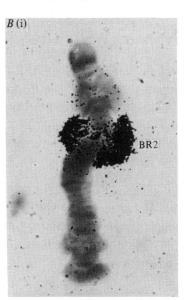
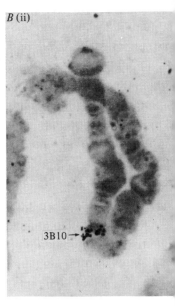

B In situ hybridisation of radioactive RNA from Balbiani ring 2 to *Chironomus* polytene chromosome IV. Phogoraphs kindly supplied by Prof. B. Daneholt (Karolinska Institutet, Stockholm).
(i) Hybridisation to the active BR2 puff site on chromosome IV in a salivary gland nucleus. Reprinted with permission from B. Lambert (1972) *Journal of Molecular Biology* 72, 65–75. Copyright 1972 by Academic Press Inc. (London) Ltd.
(ii) Hybridisation to the inactive 3B10 band on chromosome IV in a Malpighian tubule nucleus. From L. Wieslander and B. Daneholt, unpublished.

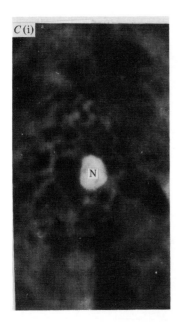 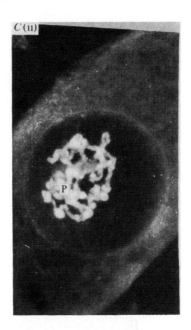

C Immunofluorescent localisation of RNA polymerases I and II in
Drosophila salivary gland polytene nuclei. Photographs kindly
supplied by Prof. E. K. F. Bautz (Molekular Genetik, University of
Heidelberg), from the thesis of M. Jamrich, University of Heidelberg, 1978. Used with permission.
(i) Nucleus stained with fluorescent-tagged antibodies against RNA
polymerase I. Staining essentially confined to nucleolus (N). Magnification ×220.
(ii) Nucleus stained with fluorescent-tagged antibodies against
RNA polymerase II. Staining confined to the polytene
chromosomes (P), particularly at expanded puff sites. Magnification ×190.

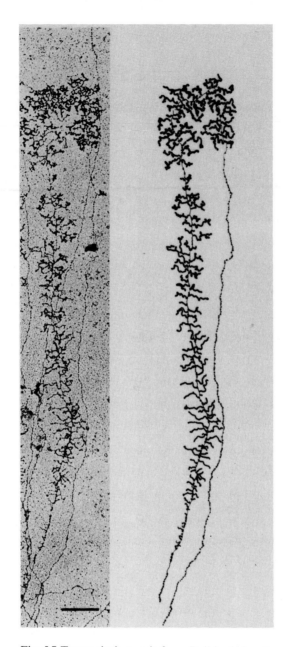

Fig. 5.7 Transcription unit from Balbiani ring 2 on chromosome IV in a *Chironomus* salibary gland nucleus. Electron migrograph of a spread preparation, by courtesy of Prof. B. Daneholt (Karolinska Institutet, Stockholm). Reprinted with permission from the copyright holders, *Cell*. From M. M. Lamb & B. Daneholt (1979) *Cell* **17**, 835–48. Note side-branches (representing nascent 75S RNA transcripts) folding into knobbed RNP structures as transcription proceeds. Bar at foot of electron micrograph (left) represents 0.5 μm.

chains. As synthesis proceeds, the nascent transcripts fold into compact RNP particles through association with nuclear proteins. Completed particles of similar size and morphology are abundant in the nucleoplasm, and some electron micrographs show them apparently squeezing through nuclear pores *en route* to the cytoplasm (Daneholt *et al.*, 1979).

The RNA synthesised at the BR2 site is somewhat unusual in two respects. Firstly, it is very large (75S or 37 kb) – much longer than would be required to encode a single copy of the fraction 2 protein. Secondly, some at least of this 37 kb RNA is transported intact into the cytoplasm, i.e. without major trimming or intron removal. Huge polysomes containing 60–100 ribosomes are found in salivary gland cytoplasm (Daneholt *et al.*, 1977; Francke *et al.*, 1982); these contain mRNA strands ranging up to 37 kb in size and complementary to BR2 DNA (Wieslander & Daneholt, 1977). Given the likelihood of some RNA degradation during isolation, this suggests that BR2 transcripts undergo little if any size alteration within the nucleus; they are apparently exported intact (albeit rather slowly) for translation in the cytoplasm.

(b) *Ecdysone-responsive puffs in* Drosophila

During the late third larval instar in *Drosophila* (prior to puparium formation), a group of new puffs appears in salivary gland nuclei in response to rising concentrations of ecdysone. Puffs which were active during the intermoult period (e.g. those encoding the larval glue proteins; see below) mostly regress in response to the hormone. The new puffs induced by ecdysone fall into two distinct classes, termed early and late (Ashburner, 1972; Ashburner *et al.*, 1973). The early puffs (e.g. 23E, 74EF and 75B) begin to appear within minutes of hormone administration, reach their maximum sizes within 1–4 hours, and then regress. By contrast, the late puffs (e.g. 22C, 63E and 82F) appear only after a lag of about 3 hours and reach their maximum sizes some 5–7 hours later, before eventually regressing. Inhibitors of protein synthesis such as cycloheximide block induction of the late puffs, implying that one or more proteins specified by the early puffs may be responsible for activating the late puffs. Since cycloheximide does not block the appearance of early puffs, these must be directly induced by ecdysone itself. Moreover, two early puffs (74EF and 75B) fail to regress in the presence of cycloheximide, which further suggests that these early puffs are normally switched off by their own protein products. Finally, if ecdysone is washed out during the first 4 hours of treatment, then several late puffs are prematurely induced, implying that ecdysone itself may normally delay puff formation at these sites. The overall system of checks and balances is shown in fig. 5.8*A* (Ashburner *et al.*, 1973), together with examples of ecdysone-induced puffing changes at the tip of the X chromosome in fig. 5.8*B*.

Walker & Ashburner (1981) have confirmed the main features of this model by using aneuploid genotypes in which the chromosomal regions bearing two of the early puffs are either increased or reduced in dosage. With extra copies of these early puff sites, the response of some (but not all) late puffs is greater and more rapid; moreover, the early puffs are active for a shorter period. The converse is true when fewer copies of the early puff sites are available, in that the period of early puff activity is prolonged and the response of the same set of late puffs is reduced. These observations confirm the dual role of early puff products in repressing their own synthesis and activating certain late puffs. Indeed, the activity of a single early puff (the *ecs* locus at 2B3) may be required to spread the effects of ecdysone to other chromosomal sites (Dubrovsky & Zhimulev, 1988). Mutations at the *ecs* locus tend to diminish the response of other ecdysone-inducible sites, whereas an increased dosage of wild-type *ecs* results in more rapid inducibility. This suggests that the *ecs*$^+$ product might be a *trans*-acting factor required for the response of other loci to ecdysone.

In situ hybridisation shows that transcripts from the early puffs 74EF and 75B appear in the RNA population of salivary gland cells only after ecdysone treatment (Bonner & Pardue, 1977b). Thus puff induction genuinely reflects the switching-on of transcription at specific

Fig. 5.8 Puffing in polytene chromosomes.

A Model for ecdysone induction of early and late puffs (modified from Ashburner *et al.*, 1973).

B Puffing patterns in *Drosophila* salivary-gland polytene chromosomes following ecdysone treatment in culture. Photographs by courtesy of Dr M. Ashburner (Dept. of Genetics, University of Cambridge) and reprinted with permission from the copyright holders, Springer-Verlag. From M. Ashburner (1972) *Chromosoma* 38, 255–81. This series of photographs shows the tip of the X chromosome; (a) control (no ecdysone); (b), (c), (d) and (e) after 1, 2, 8 and 12 hours of culture in the presence of ecdysone. An intermoult puff at site 3C regresses in culture (independent of ecdysone), while band 2B5–6 forms a small early puff in response to the hormone. This puff has regressed by 12 hr (part (e)).

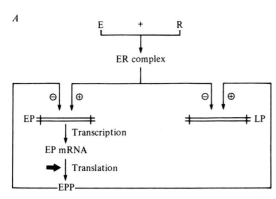

E   Ecdysone
R   Receptor protein specific
    for ecdysone
EP Early puff site(s)
LP Late puff site(s)
EPP Protein product(s)
    from early puff(s)
⊕   Activation of puffing
⊖   Inhibition of puffing
    (i.e. puff regression)
➡   Blockage by cycloheximide

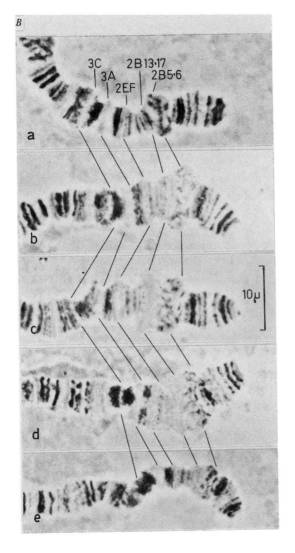

gene sites within the puffed region. Among the new RNA species synthesised in response to ecdysone is the messenger coding for dopa
decarboxylase (*Ddc*), an enzyme involved in cuticle cross-linking and
pigment production in hypodermal cells (Kraminsky *et al.*, 1980), but
also expressed in some parts of the central nervous system (particularly in catecholaminergic neurones: Konrad & Marsh, 1987). *Ddc* gene
expression peaks cyclically in response to ecdysone during larval
development and metamorphosis; however, this hormone responsiveness
is apparently confined to larval tissues and is not found in cultured
imaginal discs, where *Ddc* expression only begins some 6 hours after
ecdysone withdrawal (Clark *et al.*, 1986). Multiple DNA regions regulate the different tissue-specific patterns of *Ddc* expression in
hypodermal, neuronal and glial cells; for instance, a far-upstream
enhancer sequence is needed for neuronal but not glial expression
(Beall & Hirsh, 1987; Johnson *et al.*, 1989).

Ecdysone binds specifically to many of the chromosomal sites
whose expression it directly regulates. This has been demonstrated by
photo-crosslinking the hormone molecules to their binding sites and
then using fluorescent-tagged antibodies against ecdysone to localise
those sites on the polytene chromosomes (Gronemeyer & Pongs, 1980).
Ecdysone is bound both to early puff sites (see above), and also to
those intermoult puffs such as 68C (see below) which regress in
response to the hormone; however, specific hormone binding does not
completely explain the complex puffing changes induced by ecdysone
(Dworniczak *et al.*, 1983). A similar technique has been used to identify
the ecdysone receptor as a 130 kd cellular protein (Schaltmann &
Pongs, 1982) which shows high-affinity binding of ecdysone and rapid
accumulation of hormone within the nucleus. Essentially this system
is similar to the vertebrate examples of hormonal induction discussed
in §3.3, although there is clear evidence for repression of previously
active genes by ecdysone.

Among the puffs which regress in response to ecdysone are those
encoding the six major (and several minor) salivary-gland glue proteins (Beckendorf & Kafatos, 1976). These glue-protein (*sgs*) gene products accumulate within the salivary glads during the final larval
instar; their synthesis then ceases and the accumulated proteins are
released during puparium formation, the glue serving to attach the
pupa to its substrate. (The BR products in *Chironomus* perform a similar function.) The *Drosophila sgs* genes are present at several
chromosomal locations, including one (*Sgs*-4) which is X-linked and
shows dosage compensation (hyperexpression in XY males so as to

match the levels in XX females). This regulation can be mimicked in underproducing mutants by reintroducting a wild-type *Sgs*-4 gene plus flanking sequences into an X-chromosomal site via P-element transformation (Krumm *et al.*, 1985). The intermoult puff at 68C on chromosome 3 includes the major *sgs*-3 glue-protein gene (encoding a 1.1 kb RNA), together with two minor glue-protein genes (*sgs*-7 and -8) encoding RNAs of 0.36 and 0.32 kb (Crowley *et al.*, 1983). All three of these genes are clustered within a 5 kbp region (Meyerowitz & Hogness, 1982), the *sgs*-7 and *sgs*-8 genes lying in opposite orientations with adjacent promoters. The characteristic pattern of developmental regulation for all of these *sgs* genes has facilitated detailed analysis of the *cis*-acting DNA elements responsible, using P-element-mediated gene transfer and promoter deletion techniques. Whereas the control elements for *sgs*-5 expression lie close to the cap site (Shore & Guild, 1987), a more distant enhancer element is required for optimal expression of *Sgs*-4 (Shermoen *et al.*, 1987); this latter shows some functional redundancy (Jongens *et al.*, 1988). In the case of *sgs*-3, promoter elements close to or within the gene govern the tissue- and stage-specificity of its expression (Raghavan *et al.*, 1986), whereas a distant upstream sequence is required for maximal expression (Bourouis & Richards, 1985).

5.2.2 Imaginal discs

Much of the adult (imago) is built up during metamorphosis from reservoirs of imaginal cells which have remained diploid throughout larval life. As mentioned earlier (§5.1.2), these are present in the larva both as imaginal discs and as imaginal histoblast nests, of which only the former are considered here. Each imaginal disc comprises a single layer of ectodermal epithelial cells that will secrete a defined part of the adult cuticle during metamorphosis. Underlying this layer are adepithelial cells, including myoblasts which will give rise to the adult musculature in the region specified by each disc. There are 19 imaginal discs in *Drosophila* larvae, comprising nine pairs (bilaterally symmetrical) of head and thoracic discs plus a single fused disc which forms the posterior genital/anal structures of the adult. Three pairs of discs give rise to the labial, clypeolabral and eye/antennal structure of the adult head, while six pairs generate the dorsal prothorax, wings, halteres, first, second and third legs of the adult thorax.

As implied by this classification, each disc is able to form only a limited region of the adult exoskeleton, and is thus **determined** to follow a certain path of development during metamorphosis. By the late larval stages, this is also true for small groups of cells within a given disc, each group giving rise to one particular element of the adult structure (e.g. the different proximo-distal regions of a leg). Thus a mature imaginal disc can be viewed as a mosaic of determined and largely autonomous cell regions, each responsible for generating one element of the final cuticular pattern. Usually damage to one part of a disc will result in the absence or malformation of the corresponding adult element. However, if a sufficient period of growth elapses between damage and metamorphosis, then either reduplication or regeneration may occur (Schubiger, 1971; Bryant, 1971). By deleting particular regions of a late larval imaginal disc and observing which adult elements are lost in consequence, it is possible to build up a two-dimensional **fate map** of that imaginal disc. This shows which parts of the final cuticular structure derive from which regions of the cell sheet in that disc.

Despite their variety of predetermined fates, imaginal disc cells are typically small and cuboidal, forming an undifferentiated epithelial sheet. Only in the eye discs are there clear signs of future adult differentiation by the end of the larval stage. Furthermore, 2D gels detect few differences between the populations of prevalent proteins in imaginal disc cells, whether comparing different regions of the same disc (Greenberg & Adler, 1982) or even different types of disc (Ghysen *et al.*, 1982). This is hardly surprising, given that the differentiated phenotypes of these cells will be mostly rather similar, i.e. muscle tissue derived from adepithelial myoblasts and cuticle-secreting epidermis from the sheet of ectodermal cells. Contrasts between the structures derived from different discs (or parts of the same disc) mainly reflect the varied three-dimensional patterning of the cuticle and underlying musculature. Some of the genes controlling pattern (whose protein products are unlikely to be abundant) are discussed below in § § 5.4 and 5.5.

The stability of the determined state in imaginal disc cells is most clearly revealed by transplantation experiments, where whole discs or parts of discs are removed from a donor larva and placed in the abdominal cavity of a host larva or adult; usually host and donor carry distinctive genetic markers so that the origin of donor structures can be confirmed. In a larval host, the transplanted disc will differentiate into its appropriate (predetermined) structure at the time of host metamorphosis, i.e. under the influence of ecdysone in the absence of juvenile hormone. In an adult host, however, ecdysone concentrations are very low, and the donor disc will not reveal its determined state until retransplanted into a larval host which is then allowed to meta-morphose. Imaginal disc material can be passed from one adult host to another (serial transfer) over prolonged periods, yet still retain its original determined state. The same is true when imaginal disc frag-ments are cultured *in vitro*; again they will differentiate into ap-propriate **autotypic** structures only when ecdysone is added to the medium (Seybold & Sullivan, 1978; Edwards *et al.*, 1978).

However, rare exceptions do occur, where **allotypic** structures inap-propriate to the original disc type may develop from disc material that has been cultured for a long time *in vivo* or *in vitro*. This phenomenon is termed **transdetermination** (reviewed by Hadorn, 1978), and is observed only among dividing imaginal disc cells. Transdetermination represents a discrete change in the determined state, and affects only some cells in the disc population; these will become manifest as a patch showing allotypic characteristics surrounded by autotypic struc-tures when the disc transplant is exposed to ecdysone (or larval host metamorphosis). The boundary between autotypic and allotypic structurs is always sharp, with no regions showing intermediate characteristics. Transdetermination switches development from the expected (autotypic) path into an alternative (allotypic) pathway which is inappropriate for cells of that disc type, although it would be per-fectly normal for cells originating from a different disc. The fact that developmental fates become switched rather than merely disrupted, together with the relative frequency of transdetermination (far greater than the spontaneous mutation rate), suggest that we are not dealing with a classical mutational event. Neverthless, both determined and transdetermined states are heritable characteristics, being passed on (usually unchanged) from one cell generation to another. One plausi-ble explanation invokes pattern-controlling **selector genes**, whose activity states can be inherited from parent to daughter cells. Altering the activity state of one such gene might suffice to change the pattern from, say, haltere to wing specificity, as observed in transdeter-mination. There is a definite hierarchy in the order and probability of particular transdetermination events (fig. 5.9*A*), which implies that a limited number pattern-controlling genes is involved. These probably include the homoeotic genes, since there are close similarities between certain transdeterminations and the pattern abnormalities seen in some homoeotic mutants (§ 5.5 below).

Both determination and transdetermination involve decisions taken by groups of cells together. The main evidence for this is derived from genetic marking of clones of cells via X-ray-induced somatic

A Frequency of transdetermination events (after Haborn, 1978).

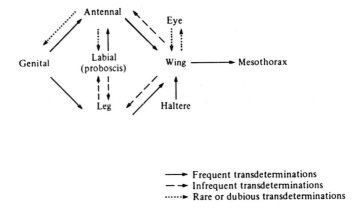

⟶ Frequent transdeterminations
– ➤ Infrequent transdeterminations
······➤ Rare or dubious transdeterminations

B Groups of cells are involved in transdetermination (modified from Alberts *et al.*, 1983).

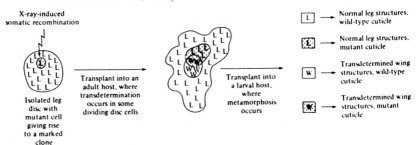

Fig. 5.9 Transdetermination.

recombination, as decribed in §5.1.3(b). Clones induced early in development will of course be larger than those induced later, because on average the marked founder cell will have passed through more division cycles in the former case. Commonly, marked cells from an early clone will produce a large patch of mutant cuticle spreading over several adjacent elements in a structure, whereas a later clone will give rise to a mutant cuticle patch contained within one of those elements. This implies that cells have become determined to form one or other subsidiary element at some stage *after* the induction of the early clone (hence some marked cells may contribute to one element, some to another), but *before* the induction of the later clone (by which time the marked founder cell will already have decided its future fate). Now consider what would happen if the founding of a marked clone were to coincide with such a determination decision. If this decision is **clonal**, i.e. involves just a single cell, then sometimes that same cell would become the founder of a marked clone, such that the *entire* adult structural element would be composed of mutant cuticle. In fact, this has never been observed; all such structural elements include wild-type as well as mutant cuticle, irrespective of the stage at which the marked clone was induced. This implies that each determination decision must be **polyclonal**, involving a group of cells which take the same decision together. Because only one of these founding cells can be genetically marked, it is impossible for mutant cuticle to occupy the whole of the final structure to which those cells give rise.

A more dramatic demonstration of this involves inducing fast-growing marked clones on a slow-growing background, such that the mutant cuticle patch is far larger than normal (see discussion of compartments in §5.4.1 below); even so, the adult structure always includes some wild-type cuticle derived from the slower-growing, non-marked founding cells. The general principle here is illustrated in fig. 5.9B, showing how a marked clone induced prior to a transdetermination event can give rise to part but not all of the transdetermined region.

Polyclonal determination decisions characterise even the earliest stages of embryogenesis; in the cellular blastoderm, a small group of cells (perhaps 10 or so) is set aside to found each imaginal disc. By the time of pupation, many thousands of cells are present in each disc, e.g. about 50 000 in a mature wing disc. Thus on average each founding cell must have passed through 10–12 division cycles during embryonic and larval development. However, mitotic rates are not uniform throughout a disc; rather, different groups of cells divide at their own characteristic rates, some groups dividing faster or ceasing division earlier than others. Programmed cell death also occurs in some regions. These local differences in mitotic pattern result in complex changes of disc morphology, converting a simple inpocketing of the early larval epithelium into a highly convoluted structure (whose shape and folding patterns are characteristic for each disc type) by the onset of pupation. During metamorphosis, ecdysone induces a change in shape among the imaginal disc cells from cuboidal to squamous epithelial; this causes each disc to become everted like a balloon (fig. 5.10), from which the final adult structure is modelled. For instance, flattening occurs during wing formation to produce an ovoid double layer of cells. Recent work has identified several ecdysone-inducible genes involved in disc morphogenesis, whose transcripts are associated with membrane-bound polysomes. The early IMP-E1 gene

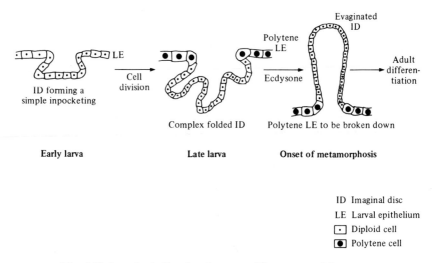

Fig. 5.10 Imaginal disc development (diagrammatic).

product is probably involved in the cellular rearrangements associated with disc morphogenesis and the formation of glial cell layers around the brain (IMP-E1 transcripts are expressed in both; Natzle *et al.*, 1988). Expression of the late IMP-L2 gene occurs first in imaginal discs and later in imaginal histoblasts, in both cases preceding the spreading and fusion of separate primordia to form the continuous adult epidermis (Osterbur *et al.*, 1988).

Metamorphosis in holometabolous insects is a period of renewed developmental activity, converting the larva into a radically altered imago adapted to a different life style. This poses the question as to whether there might be an imago-specific set of genes required for the growth/determination/differentiation of imaginal discs, but dispensable for embryonic and larval development. There are indeed several 'discless' mutants in which the larvae appear relatively normal but the imaginal discs remain rudimentary, such that metamophosis is blocked. However, these mutants prove to have defects affecting cell proliferation in general (noticeable in at least some larval tissues), rather than specifically eliminating adult pathways of development (Szabad & Bryant, 1982); imaginal discs are most obviously affected because of their rapid mitotic rate during larval life. However, some mutations in this class appear to be more selective in their effects; the *defective dorsal discs* (*ddd*) mutant, for instance, results in reduced or absent dorsal (e.g. wing and haltere) discs but has normal ventral (leg) discs in the same thoracic region (Simcox *et al.*, 1987). Chimaeric wing discs composed of a mixture of mutant and wild-type cells develop normally, suggesting that the *ddd*⁺ product may be a diffusible factor required for the growth of dorsal discs. Overall, it seems likely that the main elements of the adult body pattern are controlled by the *same* set of regulatory genes as that which is active during embryogenesis, i.e. the segmentation and homoeotic genes discussed in § §5.4 and 5.5 below. In some cases the same gene fulfils a similar role both in the embryo and in imaginal discs (e.g. the distribution of *engrailed* protein in posterior but not anterior compartments; see §5.4 and fig. 5.22), but in other cases the same gene may function differently in these two situations (e.g. the *wingless* gene product; Baker, 1988a).

5.2.3 Sex determination in Drosophila

Both *Drosophila* and humans have similar sex-chromosome constitutions, namely XY males and XX females. But there the resemblance ends. In humans, the small Y chromosome confers maleness, such that XX and XO are both female, whereas XY, XYY and even XXY are male. Rare XX males carry a translocated section of Y chromosome on another chromosome, while rare XY females have a deletion of part of the Y chromosome. By mapping these deletions and translocations, it is possible to define a 350 kbp segment of the Y chromosome which is essential for male sexual development. This region has been cloned as a series of overlapping DNA fragments, and it contains a candidate zinc-finger gene (ZFY) which may help to specify male as opposed to female gonadal development (Hodgkin, 1988b). However, the picture is complicated by a closely related X-linked gene (ZFX) and by the involvement of at least one further locus on the Y chromosome (Burgoyne, 1989). Secondary sexual characters are controlled by the sex steroid hormones (mainly androgens in males, oestrogens and progestins in females; §3.3) which are secreted by the gonads, providing a further regulatory tier.

Drosophila sex determination is radically different, in that the sexual phenotype of each cell is set autonomously by its chromosomal constitution and is *not* mediated via sex hormones. Moreover, XO individuals are male rather than female, implying that the Y chromosome does not determine maleness (though it is required for spermatogenesis). In fact, it is the *ratio* between the number of autosome sets (A) and the number of X-chromosomes which determines sex. Where this X:A ratio is 0.5 (as in diploid XY or XO individuals), male development ensues; where this ratio is 1.0 (as in XX diploids), female development results. In triploid flies with only two X chromosomes (i.e. 2X:3A, ratio 0.67), an intersex phenotype develops as a coarse mosaic of male and female cell groups. This further suggests that the X:A ratio is sensed once during early development, so establishing the future sex of each cell (together with all of its descendents). In the triploid intersexes, some cells may sense this ratio (0.67) as closer to 1.0 and hence choose the female option, while others sense it as closer to 0.5 and hence become male; the balance between these choices may be so finely set that minor environmental differences can tip it one way or the other in particular cells (Hannah-Alava & Stern, 1957).

Chimaeric flies consisting partly of male and partly of female cells (gynandromorphs) can also be generated when an XX zygote loses an X chromosome during one of the early cleavage divisions, so generating a male XO nucleus alongside one or more female XX nuclei. Normally such events are very rare, but their frequency can be increased by using a ring-X chromosome or various mutants which induce chromosome loss during these early divisions (see Hall *et al.*, 1976). Because the orientations of these nuclear divisions are not fixed, adult gynandromorphs will have different regions populated by cells showing male as opposed to female characteristics. Each early cleavage nucleus gives rise to a surface patch of cells in the blastoderm, and this in turn will later produce some part of the adult exoskeleton (the size of this will depend on how many cells in that blastoderm patch were set aside to form imaginal discs or histoblast nests). It is possible to build up a two-dimensional blastoderm fate map (see fig. 5.3) by measuring the frequency with which nearby adult structures are separated by a boundary between the male and female zones in such gynandromorphs. Structures which are normally derived from adjacent regions of blastoderm will be separated less frequently than structures formed from widely spaced founder regions. The mapping procedure involves triangulation of the percentage separation frequencies (sturt values) between any three nearby landmark structures, and then proceeding to further nearby structures until the whole blastoderm has been covered. (In fact, the sturt values for left and right halves can be counted together, since the animal is bilaterally symmetrical; see Sturtevant, 1929; Garcia-Bellido & Merriam, 1969; summarised in chapter 8 of Sang, 1984). In the present context, it is important to note that the very existence of boundaries in such male:female mosaics implies that the sexual phenotype of each cell is determined autonomously by its chromosomal makeup and is not influenced by neighbouring cells.

A second tier of control in sexual differentiation has been revealed by fate-mapping the genital disc which gives rise to the genital and anal structures of the adult. Within this disc, there is an antero-posterior arrangement of precursors for female genitalia, male genitalia and analia (Schupbach *et al.*, 1978). In gynandromorphs, a complete set of female genital structures will form only if the XX zone includes the precursor region for the female genitalia; likewise male genitalia can only form if the appropriate precursor region is composed of XO cells. If the male:female boundary runs antero-posteriorly through the genital disc, this will result in gynandromorph adults with male genital structures on one side and female structures on the other (Nothiger *et al.*, 1977). If no male nuclei are present in the precursor region for male genitalia, these cells fail to develop further; likewise, the precursor region for female genital structures does not develop unless occupied by female nuclei (Epper & Nothiger, 1982). By contrast, the anal precursors can develop along *either* male *or* female pathways, depending on whether the anal precursor region is populated by male or by female nuclei. In this respect the anal precursors follow the general rule in *Drosophila*; namely, that the phenotype of all sexually dimorphic structures (except the precursors of the genitalia, see above) is regulated by the X:A ratio in the contributing nuclei.

Table 5.1 lists a number of mutations which are either lethal to one sex (*Sxl, da*), or which alter the somatic sexual phenotype (*tra, tra-2, ix, dsx*). Note that germ cells are not affected by the latter group of mutations; thus mutant pole cells can develop into ova or sperm according to their XX or XY chromosomal constitution, as shown by transplanting them into normal embryos (Schupbach, 1982). Thus the pathway involving these genes governs sexual differentiation in somatic cells only (many bristles and other cuticular features of the adult are sexually dimorphic).

The phenotypes of double-mutant combinations can be used to establish epistatic relationships between these gene functions and so elucidate the order in which they act, although the issue is somewhat complicated by their sex-specific effects. Thus two mutations mapping close together within the the X-linked *Sxl* locus have opposite effects, one causing male lethality (*Sxl^{M1}*) and the other female lethality (*Sxl^{f1}*). One model (Cline, 1978) suggests that the *Sxl$^+$* gene product is essential for female development but lethal to males. Null mutations at the *Sxl* locus would therefore kill females but allow males to survive (*Sxl^{f1}*); conversely, a regulatory mutation resulting in constitutive *Sxl$^+$* expression would give the opposite phenotype, killing males but allowing females to develop normally (*Sxl^{M1}*). Positive autoregulation of *Sxl$^+$*

Table 5.1. Drosophila *mutations affecting somatic sexual phenotype (modified from Sang, 1984)*

Gene	Chromosome	Principal effects
Sex lethal (Sxl)	1 (X)	Different mutations at this locus are dominant male or female lethals; i.e. cause death of either all male offspring or all female offspring.
daughterless (da)	2	Maternal-effect female lethal
transformer (tra)	3	Females converted into
transformer-2 (tra-2)	2	males; males normal (*tra*) or sterile (*tra-2*)
double-sex (dsx)	3	Both males and females converted into intersexes by null mutations; other mutations at this locus convert only males or only females into intersexes
intersex (ix)	2	Females converted into intersexes; males normal

has been proposed as a mechanism for the cellular memory effect, whereby each cell 'remembers' its X:A ratio sensed once earlier in development (Cline, 1984). This is achieved via sex-specific RNA splicing (Bell *et al.*, 1988), as is the case for both *tra* and *dsx* (see below). Thus both males and females transcribe the *Sxl* gene, but in all male mRNAs the open reading frame is truncated by a stop codon in an additional exon that is removed by splicing from female mRNAs (Salz *et al.*, 1989). Thus only female-specific mRNAs can encode a full-length Sxl^+ protein. The Sxl^+ protein exhibits sequence similarity to known RNA-binding proteins, suggesting that it may interact both with its own transcripts (splicing them to female-specific mRNAs, providing the positive autoregulation inferred) and also with those of genes acting downstream in the pathway (e.g. *tra*; see below). The activity of the Sxl^+ gene in females requires both the maternal *daughterless* (da^+) gene-product (Cline, 1976), and also the zygotic *sisterless-a/-b* ($sis\text{-}a^+/\text{-}b^+$) gene-products (Cline, 1986); these functions are needed only for female development, and may play some role in activating *Sxl* and/or in sensing the X:A ratio (1.0) in female cells. Thus the progeny of mothers homozygous for *daughterless* mutations (*da/da*) are all male, because XX females cannot develop in the absence of functional da^+ product supplied by the oocyte. This defect can be rescued by injecting wild-type egg cytoplasm into mutant eggs. However, in addition to this requirement for da^+ function in the maternal germ-line to permit development of daughters, there is also an essential zygotic function for the da^+ product in somatic cells of both sexes (Cronmiller & Cline, 1987), specifically for the formation of peripheral neurones (Caudy *et al.*, 1988). As mentioned earlier, the da^+ protein contains a helix-loop-helix domain similar to that in the mammalian family of MYC-related proteins (see §3.1; Murre *et al.*, 1989), suggesting a role in transcriptional regulation (see Cline, 1989). In addition to da^+ and sis-a^+/-b^+, the liz^+ gene-product is required both maternally and zygotically for Sxl^+ activity in females (Steinmann-Zwicky, 1988). Unlike sis-a^+, however, the liz^+ function is also required for 'constitutive' *Sxl* activity in XY individuals carrying the Sxl^{M1} mutation (i.e. this is lethal to males in the presence but not in the absence of liz^+ function). Sxl^+ stands at the apex of several regulatory hierarchies in *Drosophila*, governing not only somatic sex but also dosage compensation and germ-cell differentiation (Schupbach, 1982). Only the first of these is further considered here; the remaining genes in this pathway act only on somatic sexual phenotype.

Extensive genetic studies suggest the following order of gene functions (see Belote *et al.*, 1985b and fig. 5.11 for details): $Sxl^+ > tra^+ \gg tra$-$2^+ > dsx^+ \gg ix^+ >$ terminal differentiation. The two tra^+ functions are specifically required in females but not in males, since XX tra^- and tra-2^- homozygotes are transformed into phenotypic males. Cloning of some 25 kbp of DNA from the *tra* region has revealed at least six transcription units; however, P-element-mediated transformation experiments indicate that a single 2 kbp subfragment is sufficient to supply the tra^+ function. This *tra* gene is transcribed in both males and females, but the primary transcripts are differentially spliced, producing a 1.0 kb tra^+ RNA that is female-specific and a 1.2 kb RNA that is present in both sexes (McKeown *et al.*, 1987). Notably, only the female-specific RNA contains a long open reading frame that could encode a functional tra^+ protein; the RNA species present in both sexes does not have signficant protein-coding potential and is apparently without function. Ectopic expression of the female-specific tra^+ RNA in chromosomal males causes them to develop as phenotypic females (McKeown *et al.*, 1988). Female-specific splicing of *tra* transcripts

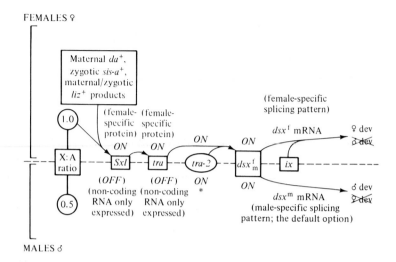

Fig. 5.11 Somatic sexual differentiation in *Drosophila*; a regulatory hierarchy (modified from Belote *et al.*, 1985b).
*Note: *tra-2* is expressed in males and its function is required for spermatogenesis. Since *tra-2* encodes an RNA-binding protein, it is possible that the *tra*$^+$ and *tra-2*$^+$ products act in conjunction to direct the female-specific splicing pathway for *dsx* transcripts (giving *dsx*^{f} mRNA). Thus *tra-2* is an adjunct rather than a primary member of the sex determination pathway, hence it is circled rather than boxed in the figure.

requires another gene product acting upstream of *tra*$^+$ (probably the *Sxl*$^+$ protein), since the female-specific *tra*$^+$ product does not autoregulate the splicing of its own RNA (McKeown *et al.*, 1988). The female-specific *tra*$^+$ products in turn regulate the *dsx* function, perhaps indirectly in conjunction with the *tra-2*$^+$ gene-product.

The *tra-2* gene has been cloned (Amrein *et al.*, 1988), and like *Sxl* it encodes a putative RNA-binding protein. P elements carrying a 3.9 kbp DNA fragment from the *tra-2* locus are able to rescue *tra-2*$^-$ mutants (Goralski *et al.*, 1989). Again, *tra-2*$^+$ transcripts (1.7 kb) are expressed (mainly in the germ line) in males as well as females, but both appear to contain identical protein-coding regions. Thus *tra-2*$^+$ transcripts are not differentially spliced to yield translatable mRNAs in females but non-translatable RNAs in males, as in the case of *Sxl* and *tra* (see above); this is consistent with the fact that the *tra-2*$^+$ function is required for spermatogenesis in males as well as for female sexual differentiation (see Table 5.1). The *tra-2*$^+$ protein may interact with *dsx* transcripts to specify their female-specific splicing mode (giving *dsx*[f] mRNA; see below), perhaps acting in conjunction with the female-specific *tra*[+] protein (since *tra-2*[+] itself is expressed in both sexes; Amrein *et al.*, 1988).

The *dsx*[+] gene appears to have two distinct functions, one of which is female-specific (*dsx*[f]) and can be activated only via the *Sxl*[+] > *tra*[+] ≫ *tra-2*[+] pathway (above). The male-specific function of *dsx*[+] (*dsx*[m]) is apparently a default or basal state of expression which is adopted whenever one (or more) of the above regulatory genes is (are) inactive. Note that *dsx*[m] is the first function required specifically for male sexual differentiation. The fact that null mutations of *dsx* (eliminating both *dsx*[f] and *dsx*[m] functions) always produce intersexes needs some explanation; in normal females *dsx*[f] activity blocks male differentiation, whereas in normal males *dsx*[m] activity prevents female development. Thus when both *dsx* functions are absent, neither pathway is blocked, hence cells simultaneously develop both male and female characteristics – resulting in the intersex phenotype (note that this is *not* a coarse mosaic of male and female regions, as in the triploid intersexes). Inactivation of *dsx*[m] but not *dsx*[f] will result in normal females while males develop as intersexes; conversely, inactivation of *dsx*[f] but not *dsx*[m] will give females with the intersex phenotype but normal males. Thus *dsx*[+] is a bifunctional switch gene whose two functions (*dsx*[f] and *dsx*[m]) each act negatively to repress the inappropriate pathway of sexual differentiation. Once again this is achieved via differential splicing of primary transcripts from the 40 kbp *dsx* locus, using sex-specific 3′ exons and different poly(A)-addition sites (Nagoshi *et al.*, 1988; Baker & Wolfner, 1988; Burtis & Baker, 1989). Notably, the sex-specific *dsx* mRNAs become detectable only from the late larval stages onwards; prior to this a non-sex-specific *dsx* RNA of unknown function predominates. In females with active *Sxl*[+]/*tra*[+]/*tra-2*[+] gene products, the *dsx*[+] transcripts are spliced in a female-specific pattern leading to *dsx*[f] mRNA(s); by contrast, in males (or in chromosomal females lacking any of these three gene products) an alternative 'default' splicing pattern is used, leading to *dsx*[m] mRNA(s). The female-specific product of the *intersex* (*ix*[+]) gene functions downstream of *dsx*[f] (or perhaps in conjunction with it), hence the intersex phenotype in *ix*/*ix* homozygotes is confined to chromosomal females. The entire sex-determination pathway is regulated largely via differential sex-specific splicing of RNAs transcribed in both sexes (see reviews by Hodgkin, 1989, and Baker, 1989).

Finally, the lowest tier of regulatory genes in each pathway (i.e. *dsx*[m] in males; *dsx*ᶠ and/or *ix*⁺ in females) must influence the activity of a wide range of terminal differentiation genes so as to produce the male or female phenotypes of all sexually dimorphic features (reviewed by Wolfner, 1988). Most such features develop only in one sex, suggesting that the sex-determination genes function within progenitor cells to direct a sex-specific developmental pathway (c.g. the male accessory glands, or the follicle cells surrounding each egg chamber in females). Although the fat body is present in both sexes, it carries out yolk-protein (yp) synthesis only in females and never in males. The yp genes are regulated by both ecdysone and juvenile hormone as well as by products of the sex determination pathway (Belote *et al.*, 1985a); notably, female-specific regulatory functions are still required during adult life for the maintenance of yp expression. An important tool in this and related studies has been the use of temperature-sensitive alleles of *tra-2*. Although *tra-2*⁺ products are expressed in both sexes, they participate in the female-specific splicing pathway for *dsx* transcripts (perhaps in conjunction with the female *tra*⁺ protein; see earlier). This means that the splicing pattern for *dsx* transcripts can be shifted from the female to male mode simply by raising the temperature from permissive (16 °C) to restrictive (29 °C) levels and thereby inactivating the *tra-2*ᵗˢ product. Adult XX somatic tissues can thus be switched from the female (16 °C) into the male (29 °C) pathway, and back again after cooling. In the case cited, yp gene expression ceases whenever the *tra-2*ᵗˢ product is inactivated, but is resumed on lowering to the permissive temperature, implying a continuing dependence on sex-determination functions. By contrast, the patterns of gene expression in sex-specific somatic cell-types (e.g. male accessory gland, female follicle cells) cannot be altered by changing *tra-2*ᵗˢ activity in adult flies via such temperature shifts; this indicates that the sex-determination genes function earlier in the specification and development of these cell types (see Wolfner, 1988).

The genetic control of sex-determination in *Drosophila* differs mechanistically from that in *C. elegans* (see § 4.5 and fig. 4.11). Although insects (and even dipterans) display a bewildering variety of sex-determination mechanisms, most of these turn out to be variants on a common pattern related to that in *Drosophila* (Nothiger & Steinmann-Zwicky, 1985). Invariant features of this pattern include a **primary signal** (the X:A ratio in *Drosophila*) controlling a **key gene** (cf. *Sxl*⁺) which in turn regulates a **genetic double switch** (cf. *dsx*⁺).

5.3 Dorso-ventral polarity and neurogenesis

5.3.1 D/V polarity in Drosophila

Many maternal gene-products are essential for normal *Drosophila* embryogenesis, yet few of these seem to be localised within precise subregions of the zygote; still fewer direct the development of only one particular tissue or structure later on (as in ascideans: § 2.6.1). Exceptional on both counts are the posterior pole factors (see § 5.1.2 above), which are segregated into the pole cells and essential for germ-cell formation. Many maternal mutations affecting embryonic pattern in *Drosophila* cause *global* alterations to that pattern, as would be predicted if the maternal determinants were distributed as gradients across the egg (Nusslein-Volhard, 1979). Thus if one part of the overall pattern is deleted, neighbouring elements may also be reduced, while more distant elements expand to fill the gap or even become duplicated. Maternal mutants affecting antero-posterior (A/P) polarity show this global effect particularly clearly, since most metameric units in the A/P sequence acquire distinctive identities (both morphologically and in terms of gene activity) during later embryonic development (see § § 5.4 and 5.5). The same is true for maternal mutants affecting D/V polarity, except that here we are dealing with a series of tissue types within the germ-band region. From dorsal to ventral, these are: (i) the amnioserosa (an extra-embryonic structure formed from the dorsalmost strip of cells); (ii) dorsal epidermis; (iii) ventral epidermis and nervous system (the ventral neurogenic region; see § 5.3.2 below); and (iv) mesoderm (formed from the ventralmost strip of cells which invaginate during gastrulation).

Mutations in some 20 genes (both maternal and zygotic) alter the embryonic D/V pattern (reviewed by Anderson, 1987; Levine, 1988; these fall into two distinct categories (table 5.2). **Dorsalising** mutations reduce or abolish ventral structures and cause a corresponding expansion of dorsal and lateral structures (e.g. a tube of dorsal epidermis with no mesoderm or ventral nervous system). Conversely, **ventralising** mutations delete or reduce the dorsal tissues and expand the ventro-lateral components. Importantly, these mutants do not cause extensive cell death in the region affected (i.e. V or D structures are not simply lost), but rather produce a general shift in the positional identities of cells towards a more dorsal or more ventral fate.

The egg itself shows structural asymmetry along the D/V axis, being more curved ventrally and flatter dorsally (fig. 5.12), with special

Table 5.2. *Genes affecting D/V polarity in* Drosophila *(modified from Anderson, 1987)*

Maternal-effect mutations		Zygotic mutations
Affecting egg shape and embryo polarity	Affecting embryo polarity only	
D *fs(1)K10*	*dorsal (dl)*	*twist (twt)*
	windbeutel (wbl)	*snail (sna)*
	gastrulation-	
	defective (gd)	
	nudel (ndl)	
	pipe (pip)	
	tube (tub)	
	snake (snk)	
	easter (ea)	
	pelle (pll)	
	spatzle (spz)	
	Toll (Tl)	
V *gurken (grk)*	*cactus (cac)*	*decapentaplegic (dpp)*
(torpedo (top))*		*zerknullt (zen)*
		twisted gastrulation
		(twg)
		tolloid (tld)
		shrew (srw)

All genes are classed according to the phenotypes of null mutants as dorsalising (D, upper part) or ventralising (V, lower part).
    * required in somatic follicle cells but not in the germ line.
All other gene products in the first two columns are required in the oocyte, and those in the third column at various stages during early embryogenesis.

chorionic appendages confined to the dorsal side. This asymmetry presumably reflects the spatial organisation of the egg chamber and chorion-secreting follicle cells during oogenesis (§5.1.2). Three of the maternal D/V mutations alter egg shape as well as embryonic polarity; *fs(1)K10⁻* mutant mothers produce dorsalised eggs and embryos, whereas both *torpedo⁻* (*top/top*) and *gurken⁻* (*grk/grk*) mutant mothers produce ventralised eggs and embryos. The wild-type *fs(1)K10⁺* gene encodes an oocyte-specific nuclear protein which may bind to DNA and perform a regulatory function during oogenesis (Prost *et al.*, 1988). Studies with germ-line mosaics (Schupbach, 1987) suggest that the *torpedo⁺* product is required in the somatic (follicle cell) component of the ovary, while the *gurken+* function is required in the germ-line component (nurse-cell/oocyte complex). Thus somatic and germ-line cells cooperate during oogenesis in order to establish the normal D/V pattern of the egg shell and embryo. The *top+* function is also required after fertilisation, since the zygotic lethal, *faint little ball*, is allelic to *top*; a general function in cell communication is implied by sequence relationships between the *top+* product and the mammalian EGF receptor (Price *et al.*, 1989). Other maternal effect mutations affecting embryonic D/V (and A/P) polarity act autonomously within the germ line; that is, embryos derived from mutant oocytes always manifest the mutant phenotype even if they were surrounded by wild-type follicle cells in the ovary (Schupbach & Wieschaus, 1986).

Apart from the three cases mentioned above, maternal D/V mutations have no effect on egg shape, and most cause embryo dorsalisation (the dorsal class genes). Amongst those maternal mutants which do give ventralisation, most are attributable to dominant mutations at loci whose null (recessive) phenotypes are dorsalised (e.g. *Toll^D*, see below). Only in the case of *cactus* does a maternally acting null mutation cause ventralisation of the embryonic pattern. So far, 11 maternal genes have been assigned to the dorsal class (table 5.2), implying that their wild-type functions are all needed to establish ventral pattern and that the dorsal pathway may be a 'default' option followed when one (or more) of their products is deficient.

For seven of these dorsalising mutants, the defect can be rescued at least partially by injecting wild-type (wt) embryo cytoplasm into zygotes or early embryos derived from homozygous mutant mothers. For instance, *dl⁻* embryos (from a *dl/dl* mother) will develop some ventral structures if injected with wt cytoplasm. Rescue can also be achieved by injecting cytoplasm from embryos which are deficient for a different dorsal-class product (e.g. from *pll/pll* or *ea/ea* mothers). This shows that the rescuing activity is **locus-specific** and probably represents the wt product of the gene in question; in the case cited, *dl+* product would be available from the transplanted (donor) cytoplasm while *pll+* (or *ea+*) activity would be supplied by the recipient embryo. Wild-type embryonic poly(A)+ RNA can partially rescue mutant embryos lacking any one of six maternal dorsal-class gene-products. The injected RNA must therefore include surviving maternal transcripts from the wt gene, which can provide sufficient functional protein to effect rescue (Anderson & Nusslein-Volhard, 1984). In some cases, the ability of maternal RNA to effect rescue changes during the course of early development. For instance, wt embryonic RNA cannot

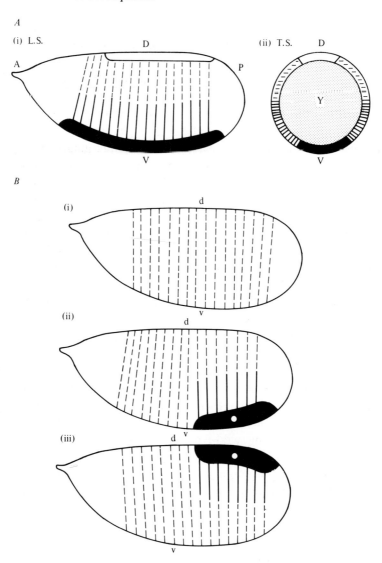

Fig. 5.12 Dorso-ventral polarity in *Drosophila* (modified from Anderson, 1987).

A D/V polarity (blastoderm stage embryo). (i) L.S.; (ii) T.S.; D, dorsal; V, ventral; A, anterior; P, posterior; [⠿], yolk. *Presumptive tissues:* [▭], amnioserosa; [⦙⦙], dorsal epidermis; [▥], ventral neurogenic region (precursors of ventral nervous system and ventral epidermis); [■] mesoderm.

rescue *pll⁻* embryos beyond the time of pole-cell formation, whereas wt cytoplasm can rescue them right up to the late blastoderm stage (Muller-Holtkamp *et al.*, 1985). This implies that *pll⁺* mRNA is normally translated at an early stage, and thereafter only the *pll⁺* protein can effect rescue. In *tub⁻*, *ea⁻* or *snk⁻* embryos, the effectiveness of cytoplasmic rescue is independent of the site of injection. For *pll⁻* and particularly *dl⁻* embryos, however, there is a pronounced position dependence. Thus wt cytoplasm injected ventrally (as judged by egg shape) elicits rescue more efficiently than a similar dorsal injection. Furthermore, cytoplasm from the ventral side of wt donor embryos is more effective than dorsal cytoplasm in rescuing *dl⁻* (or *pll⁻*) embryos. These position-dependent effects become more pronounced during early donor development, implying that active *dl⁺* and *pll⁺* products become concentrated ventrally during normal embryogenesis (Santamaria & Nusslein-Volhard, 1983; Muller-Holtkamp *et al.*, 1985). However, for all of the mutants described so far, rescue experiments reveal an innate D/V polarity corresponding to egg shape. Thus even if the rescuing injection is made dorsally, ventral structures will develop only on the expected ventral side (in this case, opposite to the site of injection).

The *Toll⁺* function is strikingly atypical in this respect (Anderson *et al.*, 1985a,b). *Toll⁻* embryos (from *Tl/Tl* null-mutant mothers) have no innate D/V polarity, in that wt cytoplasm causes rescued ventral structures to appear only near the site of injection (whether dorsal or ventral, as judged by egg shape; fig. 5.12). This implies that the active *Toll⁺* product may be a primary determinant of D/V polarity, a conclusion reinforced by the existence of dominant ventralising (*Tollᴰ*) mutations at this locus. Among the other maternal dorsal-group genes,

Fig. 5.12 (*cont.*)

B Role of *Toll⁺* function in D/V polarity.
(i) *Toll⁻* embryo (from *Tl/Tl* null mutant mother), comprising a tube of dorsal epidermis with no ventral structures. d, 'dorsal' side of egg (determined from egg shape); v, 'ventral' side of egg (determined from egg shape).
(ii)/(iii) Local rescue of ventral structures following injection of wild-type cytoplasm into *Toll⁻* embryos. ▆ ▆ , site of injection (future mesoderm).
(ii) Ventral tissues formed on 'ventral' side of egg following a 'ventral' injection.
(iii) Ventral tissues formed on 'dorsal' side of egg following a 'dorsal' injection.

only *easter* has dominant ventralising (*ea*^D) alleles (see below; Chasan & Anderson, 1989). As we have seen previously, mutants involving developmental switch genes (e.g. *lin-12* in *C. elegans* or *Sxl* in *Drosophila* sex determination) frequently fall into two classes with opposite phenotypes, one given by recessive mutations abolishing wild-type gene activity, the other due to dominant regulatory mutations causing constitutive or ectopic expression of the gene in cells where its activity is inappropriate. However, the ventralised *Toll*^D phenotype is not simply due to overexpression of the *Toll* gene, nor is the wild-type *Toll*^+ product localised ventrally (e.g. in a ventral-to-dorsal gradient).

This is shown clearly by the fact that *Toll*^− embryos can be rescued equally well by cytoplasm from any region of a wild-type embryo. Thus *Toll*^+ product must be available throughout the embryo and is not distributed asymmetrically (compare with the rescue of *dl*^− or *pll*^− embryos described above). One possible resolution of this paradox is that the maternal *Toll*^+ gene-product is present throughout the egg in an inactive form, but is activated only locally (presumably in the ventral region) by other dorsal-group gene-products (see Woodland & Jones, 1986; Anderson, 1987). The *Toll*^D ventralised phenotype may be due to an altered *Toll* protein which is active indiscriminately, rather than requiring local activation. This view is supported by the phenotypes of embryos derived from double-mutant mothers carrying a *Toll*^D allele in combination with a homozygous dorsal-group mutation (Anderson *et al.*, 1987). In most such combinations, the embryos develop some ventral structures, implying that these dorsal-group gene-products normally interact with (activate) the *Toll*^+ product. However, this is not the case for *dl*, since *dl*^− embryos from *Toll*^D mothers are fully dorsalised, suggesting that the *dl*^+ product acts downstream of *Toll*^+ rather than participating in the local activation process. As mentioned earlier, there are also dominant ventralising *ea*^D alleles of the *easter* gene. Synthetic *ea*^+ transcripts injected after fertilisation can rescue *ea*^− mutants completely (Chasan & Anderson, 1989). Overall, *ea*^+ and *Toll*^+ products may act together in a cyclical fashion to effect the local activation inferred above. Neither product is localised in the egg, and the interaction between them is necessary only after fertilisation in order to specify ventral development.

Further elucidation of the genetic pathway controlling D/V pattern has come from recent molecular analysis of several maternal dorsal-group genes. The *Toll* locus encodes a large (5.3 kb) ovarian transcript, whose expression is sufficient to rescue *Toll*^− embryos, as shown by P-element-mediated transformation studies (Hashimoto *et al.*, 1988).

From sequence analysis it is inferred that the *Toll*+ product is an integral membrane protein with both a cytoplasmic and a large extracellular domain. It is unclear where this protein could be membrane-bound in precellular embryos, nor how it could be 'activated' so as to function locally as a primary determinant of ventral pattern elements, although various suggestions have been made such as proteolytic cleavage. In this context, it is noteworthy that both the *snake*+ and *easter*+ products are serine proteases, related in the former case to factors involved in the vertebrate blood-clotting cascade (deLotto & Spierer, 1986; see also Chasan & Anderson, 1989); it will be interesting to see whether other dorsal-group genes also encode proteases which might help to activate the *Toll*+ product. The *dorsal* (*dl*) locus comprises an 8 kbp gene encoding a 2.8 kb RNA which is transcribed only in ovaries, but which remains present in embryos until the blastoderm stage (Steward *et al.*, 1984). The *dl* cDNA sequence is related to that of the viral oncogene v-*rel* and its avian and human c-*rel* counterparts, all of which encode nuclear (DNA-binding?) proteins of unknown function (Steward, 1987). Although *dl*-rescuing activity becomes progressively localised in ventral regions during early embryogenesis, this regulation is not achieved at the level of transcription or RNA localisation; *in situ* hybridisation studies show that *dl*+ RNA is distributed uniformly in the egg and embryo (Steward *et al.*, 1985). However, during early embryogenesis, the *dl*+ protein shows a ventral-to-dorsal gradient of nuclear localisation among the peripheral nuclei of the syncytial and cellular blastoderm stages (Steward *et al.*, 1988; Roth *et al.*, 1989; Rushlow *et al.*, 1989; Steward, 1989; Hunt, 1989). Both the ventral localisation of nuclear *dl*+ protein and its action downstream of *Toll*+ (see above) suggest that the *dl*+ function acts late in the maternal pathway controlling D/V cell identities (Levine, 1988), perhaps as the D/V morphogen.

By analogy with A/P polarity (§ 5.4.2), it is likely that the maternal D/V gene-products interact to generate a system of positional information in the early embryo, to which various zygotic genes then respond with spatially restricted patterns of expression. Such genes might be switched on or off by spatial cues such as threshold amounts of particular maternal gene products. Among the zygotic D/V genes, two (*twist* and *snail*) give dorsalised mutant phenotypes. Both *twt*− and *sna*− embryos fail to form a ventral furrow or to develop mesoderm, suggesting that the wild-type products of these genes are required specifically in the ventralmost regions (i.e. the mesodermal precursors). Indeed, transcripts of the *twt*+ gene are localised initially along the ventral furrow and later in the invaginating mesoderm of normal gastrulating embryos (Thisse *et al.*, 1987). Expression of the nuclear *twt*⁺ protein is also found in the anterior and posterior midgut invaginations (endodermal primordia) until the late blastoderm stage, consistent with the fact that *twt*⁻ mutant embryos lack all internal organs including endodermal as well as mesodermal derivatives (Thisse *et al.*, 1988). The expression of *twt*⁺ RNA is abolished in dorsalised *dl*⁻, *pll*⁻, *ea*⁻ or *Tl*⁻ embryos, implying that the wild-type products of all of these genes serve directly or indirectly to activate the *twt*⁺ gene (Thisse *et al.*, 1987). A close genetic interaction between *dl* and *twt* has been described (Simpson, 1983); whereas heterozygotes for mutations in either gene singly develop normally, double heterozygotes (*dl*/+, *twt*/+) display a *twt*⁻-like phenotype lacking ventral structures, implying that a reduced dose of *dl*⁺ product results in lower *twt*⁺ activity. The *sna* gene has been cloned, and the predicted sequence of its protein product includes five zinc-finger motifs likely to be involved in DNA-binding (cf. the *Xenopus* TFIIIA protein, see § § 2.4.2 and 5.6.2 below; Boulay *et al.*, 1987). The *single-minded* (*sim*) gene is expressed more laterally in two strips of cells flanking the mesoderm, i.e. the ventral neurogenic regions. Since *sim*⁻ mutants show losses of both neural and epidermal tissue along the ventral midline (after mesoderm invagination), the *sim*⁺ function is more likely to be involved in D/V patterning than in neurogenesis (Thomas *et al.*, 1988). The *sim* gene encodes a nuclear protein which shows some homology to the *period*⁺ gene-product involved in controlling the periodicity of biological rhythms in *Drosophila* (Crews *et al.*, 1988). The *sim* gene is one member of the spitz group (not shown in Table 5.2), which all have related mutant phenotypes affecting the ventral neurogenic region; this group also includes the maternal *sichel* and zygotic *spitz, Star, pointed* and *rhomboid* genes (Mayer & Nusslein-Volhard, 1988).

Mutations affecting the other zygotic D/V genes produce the opposite (ventralised) phenotype, typically deleting or reducing the dorsalmost tissues. Thus *zen*⁻ mutants eliminate the amnioserosa, while some *dpp*⁻ mutants delete the dorsal epidermis. Both *zen* and *dpp* genes have been cloned and characterised in molecular terms. The *dpp* locus is complex, but part of it encodes a protein with homology to vertebrate transforming growth factor-β (Padgett *et al.*, 1987). A mutation (*dpp*ᵈⁱˢᶜ) affecting this part of the locus causes massive local cell death in those parts of imaginal discs which normally produce distal structures (Bryant, 1988). This defect is non-autonomous, since it can be rescued by coculture of mutant with wild-type disc material, implying that one product of the *dpp* locus is indeed a secreted growth factor. *In situ* hybridisation of a *dpp* probe to wild-type embryos reveals that *dpp*⁺ transcripts are localised initially in a longitudinal dorsal stripe at the early syncytial blastoderm stage; during germ-band elongation *dpp*⁺ RNAs are expressed throughout the region which will give rise to the dorsal epidermis (this is notably lacking in at least one *dpp*⁻ mutant); later still, the *dpp*⁺ transcripts become restricted to two narrow stripes on either side of the dorsal midline, perhaps reflecting a further sub-division of the D/V axis (St Johnston & Gelbart, 1987).

The *zen* locus lies in the Antennapedia complex of homoeotic genes (§ 5.5 below), and comprises two small linked genes (z1 and z2) each containing a homeobox sequence (see § § 5.5 and 5.6 below). Both z1 and z2 show similar patterns of expression, yet the proteins they encode are highly divergent apart from their homeodomains. P-element-mediated gene transfer shows that the z1 gene alone is sufficient to provide *zen*⁺ function, implying that the z2 gene may be dispensable (Rushlow *et al.*, 1987a). In the blastoderm embryo, *zen*⁺ transcripts accumulate transiently in a dorsal stripe (narrowing from about 30% of embryo circumference initially to about 10% by gastrulation), which represents the precursors of amnioserosa and dorsalmost epidermis (Doyle *et al.*, 1986). This dorsal-on/ventral-off pattern of expression is readily detectable using antibodies against the *zen*⁺ protein (Rushlow *et al.*, 1987b). In dorsalised embryos lacking maternal products from one of the dorsal-group genes, zygotic *zen*⁺ expression extends into ventral regions. Thus one or more of these maternal genes might encode a repressor which normally prevents *zen*⁺ expression ventrally. Conversely, maternal ventralising mutations (*cac/cac* or *Toll*ᴰ mothers) virtually abolish embryonic *zen*⁺ expression. This is the effect predicted for *Toll*ᴰ, but also suggests a possible role for the wild-type *cac*⁺ gene product, which may restrict the repressor activity to ventral regions, so permitting dorsal expression of *zen*⁺ products. The ventralised phenotype of *cac*⁻ embryos might therefore reflect a failure to activate *zen* (and other genes?) dorsally. Thus *Toll*⁺ may promote and *cac*⁺ inhibit the ventral localisation of the nuclear *dl*⁺ morphogen (Hunt, 1989), which itself might repress zygotic *zen* expression.

5.3.2 Neurogenesis

The central nervous system (CNS) in *Drosophila* originates from two separate regions of the blastoderm fate map (fig. 5.3). Within the germ band, the ventral neurogenic region contributes the neuroblasts that give rise to the ventral nerve cord (see below), while the brain is derived from neuroblasts contributed by the procephalic neurogenic region lying anterior to the germ band. In both cases, some of the cells within the neurogenic region migrate inwards and differentiate as neuroblasts, while the remainder are directed towards non-neuronal fates.

The pattern of neurogenesis in *Drosophila* is closely related to that found in more primitive insects such as the grasshopper *Schistocerca* (Thomas *et al.*, 1984), where the larger size of cells and slower pace of development have helped to elucidate details of the morphological events and cell interactions involved in neurogenesis (Doe & Goodman, 1985a, b; Doe *et al.*, 1985). The account below refers specifically to *Schistocerca* (as does fig. 5.13), but the main principles are probably also valid for *Drosophila*. Within the ventral neurogenic ectoderm, some 20–25% of the cells become neuroblasts by first enlarging and then moving inwards in three waves. These neuroblasts will give rise to neurones via stereotyped cell lineages. This can occur either (i) directly – where an MP neuroblast divides once to give a pair of neurones, or else (ii) indirectly – where each neuroblast generates a chain of ganglion mother cells (GMCs) that each divide once to give a pair of neurones (fig. 5.13). In becoming a neuroblast, each inward-migrating cell inhibits its immediate neighbours from doing the same; instead, the latter adopt a variety of non-neuronal fates. The available options include several different types of CNS supporting cell (e.g. sheath, cap or glial cells), or cell death. In *Drosophila*, where presumptive neuroblasts are mixed in with precursors of ventral epidermis, the primary choice is between neural and epidermal fates. If a presumptive neuroblast (enlarged cell) is destroyed by laser irradiation, it is normally replaced by one of its neighbours, which is thereby switched from a non-neuronal to a neuroblast fate (Doe & Goodman, 1985a). This implies that the choice between these fates is stochastic rather than predetermined, i.e. each cell in the ventral neurogenic region has a certain probability (0.2–0.25) of becoming a neuroblast.

The above description of events in *Schistocerca* can help us to interpret the phenotypes of several mutants affecting neurogenesis in *Drosophila*. Loss-of-function mutations at any of the eight (or more) neurogenic genes produce a common phenotype, characterised by an excess of ventral neural tissue and a corresponding lack of ventral epidermis. Nevertheless, the ventral neurogenic region contains a normal number of cells initially, while both mesoderm and dorsal epidermis are unaffected. Similar abnormalities also occur in the procephalic neurogenic region, again producing excessive amounts of neural tissue

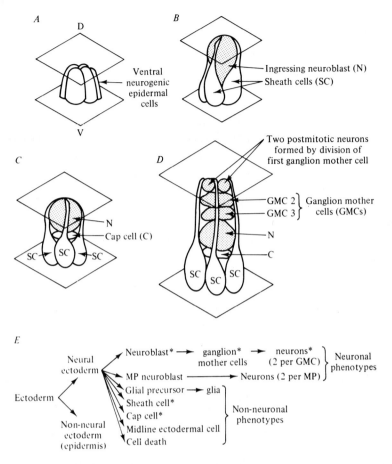

Fig. 5.13 Insect neurogenesis (from data on the grasshopper *Schistocerca*; see Doe & Goodman, 1985a).

A–D Generation of neurones and supporting cells (phenotypes marked * in part *E*). ☐, non-neuronal phenotypes (sheath and cap cells); ▨, neuronal phenotypes (neuroblasts, ganglion mother cells and neurones).

E Developmental choices available to ectodermal cells.

at the expense of cephalic epidermis. The simplest interpretation of these mutants is that too many cells within the neurogenic regions decide to become neuroblasts, while an insufficient number become epidermal dermatoblasts. The fact that each developing neuroblast normally inhibits neighbouring cells from doing likewise (at least in *Schistocerca*), immediately suggests that the mutant phenotype in *Dro-*

sophila may result from some breakdown in this inhibitory mechanism, such that more and more cells opt to follow the neuroblast route. In any case, the wild-type neurogenic functions are required to ensure epidermal development; in their absence the neuroblast pathway is followed as a default option by most cells in the neurogenic regions. The end result is neural hyperplasia and a deficiency of epidermis (graphically if exaggeratedly described as 'all brain, no skin'), hereafter referred to as the hyperneuralised phenotype.

Bearing this in mind, let us now turn to the mutants themselves. Homozygous mutations at two maternally acting loci, *almondex* (*amx*) and *pecanex* (*pcx*), produce the hyperneuralised phenotype in all offspring; moreover, both *amx*⁻ and *pcx*⁻ embryos can be partially rescued by injections of wild-type egg cytoplasm (LaBonne & Mahowald, 1985). Six further neurogenic genes are all expressed zygotically, although in some cases there is also a maternal component of expression (Jimenez & Campos-Ortega, 1982). These six are *Notch* (*N*), Delta (*Dl*), *Enhancer of split* (*E(spl)*), *neuralised* (*neu*), *big brain* (*bib*) and *master mind* (*mam*); homozygous null mutations at any of these six loci again produce the hyperneuralised phenotype (Lehmann *et al.*, 1981, 1983).

Genetic mosaics have been used to investigate the functions of several neurogenic genes. Although the *Notch*⁺ gene product seems to act locally (Hoppe & Greenspan, 1986), it is not cell autonomous (Technau & Campos-Ortega, 1987). This was shown by transplanting single marked *Notch*⁻ cells into wild-type hosts; the resultant clones of mutant cells include both neural and epidermal derivatives. Similarly, mutations in the *bib, mam, neu, Dl* and *amx* genes also act non-autonomously, in that neighbouring wild-type tissue can direct at least some cells in a mutant clone towards an epidermal rather than neural fate. However, null *E(spl)*⁻ mutations do show cell autonomy; in this case mutant clones develop only into neural tissue and never into epidermis (Technau & Campos-Ortega, 1987). One possible interpretation is that the non-autonomous neurogenic genes (i.e. *N, Dl*, etc.) may together provide the source of some cell-surface or intercellular signal, whereas *E(spl)* might encode a receptor for that signal. Cells carrying mutations in any of the non-autonomous genes would still retain the *E(spl)*⁺ receptor function and hence could respond locally to signal produced by nearby wild-type cells, so allowing some mutant cells to become epidermis. Cells lacking the *E(spl)*⁺ receptor, by contrast, would be incapable of responding to the signal and hence could never form epidermal derivatives. However, this is only the simplest of several possible models; the size and complexity of the *E(spl)* locus (large DNA deletions are required to give the null hyperneuralised phenotype) may imply a more complicated version.

A close interaction between the *E(spl)*+ and *Dl*+ functions has been inferred from genetic studies, since double heterozygotes for null mutations at both loci show the hyperneuralised phenotype, whereas heterozygotes for either mutation singly develop more or less normally (Lehmann *et al.*, 1983). This might mean that a reduced dose of *Dl*+ product can depress the response of the *E(spl)*+ function. Maternal D/V mutations (see § 5.3.1) affect the severity of the hyperneuralised phenotype by altering the extent of the ventral neurogenic territory. Thus dorsalised (e.g. *dl*−) mutants reduce or abolish the null *Notch*− phenotype, because there is little or no ventral neurogenic tissue to become hyperneuralised; by contrast, in ventralised *Toll*^{D} mutants the hyperneuralised *Notch*− phenotype affects the entire epidermis (Campos-Ortega, 1983).

Molecular data are available (so far) on three of the neurogenic genes, i.e. *Notch, Delta* and *E(spl)*. *Notch*− mutations all map within a 40 kbp region of DNA that gives rise to a single predominant RNA species of 10.5 kb. The gene spans all 40 kbp and includes at least eight exons, most or all of which are present in the spliced 10.5 kb RNA (presumably the *Notch*+ messenger; Yedvobnick *et al.*, 1985). Null mutations within the *Notch* locus give the hyperneuralised phenotype, but other classes of mutation show more restricted effects on later development, e.g. in the wing or eye. The major 10.5 kb *Notch*+ RNA is accumulated maximally during the period of neurogenesis, although it is also detectable at earlier and later stages (Yedvobnick *et al.*, 1985), particularly in the early pupa and in adult females. This last probably reflects the accumulation of maternal *Notch*+ transcripts in the oocytes. *In situ* hybridisation suggests that *Notch*+ RNA is not confined to the neurogenic regions, but rather is present in many other embryonic tissues during early neurogenesis (Yedvobnick *et al.*, 1985; Hartley *et al.*, 1987). The complete cDNA sequence of the 10.5 kb *Notch*+ RNA predicts a protein of 2703 amino acids, and includes two main types of repeating sequence. One involves 93 bp of repeated CAG and CAA triplets encoding polyglutamine; this *opa* or M repeat is also found in many other developmentally regulated transcripts (Wharton *et al.*, 1985a). Even more striking is the N-terminal half of the predicted protein, which contains 36 tandem repeats of a 40-amino-acid sequence related to vertebrate epidermal growth factor (EGF), and also to the *lin-12* product of *C. elegans* (Wharton *et al.*, 1985b; Greenwald, 1985;

Yochem *et al.*, 1988; see also Akam, 1986, and Bender, 1985). Other features of the predicted *Notch⁺* polypeptide suggest it to be a transmembrane protein, e.g. a characteristic sequence of hydrophobic amino acids that probably spans the membrane. If so, the EGF-like repeats would lie in the extracellular domain, where they might become cleaved off by proteolysis (EGF itself is also cleaved from a much larger precursor protein). There are several point mutations within the *Notch* gene affecting these EGF-like repeats, including one (*split*) which interacts specifically with the *E(spl)⁺* product (Hartley *et al.*, 1987). All of these features are consistent with (and partly underlie) the signalling role proposed for *Notch⁺* and other products in the neurogenic pathway (Technau & Campos-Ortega, 1987).

The *Dl* locus includes some *Notch*-related sequences corresponding to several of the EGF-like repeats (Knust *et al.*, 1987a). Within this cloned region, the *Dl* gene spans some 25 kbp of DNA, producing several developmentally regulated transcripts. One *Dl⁺* cDNA sequence predicts a protein of 880 amino acids with typical transmembrane features and nine tandem EGF-like repeats (Vassin *et al.*, 1987). These *Dl⁺* transcripts are detected initially in all potential neurogenic cells, but later become concentrated in those adopting the neural fate. Thus *Dl⁺* products might help to inhibit neighbouring cells from also following a neural path (but see Kopczynski & Muskavitch, 1989).

The *E(spl)* locus appears to be extremely complex: deletions of at least 34 kbp are required to give the null hyperneuralised phenotype. Some 36 kbp of DNA from the wild-type *E(spl)* locus have been cloned, but this region gives rise to at least 11 major transcripts, not all of which are necessarily required for wild-type *E(spl)⁺* function (Knust *et al.*, 1987b). However, the spatial distributions of four of these transcripts suggest a role complementary to that of *Dl⁺* in the epidermal/neural choice. Although these *E(spl)⁺* transcripts are also present throughout the neurogenic regions initially, they later become mainly confined to to the future epidermal cells. This distribution is clearly consistent with the proposed receptor functions(s) of the *E(spl)⁺* product(s). However, a further transcript from the *E(spl)⁺* locus shows the converse distribution pattern, becoming confined to the future neural tissue (Hartley *et al.*, 1988). There is evidence that this particular transcript may be crucial for *E(spl)⁺* function (Preiss *et al.*, 1988), and moreover it encodes a protein related to the β-subunit of mammalian G proteins, which have a clearly established intracellular function in signal transduction. Thus the exact roles of $N⁺$, $Dl⁺$ and $E(spl)⁺$ products in the neurogenic pathway remain to be clarified.

It is also unclear how the remaining maternal (*amx, pcx*) and zygotic (*bib, mam, neu*) neurogenic gene products will fit into this hierarchy. Yet other genes may be required for neuroblast segregation; thus the appearance of at least one class of neuroblasts is blocked by a deficiency in the *achaete-scute* complex (AS-C: Cabrera *et al.*, 1987b). The transcripts of four homologous AS-C genes are expressed in partially overlapping patterns correlated with neuroblast segregation, though none of them can be detected in differentiated neurones (Cabrera *et al.*, 1987b; Alonso & Cabrera, 1988; see also review of AS-C functions by Ghysen & Dambly-Chaudière, 1988).

As mentioned earlier, stereotyped neuroblast cell lineages give rise to the neurones of the ventral nerve cord. The individual identities of these neurones depend upon the position and time of ingression of their neuroblast ancestor, and also upon local cell interactions (e.g. between sister neurones derived from the same GMC; Doe & Goodman, 1985b). Genes involved in determining the identities of specific neurones include at least two members of the pair-rule class (*ftz* and *eve*: see § 5.4.2 below). Several further gene functions have been shown to establish the identities of particular sensory structures in adult flies. For instance, null mutations at the *cut* locus convert the external sensory organs (humidity and touch receptors) into additional chordotonal organs (e.g. internal stretch receptors; see Blochlinger *et al.*, 1988). Similarly, the *gust-B* mutation converts the single sugar-sensitive neurone in each sensilla into an additional salt-sensitive neurone (Arora *et al.*, 1987). In neither case is the number of organs or neurones altered; rather, there is a switch of identity from one class to another, implying that the wild-type gene function is required to make the distinction between them (cf. homoeotic mutants: § 5.5 below). In the case of *cut*, the wild-type gene-product is a large homeodomain protein (see § 5.6), which is detectable in the cell nuclei of external sensory organs but not in chordotonal organs; hence abolition of *cut*[+] function converts the former into the latter (Blochlinger *et al.*, 1988).

Perhaps the most exciting recent studies on adult sensory structures are those involving the compound eye (see Tomlinson, 1988; Rubin, 1989); each of its many ommatidia comprises a stereotyped pattern of eight photoreceptor cells (R1–8) plus various non-neuronal cells (e.g. cone cells with a lens function). These R cells are recruited in the order R8 (first) →R2/R5 →R3/R4 →R1/R6 →R7 (last). Mutations in the *sevenless* (*sev*) gene specifically eliminate R7 photoreceptors, converting them into additional cone cells. The wild-type *sev* gene produces an 8.2 kb RNA encoding a large transmembrane protein; this includes an essential tyrosine kinase domain, two membrane-spanning domains and an extracellular domain (Hafen *et al.*, 1987; Banerjee *et al.*, 1987a; Basler & Hafen, 1988) – all features suggesting a possible receptor function. However, the *sev*[+] protein is not confined to R7 cells, but rather occurs on the apical surfaces of all types of photoreceptor, as well as deeper in the tissue where several cells contact R8 (Tomlinson, A. *et al.*, 1987; Banerjee *et al.*, 1987b). The recently described *bride of sevenless* (*boss*) gene encodes a putative ligand product which might be recognised by the *sev*[+] receptor (Reinke & Zipursky, 1988). Notably, expression of the *boss*[+] product is required in R8 in order for a neighbouring cell to differentiate into R7, whereas the development of R1–6 and of R8 itself is independent of *boss*[+] function.

Different types of photoreceptor also show selective expression of the four available RH opsin genes (encoding protein moieties of the visual pigments). Thus RH1 (the *nina-E*[+] gene product) is expressed in R1–6, while R7 expresses either RH3 or RH4; by contrast, RH2 is expressed in the dorsal ocelli (accessory adult visual organs; Pollock & Benzer, 1988). P-elements carrying an RH1 promoter fused to the RH2 structural gene and introduced into *nina-E*-deficient flies, result in the targeted misexpression of RH2 in R1–6, with a consequent alteration of visual function (Feiler *et al.*, 1988). Another function required for the normal development of eye pattern is the homeobox-containing gene *rough* (Saint *et al.*, 1988). Analysis of somatic mosaics suggests that *rough*[+] function is required specifically in the R2 and R5 cells of each ommatidium in order to recruit R3/R4; *rough*[–] mutants can be rescued by P-elements with 8.6 kbp of DNA including the 4.3 kbp *rough* transcription unit (Tomlinson *et al.*, 1988). Finally, the *glass* gene encodes a zinc-finger protein (see §5.6) which is required for photoreceptor cell identity in all R cells (Moses *et al.*, 1989).

5.4 Metamerism and antero-posterior pattern

5.4.1 Metameric units; segments, compartments and parasegments

Segments are the visible morphological units into which the insect body (both larval and adult) is divided. In *Drosophila*, several segments (up to six; Struhl, 1981a) are fused to form the head; the three thoracic segments (T1–T3) are distinguished in the larva by their denticle belts and triradiate Keilin's organs, and in the adult by a pair of legs each plus wings on T2 and halteres on T3; finally, the abdomen comprises eight discrete segments (each showing a distinctive pattern of larval denticle belts) plus a ninth and tenth fused together. The one-to-one

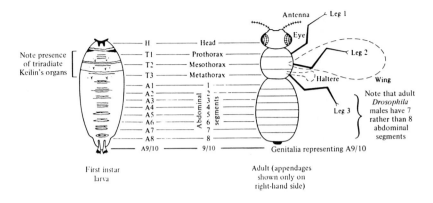

Fig. 5.14 Segment patterns of larval and adult *Drosophila*.

correspondence between larval and adult segments (see fig. 5.14) implies that the arrangement of imaginal discs and histoblast nests in the larva must be related to the basic segmental pattern, even though there is no one-to-one correspondence betwen discs and segments (see §5.2.2). From this, it would be natural to assume that segments must be a fundamental feature of organisation in the *Drosophila* embryo, yet there is now considerable evidence to suggest that this is not the case (reviewed by Martinez-Arias & Lawrence, 1985). In order to understand the alternative parasegmental model, it is first necessary to explain what compartments are and to outline their role in development.

Compartment borders were initially defined as lines of clonal restriction within imaginal discs, as revealed by plotting the boundaries of genetically marked clones arising from X-ray-induced somatic recombination (see §5.1.3). Operationally, this means that cells belonging to one compartment do not cross over its borders nor mingle with cells from neighbouring compartments during normal development (Crick & Lawrence, 1975). Patches of mutant cuticle derived from marked clones normally have irregular outlines, but wherever such a clone abuts on a compartment border, that edge is unusually regular and reproducible. However, each marked clone usually contributes only a small region of cuticle, hence a large number of instances would be required to establish the position of a given compartment border (fig. 5.15). In some cases, the border corresponds to an obvious structural discontinuity, such as the line of distinctive non-proliferating cells which separates the dorsal and ventral compartments in the wing disc (Brower *et al.*, 1982; O'Brochta & Bryant, 1985). However, no such

A   X-ray induction of marked *mwh/mwh* clones in posterior compartments of wing discs

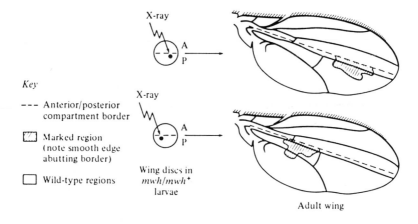

Key

--- Anterior/posterior
     compartment border

▨ Marked region
     (note smooth edge
     abutting border)

☐ Wild-type regions

Wing discs in
mwh/mwh⁺
larvae

Adult wing

B   X-ray induction of marked fast-growing *M⁺ mwh/M⁺ mwh* clone in posterior
compartment of wing disc

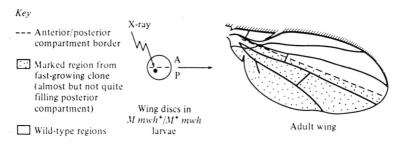

Key

--- Anterior/posterior
     compartment border

▨ Marked region from
     fast-growing clone
     (almost but not quite
     filling posterior
     compartment)

☐ Wild-type regions

Wing discs in
M mwh⁺/M⁺ mwh
larvae

Adult wing

Fig. 5.15 Anterior and posterior compartments in the *Drosophila* wing disc (after Crick & Lawrence, 1975).

discontinuity distinguishes anterior from posterior compartments, even though the boundary between them is established very early in development (see below).

A dramatic illustration of compartment borders can be made in *Drosophila* heterozygotes carrying one copy of the dominant *Minute* (*M*) mutation, which results in a much slower rate of cell division. If these flies are also heterozygous for a closely linked recessive cuticle marker such as *mwh* (see fig. 5.5) in the combination *M⁺.mwh/M.mwh⁺*, then X-ray-induced somatic recombination will sometimes generate a clone of fast-growing cells producing mutant cuticle in the adult wing (i.e. the progeny of a homozygous *M⁺.mwh/M⁺.mwh* founder cell). Surrounding cells of the original genotype will be slow-growing (*M* being dominant) but will produce wild-type wing cuticle (*mwh* being recessive). Since the *M* mutation is lethal when homozygous there will be no contribution of $M.mwh^+/M.mwh^+$ cells. The net result of this genetic trick is to generate fast-growing clones of marked cells on a slow-growing unmarked background (Garcia-Bellido *et al.*, 1973; Morata & Ripoll, 1975). Such a marked clone may almost fill a compartment, but never does so completely (see fig. 5.15); there is always a small region of wild-type cuticle also present. This implies that each compartment must be founded by a group of cells, some of which were slow-growing. Thus the binary determination decisions which give rise to the different compartments within a disc are **polyclonal**, as discussed previously in §5.2.2. Most importantly, such fast-growing clones never transgress the compartment borders once established, nor is the final size of the compartment altered (Garcia-Bellido *et al.*, 1973; Crick & Lawrence, 1975).

By inducing marked clones at successively later times during development and observing which compartments include regions of marked cells, it is possible to determine the order in which compartment borders are established. This is illustrated diagrammatically in fig. 5.16 for the imaginal discs of the second thoracic segment (see Lawrence & Morata, 1976). In the blastoderm, the future imaginal cells of both wing and second-leg discs are grouped together. Marked clones induced at this early stage are already restricted to *either* the anterior (*a*) or posterior (*p*) compartments of both wing and leg. Clones induced after the separation of leg and wing discs in the later embryo will of course produce marked regions in either wing or second leg, but not in both. The previous distinction between *a* and *p* compartments is still respected, i.e. a single marked clone will be confined within one of the four compartments so far established, namely, wing *a*, wing *p*, leg *a* or leg *p*. Still later, during larval development, marked clones in the wing disc become further confined within either dorsal or ventral compartments, and finally within either proximal (future notum) or distal (future wing blade) compartments. Thus a marked clone induced in the late larval wing disc will occupy part of just one out of the eight wing compartments (see fig. 5.16). In all cases, previous compartment assignments are respected, and the determinative decisions (involving binary choices between pairs of options) are always made by a polyclonal group of cells.

As mentioned earlier, the border between dorsal and ventral compartments in the wing disc is marked by a line of non-dividing cells which corresponds to the future wing margin. Brower *et al.* (1984) have

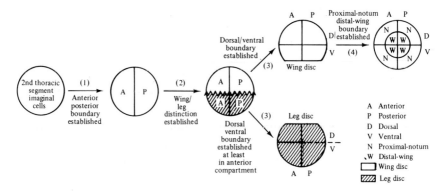

Fig. 5.16 Compartment determination among imaginal cells of the second thoracic segment (after Lawrence & Morata, 1976).

prepared monoclonal antibodies against several cell-surface glyco-proteins from the wing disc; one such protein becomes confined to the dorsal compartment during later larval development, while another becomes restricted to the ventral compartment. The time at which these proteins become segregated to different compartments in the wing disc coincides approximately with the definitive establishment of the dorsal/ventral compartment boundary (Brower *et al.*, 1985). Similar considerations may also apply in the case of the proximal/distal boundary in the wing disc, which is also established late and involves a structural discontinuity (that between notum and blade in the adult wing). On both counts, these compartmental restrictions seem to be different in kind from the *a/p* compartment boundary established in the early embryo (see Brower, 1985). In fact, the *a/p* distinction is fundamental in the metameric organisation of *Drosophila*, subdividing not only the imaginal discs and their adult derivatives, but also the embryonic, larval and adult segments. The distinction between *a* and *p* compartments requires *engrailed⁺* function in the latter (§5.4.2c). Recent data suggests a graded requirement for the *Distal-less⁺* (*Dll⁺*) gene-product to establish distal cell identities (Cohen & Jurgens, 1989).

A chain of *a* and *p* compartments can be grouped pairwise in either of two repeating patterns, one of which (*a,p/a,p/*, etc.) corresponds to the segments visible in later development, while the other (*/p,a/p,a/p*, etc.) does not. The */p,a/* repeat in the latter pattern has been termed the **Parasegment**, comprising the posterior compartment of one segment plus the adjacent anterior compartment of the next segment (see fig. 5.17). There are now strong reasons for believing that parasegments, rather than conventional segments, are the fundamental units of metameric organisation in the early *Drosophila* embryo (Martinez-Arias & Lawrence, 1985). When the embryonic germ band elongates, a series of shallow transverse grooves appears; these were originally thought by Poulson (1950) to delineate the future segments. However, several morphological criteria (in particular the position and number of tracheal pits) suggest that in fact these grooves define parasegmental units. Further evidence comes from the patterns of expression shown by many of the genes which govern the number and identity of metameric units; most of these are active in parasegmental rather than segmental regions. However, the posterior compartments are much narrower than the anterior compartments (about one as opposed to three cells wide when first established), which means that segmental and parasegmental expression patterns would be only slightly offset from one another. But in several cases where domains of gene expression in the blastoderm have been mapped on a cell by cell basis, the boundaries observed correspond to parasegments rather than to segments (see § § 5.4.2 and 5.5 below).

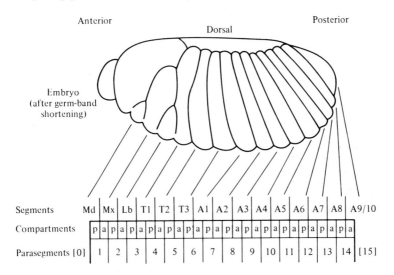

Fig. 5.17 Segments, parasegments and compartments in *Drosophila* (redrawn from Martinez-Arias & Lawrence, 1985).
Md, mandibular segment ⎫
Mx, maxillary segment ⎬ posterior head (or gnathal) segments;
Lb, labial segment ⎭
T1–T3, thoracic segments; A1–A9/10, abdominal segments; p, posterior compartment; a, anterior compartment; 1–14, definitive parasegments; [0] and [15] are designated pseudoparasegments.

The above considerations (*a* and *p* compartments, parasegments, etc.) probably apply to the central nervous system as well as to the cuticle-secreting epidermis (both are of ectodermal origin; see Lawrence, 1985). During germ-band shortening the internal mesoderm tissues shift slightly relative to the outer epidermis, such that segmental units of the latter come to overlie parasegmental units of mesoderm. These mesodermal units are not subdivided into *a* and *p* compartments like the ectoderm; thus *engrailed⁻* mutant clones induced in mesodermal tissues have little or no effect (Lawrence & Johnston, 1984a), in contrast to their drastic effects on ectoderm (see §5.4.2 below). Note that the mesoderm is itself split into visceral and somatic layers which may show different patterns of internal spatial organisation (see discussion of *Ubx⁺* function in § 5.5, below). Finally, the endoderm arises from invaginations of tissue outside the germ band (fig. 5.3); it is neither subdivided into metameric units nor affected by mutations which alter pattern within the germ band (Lawrence, 1985).

5.4.2 Genetic regulation of A/P pattern and metamerisation

The basic principles which underlie the regulation of D/V pattern (§5.3.1) also hold true for A/P pattern, though the latter is considerably more complex. In general terms, various maternal determinants are differentially distributed along the A/P axis of the egg; between them these generate a system of A/P positional coordinates to which several zygotic genes respond by becoming active within defined A/P zones. But whereas a single system of interacting maternal products suffices to define the D/V axis, three interdependent maternal systems are required to define anterior, posterior and terminal regions along the A/P axis. In the early embryo, this maternal information is used to activate the zygotic gap genes, each of which is expressed within broad A/P regions. Subsequently, the germ band is divided into smaller metameric units through activation of the pair-rule genes (expressed in 7-stripe patterns with double-segment periodicity) and segment polarity genes (expressed in 14-stripe patterns with single-segment periodicity). Thus the A/P axis first becomes regionalised (maternal and gap genes), and then the central germ band (fig. 5.3) becomes subdivided into metameric units (pair-rule and segment-polarity genes). The activities of these 'segmentation' genes together define the number, polarity and size of the metameric units (Nusslein-Volhard & Wieschaus, 1980), while the distinctive identities of such units are largely controlled by the homoeotic genes (§5.5

below). Since the spatial domains of homoeotic gene expression are specified by maternal, gap and pair-rule products, these systems are all interlinked in a complex regulatory hierarchy or network (reviewed by Akam, 1987; Scott & Carroll, 1987; Ingham, 1988). Strikingly, many of the general and detailed features of this hierarchy can be simulated by a computer model based on reaction-diffusion mechanisms and mutual inhibition, in order to generate discrete cell-states coexisting side by side (Meinhart, 1986).

(a) *Maternal control of A/P pattern*

Although the *Drosophila* larva appears segmented throughout much of its length, these structures all arise from the germ band representing about half of the blastoderm fate map (fig. 5.3). Regions anterior to the germ band give rise to anterior head and 'acron' structures (which are largely involuted in the larva), while regions posterior to it produce telson structures (fig. 5.18*A*). As defined by Nusslein-Volhard *et al.*, (1987), the telson includes most structures posterior to A8 (such as posterior gut, Malpighian tubules and the fused primordia of A9/10) but not the pole cells, whereas the 'acron' comprises the labrum and much of the cephalopharyngeal skeleton but not the antennal or maxillary structures. Other definitions restrict both terms to a subset of the structures mentioned, but they are used here in the broader sense throughout. Although these distinctions may seem academic or even arbitrary in terms of adult structure, they appear to be of fundamental importance in the organisation of the egg and early embryo, since a separate set of terminal gene functions is required to define both 'acron' and telson (see below).

Maternal-effect mutations can alter the embryonic A/P pattern in one of three distinct ways: (i) by deleting anterior (head/thoracic) elements, and in some cases replacing the acron with a duplicated telson (fig. 5.18*B*); (ii) by deleting posterior (abdominal) structures, and sometimes also the pole cells, but without affecting the telson (fig. 5.18*C*); or (iii) by deleting both acron and telson together (fig. 5.18*D*). Members of these three classes are termed anterior, posterior and terminal mutants respectively; they are listed in table 5.3.

The anterior-class functions present what is so far the clearest molecular paradigm of how maternal genes control A/P polarity. Null *bcd⁻* embryos show severe deletions of head and thorax plus a transformation of the acron into a second telson (fig. 5.18*B*). This pattern of defects can be phenocopied simply by removing cytoplasm from the anterior pole (AP) of a wild-type egg, implying that anterior deter-

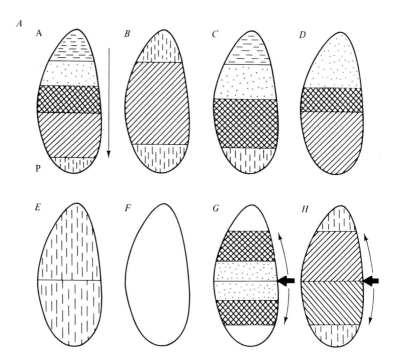

Fig. 5.18 Schematic representations of some maternal A/P mutant phenotypes (after Nusslein-Volhard *et al.*, 1987).
Key to cuticular structures observed: [⊟⊟], acron; [∴∴], head; [▨▨], thorax; [⧄⧄], abdomen; [▥▥], telson; ◀, site of injection; A ⟶ P, polarity of embryonic structures within germ band.

A Wild-type embryo.
B bcd⁻ embryo (head and thorax deleted; acron transformed to a second telson).
C osk⁻ embryo (abdomen deleted); similar phenotype for other mutants in posterior class.
D tor⁻, trk⁻ or *tsl⁻* embryo (acron and telson both deleted).
E bcd⁻/osk⁻ double mutant (polarity neutral), comprising two juxtaposed telsons (head, thorax and abdomen deleted; acron transformed to second telson).
F bcd⁻/osk⁻/tsl⁻ triple mutant (no differentiated cuticular structures present).
G Central head embryo formed by injecting wild-type anterior pole cytoplasm into centre of a *bcd⁻/osk⁻* or *bcd⁻/osk⁺* embryo.
H Central abdomen embryo formed by injecting wild-type posterior pole cytoplasm into centre of a *bcd⁻/osk⁻* (but not *bcd⁺/osk⁻*) embryo.

Table 5.3. *Maternal mutations affecting A/P pattern (modified from Nusslein-Volhard et al., 1987)*

Gene	Phenotype of deficient embryo	Comments (see text)
Anterior class		
bicoid (bcd)	Head/thorax deleted; acron transformed to 2nd telson	Localised anterior signal/determinant
exuperantia (exu) *swallow (swa)*	Weaker anterior deletions	Affect localisation of the *bcd*+ product
bicaudal (bic) *Bicaudal C (BicC^D)* *Bicaudal D (BicD^D)*	Large anterior deletion/ duplication of posterior abdomen and telson	*BicD^D* has anterior duplication of the posterior activity
Posterior class		
oskar (osk) *vasa (vas)* *valois (val)* *staufen (stau)* *tudor (tud)*	Deletion of abdomen and pole plasm, but no effect on telson; *stau*− also causes anterior deletions	All 5 are involved in localising the posterior activity in posterior pole cytoplasm
nanos (nos) *pumilio (pum)*	Deletion of abdomen, but no effect on pole plasm or telson	*nos*+ product may be localised posterior signal; *pum*+ product required for signal transmission
Terminal class		
torso (tor) *torsolike (tsl)* *trunk (trk)*	Deletion of both acron and telson	

minants are normally localised at the AP. Transplantation experiments reveal that wild-type AP cytoplasm can rescue head and thorax development when injected at any site into *bcd*− embryos (Frohnhofer & Nusslein-Volhard, 1986; example in fig. 5.18*G*). The rescuing activity (presumably *bcd*+ product) declines rapidly away from the anterior end in wild-type embryos, suggesting a gradient distribution; moreover, the rescuing ability of AP cytoplasm is dependent on the dosage of wild-type *bcd*+ genes in the donor (3 copies > 2 > 1). These studies all imply that the *bcd*+ product is a primary determinant of anterior development (Frohnhofer & Nusslein-Volhard, 1986).

The *bcd* gene has now been cloned (Frigiero *et al.*, 1986), and contains both a homeobox and other sequences related to the *paired* pair-rule gene. Normally, *bcd*⁺ RNA is synthesised by nurse cells during oogenesis, but remains sharply localised at the anterior pole of the egg (where the ring canals enter the oocyte; Belerth *et al.*, 1988; fig. 5.19). This accords with the interpretation proposed earlier for the *dicephalic* mutant (§5.1.2), which has two such entry points. Sequences responsible for anterior localisation of *bcd*⁺ mRNA are located in the 3' untranslated region of this RNA (Macdonald & Struhl, 1988). During preblastoderm stages of embryogenesis, the *bcd*⁺ protein product forms a roughly exponential gradient, declining steeply away from the AP (Driever & Nusslein-Volhard, 1988a; fig. 5.19). Changes in the shape of the *bcd*⁺ protein gradient protein gradient are directly reflected in the altered embryonic fate map; extra copies of the *bcd*⁺ gene cause an anterior shift of pattern, while fewer copies result in a posterior shift (Driever & Nusslein-Volhard, 1988b). The *bcd*⁺ protein product binds via its homeodomain to DNA (consensus TCTAATCCC); several such target sequences occur upstream of the zygotic gap gene *hunchback* (three between −50 and −300) which is positively regulated by *bcd*⁺

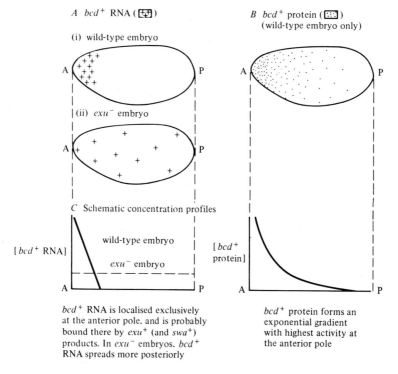

A bcd⁺ RNA (▦)

(i) wild-type embryo

(ii) *exu*⁻ embryo

B bcd⁺ protein (▦)
(wild-type embryo only)

C Schematic concentration profiles

[*bcd*⁺ RNA]

wild-type embryo

exu⁻ embryo

[*bcd*⁺ protein]

bcd⁺ RNA is localised exclusively at the anterior pole, and is probably bound there by *exu*⁺ (and *swa*⁺) products. In *exu*⁻ embryos, *bcd*⁺ RNA spreads more posteriorly

bcd⁺ protein forms an exponential gradient with highest activity at the anterior pole

Fig. 5.19 Localisation of *bcd*⁺ gene products in preblastoderm embryo.

protein (Driever & Nusslein-Volhard, 1989). *bcd⁻* embryos can be rescued by fusion proteins containing foreign activation domains fused to the *bcd⁺* DNA-binding domain (Driever *et al.*, 1989a).

Two other anterior-class mutants (*exu⁻* and *swa⁻*) resemble weak *bcd⁻* alleles in producing milder deletions of anterior structures. However, removal of AP cytoplasm scarcely alters the phenotype of *exu⁻* or *swa⁻* embryos, but greatly increases the severity of weak *bcd⁻* phenotypes; *exu⁻* and *swa⁻* embryos also show an expansion of posterior head and thoracic regions, only the anterior head being deleted. This suggests that *bcd⁺* activity is not localised anteriorly in *exu⁻* or *swa⁻* embryos, but rather extends into more posterior regions than normal (Frohnhofer & Nusslein-Volhard, 1987). This has been confirmed directly by studies of *bcd⁺* RNA localisation in *exu⁻* or *swa⁻* embryos; both show lower concentrations of *bcd⁺* RNA spread throughout much of the embryo (Belerth *et al.*, 1988: fig. 5.19). Plausibly, the *exu⁺* and *swa⁺* products (cytoskeletal components?) might serve to anchor *bcd⁺* mRNA via its 3′ localisation sequences at the anterior end of the egg.

BicD^D is a dominant (gain of function) mutation which also causes deletion of anterior structures but replaces them with a mirror-image duplication of both telson and abdomen (Mohler & Wieschaus, 1985). There is evidence to suggest that a second centre of posterior activity is present at the AP in *BicD^D* embryos, rather than altered *bcd⁺* localisation; other bicaudal-type mutants may involve similar abnormalities.

The fact that *bcd⁻* mutants show a telson duplication in place of the acron does not imply any additional posterior activity, since both acron and telson are separately controlled by the maternal terminal-class genes. Both structures are specifically deleted in *tor⁻*, *trk⁻* or *tsl⁻* embryos (fig. 5.18*D*). These terminal gene products on their own direct telson formation, but in conjunction with high *bcd⁺* activity they direct acron development instead. Thus a lack of *bcd⁺* activity at the anterior pole results in replacement of the acron by a second telson, because of the failure of this *bcd⁺* switch function (Nusslein-Volhard *et al.*, 1987). There is some overlap between the posterior and terminal functions, since *tor⁻* mutants show defects in A7 as well as the telson. The maternal terminal genes probably influence embryonic pattern via the zygotic gap gene *tailless* (*tll*), which may become activated locally in the terminal regions. Thus *tll⁻* embryos delete acron and telson structures which are a subset of those lost in *tor⁻*, *trk⁻* or *tsl⁻* mutant embryos (Strecker *et al.*, 1986, 1988; Mahoney & Lengyel, 1987).

In the case of *tor*, dominant gain-of-function (*tor^D*) alleles cause a converse phenotype in which the segmented germ band is deleted, the embryo consisting of an acron juxtaposed next to an enlarged telson;

the expression of zygotic segmentation genes such as *ftz* is abolished in such *tor*[D] mutants (Klingler *et al.*, 1988; Strecker *et al.*, 1989). However, in double mutants which also lack the zygotic *tll*[+] function, segmentation of the germ band is restored in spite of the *tor*[D] mutation. This demonstrates that the maternal *tor*[+] function affects embryonic pattern through zygotic *tll*[+] activity; the *tor*[D] mutations may cause inappropriate expression of *tll*[+] in the central regions of the embryo, thereby suppressing the segmentation functions normally active within the germ band (Klingler *et al.*, 1988). However, neither *tor*[+] mRNA nor its membrane-bound protein product is confined to the ends of the early embryo, implying local activation of the protein in the terminal regions (Sprenger *et al.*, 1989; Casanova & Struhl, 1989). The sequence of the *tor*[+] products suggests transmembrane protein with a tyrosine kinase domain, related to vertebrate growth-factor receptors. The *l(1) pole hole* gene encodes a serine/threonine kinase which is itself activated locally by the *tor*[+] product in the terminal regions (Ambrosio *et al.*, 1989). The A/P pattern of mutant embryos can also be determined by monitoring the blastoderm expression patterns of several zygotic genes (including *ftz*; also *Dfd* and *cad*, expressed respectively at the anterior and posterior ends of the germ band; see §5.5). This approach confirms the localised effects of *tor*[−] and *trk*[−] mutations (largely confined to acron and telson), whereas both *bcd*[−] and *exu*[−] embryos show longer-range alterations in such gene expression patterns, consistent with the morphological changes (Mlodzik *et al.*, 1987).

The maternal posterior-group functions show a more complex pattern of interactions. Embryos deficient for any of these seven gene products show a common mutant phenotype (fig. 5.18*C*) in which the entire abdomen is deleted. (A similar phenotype, but retaining segment A1, is given by mutations in the zygotic gap gene *knirps*.) Five of these maternal mutants also fail to develop pole cells (e.g. *tud*[−], see §5.1.2 and table 5.3), termed the grandchildless-*knirps* phenotype (Carroll *et al.*, 1986b). Thus wild-type *vasa*[+] products are required specifically for the development of the female germ line (Lasko & Ashburner, 1988; see §5.1.2). However, the other two mutants (*nos*[−] and *pum*[−]; table 5.3) form pole cells normally. Six of the seven mutants can be rescued by transplantation of cytoplasm from the posterior pole (PP) of wild-type embryos (the *tud*[−] phenotype is too variable for this to provide a reliable assay). The best studied example is *osk*; *osk*[−] embryos can be rescued only by PP cytoplasm from wild-type donors. If this is injected into the PP of recipient *osk*[−] embryos, pole-cell formation is rescued but abdominal segmentation remains defective; conversely, injections at a more anterior site rescue abdominal segmentation but not pole-cell formation (Lehmann & Nusslein-Volhard, 1986). Injection of wild-type PP cytoplasm into the anterior tip of recipient embryos has only a slight 'posteriorising' effect on their A/P pattern (Nusslein-Volhard *et al.*, 1987), implying that *bcd*+ activity may inhibit the effects of posterior determinants. In *bcd⁻/osk⁻* double-mutant embryos (which are polarity-neutral and normally comprise two juxtaposed telsons; fig. 5.18*E*), central injections of wild-type PP cytoplasm produce a striking bicaudal phenotype in which the anterior abdominal segments flank the injection site while posterior abdominal segments adjoin both telsons (fig. 5.18*H*). If the *osk*+ function were a direct determinant of posterior pattern, one would predict the opposite polarity, i.e. with the most posterior abdominal structures flanking the central injection site (compare fig. 5.18*G* and *H*). In addition to the grandchildless-*knirps* defects, *stau⁻* embryos also show some anterior deletions, due apparently to an effect on *bcd*+ product distribution; this may imply a localisation function for the *stau*+ product.

Finally, what of the two posterior-class mutants that do not affect pole-cell formation? In the case of *pum*, PP cytoplasm from *pum⁻* embryos can rescue abdominal segmentation in embryos deficient for any of the posterior-class functions, but only when injected into the prospective abdomen (Lehmann & Nusslein-Volhard, 1987a). This exceptional feature suggests that the *pum*+ product is required to transport a posterior signal away from its PP location into the abdominal area where its function is needed.

This leaves us with *nanos* (*nos*) as the only remaining candidate which might encode the posterior signal itself. Lehmann has recently reported experiments which support such a role (quoted in North, 1988; Sander & Lehmann, 1988), using transplants of mutant nurse-cell cytoplasm to rescue posterior-class deficiencies. PP cytoplasm from *osk⁻* embryos cannot rescue the *osk⁻* phenotype, yet nurse-cell cytoplasm from *osk⁻* egg chambers can do so effectively. The same is true for the other grandchildless-*knirps* mutants, but not for *nos⁻* mutants – whose nurse-cell cytoplasm fails to rescue mutant embryos lacking any of the posterior-class products. This is consistent with the proposed signalling role of *nos*+, although a localised posterior distribution for *nos*+ products remains to be demonstrated. Recent evidence suggests that the posterior signalling role of *nos*+ may function indirectly via removal of the maternal *hunchback* (*hb*+) product from the posterior end of the egg (see below). Within the grandchildless-

knirps group of mutations, the embryonic pattern defects can be rescued interchangeably by mutant nurse-cell cytoplasm. From this one might infer that the wild-type products of these five genes cooperate to provide a source for the posterior (*nos*⁺?) signal, e.g. transporting it from the nurse cells and localising it correctly at the PP. The same five genes (but not *nos*⁺) also have an important role in the formation of pole cells; notably, the *vasa*⁺ protein is localised in these cells from the time of their formation. If any of these five maternal gene-products is lacking, no pole or germ cells develop in the embryo, hence the grandchildless aspect of all five mutant phenotypes (§ 5.1.2).

Before leaving the maternal A/P genes, it is worth mentioning that two genes with essential zygotic functions also show a graded maternal component of expression; these are the gap gene *hunchback* (*hb*) and the homeobox gene *caudal* (*cad*). The maternal products of both genes become distributed in a gradient pattern during early (preblastoderm) embryogenesis; *hb*⁺ product levels are high in the anterior half of the egg but decline to zero at the posterior end (Tautz, 1988), while the *cad*⁺ gradient runs from high posteriorly to low anteriorly (Mlodzik *et al.*, 1985; Macdonald & Struhl, 1986). There are two promoters driving a single protein-coding region in both the *cad* and *hb* genes; in each case, only one of these promoters is used for maternal expression, the other (*cad*; Mlodzik & Gehring, 1987) or both (*hb*; Schroder *et al.*, 1988) being used zygotically. In *cad*⁻ embryos lacking the maternal gradient there is some disturbance of the global A/P pattern (Macdonald & Struhl, 1986); the maternal *cad*⁺ product also activates zygotic *ftz* transcription in the posterior half of the embryo (Dearolf *et al.*, 1989). Similarly, the zygotic *hb*⁻ gap phenotype becomes more severe in the absence of maternal *hb*⁺ product (comparing *hb/hb* with *hb/*⁺ germ cells; see below).

A more central role for the maternal *hb*⁺ product has been inferred recently from its interaction with the *nos*⁺ posterior signal. The *nos*⁻ mutant phenotype (no abdomen) can be rescued completely by eliminating the maternal gradient of *hb*⁺ product; thus *hb*⁻/*nos*⁻ eggs can develop into normal embryos (*nos*⁻ effect suppressed) provided they are fertilised by wild-type sperm so as to provide the essential zygotic *hb*⁺ function (Irish *et al.*, 1989a; Hülskamp *et al.*, 1989). This suggests that the *nos*⁺ product may in fact serve to eliminate maternal *hb*⁺ products from the posterior end of the egg (giving the maternal gradient described above). Such a function would be abolished in eggs from *nos*⁻ mutant mothers, resulting in the inappropriate presence of maternal *hb*⁺ products in the posterior regions of such eggs. The *nos*⁺ control mechanism must act post-transcriptionally to govern the distribution of maternal hb^+ products, since these are expressed in both nos^+ and nos^- eggs. Heat shock can be used to induce overexpression of maternal hb^+ products in eggs carrying a P-element construct with a heat-shock promoter fused to the hb coding sequence. The net effect is to swamp the nos^+ control with high levels of hb^+ products throughout the egg, and the resultant embryonic phenotype is closely similar to that of maternal nos^- mutants (Struhl, G., 1989). Again, this suggests that the wild-type function of nos^+ is to remove maternal hb^+ products from posterior regions of the egg. This is an example of double negative control, with the posterior nos^+ signal eliminating maternal hb^+ products from the posterior end of the egg, since these would otherwise suppress abdominal segmentation.

(b) *The zygotic gap genes*

Mutations in the zygotic gap genes delete broad regions of the embryonic A/P pattern, in each case spanning several contiguous segments. The extent of these deletions is outlined in Table 5.4 for each of the five candidate genes in this class, namely, *hunchback* (*hb*, affecting thorax and posterior head, plus A7/8), *Kruppel* (*Kr*, affecting thorax and anterior abdomen, plus some telson structures), *knirps* (*kni*, affecting most of the abdomen), *giant* (*gt*, affecting head and posterior abdomen), and *tailless* (*tll*, affecting non-segmented acron and telson structures only).

These genes may be among the immediate zygotic targets which are regulated directly by the maternal A/P determinants discussed above. As mentioned earlier, mutations in the *tll* gap gene cause deletions of both acron and telson structures, similar to those lost in the maternal terminal mutants (tor^-, trk^- or tsl^- embryos). Thus in the early embryo, the tll^+ function may become activated locally in response to the products of maternal terminal genes (see discussion of tor^D mutants in (a) above). Mutations in the *kni* gap gene cause abdominal deletions similar to those in embryos lacking maternal posterior-class products, although A1 is retained in the former but not the latter; as discussed earlier, the response of the kni^+ function to maternal posterior determinants is likely to be an indirect one (involving the removal of maternal hb^+ products from posterior regions of the egg). The wild-type *hb* gene itself is active zygotically in an anterior domain, suggesting positive regulation by bcd^+ activity. Tautz (1988) has confirmed this by showing that zygotic hb^+ expression depends on the bcd^+ function. There is also direct evidence that the bcd^+ protein acts as a positive transcription factor binding to sequences upstream of the hb^+ cap site

Table 5.4. *Zygotic gap genes (modified from Akam, 1987)*

Gene	Segments lost in mutant	Regulation	
		+ve	−ve
hunchback (hb)	p. head to T3 + A7/8 (*hb/+* gc)	*bcd*	*Kr,*
	a. head to A2/3 + A7/8 (*hb/hb* gc)		(MPG)
Kruppel (Kr)	T1 to A5 inclusive (also	?	*bcd, hb,*
	Malpighian tubules)		*kni,*
			MPG
knirps (kni)	A1/2 to A7 inclusive	low *Kr*	*tll*
giant (gt)	Some head structures + A4 to A8	?	?
tailless (tll)	Acron and telson structures only	MTG	?

Abbreviations: a., anterior; p., posterior; gc, germ cell; MPG, maternal posterior genes; MTG, maternal terminal genes.

and activating its zygotic expression (Driever & Nusslein-Volhard, 1989). Both high- and low-affinity *bcd*$^+$-protein-binding sites are present in the proximal *hb* promoter; fusion constructs regulated by high-affinity sites are expressed throughout the anterior half of the embryo, whereas those regulated by low-affinity sites are expressed only in the anteriormost quarter (Driever *et al.*, 1989b). During the late blastoderm stage, *hb*$^+$ products disappear from the most anterior regions of the embryo, perhaps implying negative regulation by the *tll*$^+$ function. *Kr*$^+$ normally acts within a central zone, implying negative regulation by both posterior and anterior determinants. Again this has been demonstrated directly, since the *Kr*$^+$ protein domain spreads anteriorly in *bcd*$^-$ embryos but posteriorly in embryos lacking posterior-class functions (Gaul & Jackle, 1987). Maternal A/P determinants therefore provide spatial cues which define the zones of gap-gene activity, though clearly this regulation is complex.

Both *Kr* and *hb* genes have been cloned (Preiss *et al.*, 1985; Jackle *et al.*, 1986), as have the *kni* and *gt* genes more recently (Nauber *et al.*, 1988; Mohler *et al.*, 1989). *Kr*, *hb* and *kni* all encode nuclear proteins with DNA-binding zinc-finger structures (cf. those in TFIIIA and steroid receptors: Rosenberg *et al.*, 1986; Gaul *et al.*, 1987; Tautz *et al.*, 1987; Tautz, 1988; Nauber *et al.*, 1988; Stanojevic *et al.*, 1989; Treisman & Desplan, 1989). The case of *Kr* is the simplest, as there is no maternal expression of this gene to complicate the issue. Wild-type embryos injected with antisense *Kr* RNA develop as phenocopies of the *Kr*$^-$ gap mutant (Rosenberg *et al.*, 1985). Zygotic transcripts from the *Kr*$^+$ gene are first detected in the central regions of the syncytial blastoderm, and by the cellular blastoderm stage there is a sharply defined band of Kr^+ transcripts spanning the region from about T2 to A1 in the prospective germ band (Knipple *et al.*, 1985). This is considerably smaller than the group of segments deleted in null Kr^- embryos (T1 to A5). Somewhat later, a smaller zone of Kr^+ transcripts appears near the posterior pole, perhaps correlated with a requirement for Kr^+ in the development of Malpighian tubules. Indeed, the Kr^+ function acts as a homoeotic switch to distinguish Malpighian tubule cells from hindgut (Harbecke & Janning, 1989). By gastrulation the Kr^+ transcript pattern becomes considerably more complex; still later, Kr^+ protein is detectable in several tissues including the amnioserosa, muscle precursors and parts of the nervous system (Gaul *et al.*, 1987).

In blastoderm embryos, the *kni* gene is expressed antero-ventrally and in two circumferential zones, one posterior and one anterior to the Kr^+ transcript domain (Rothe *et al.*, 1989); the posterior zone clearly correlates with the known requirement for kni^+ function in the abdomen (table 5.4). This posterior zone of kni^+ transcripts is abolished in *nos*$^-$ embryos (developing from *nos/nos* eggs). This is probably an indirect effect; the inappropriate presence of maternal hb^+ products in posterior regions of *nos*$^-$ eggs might prevent the activation of *kni* transcription there. The kni^+ gene-product is a member of the steroid receptor superfamily (see also §3.3.1); it may well act as a transcriptional regulator, although the nature of its activating ligand (if any) remains unknown. A closely similar gene known as *knirps-related* (*knrl*) has been cloned on the basis of its partial homology to the human retinoic acid receptor gene (Oro *et al.*, 1988); maternal $knrl^+$ transcripts are uniformly distributed but zygotic transcripts accumulate in spatially restricted zones similar to those for kni^+ transcripts. A third member of this group is the *egon* gene, expressed only in late embryonic gonads (Rothe *et al.*, 1989). Zygotic kni^+ expression is repressed by tll^+ activity (hence the kni^+ domain extends posteriorly in tll^- mutants); by contrast, kni^+ expression is enhanced by low levels of Kr^+ protein extending through the kni^+ domain (though the kni^+ and Kr^+ transcript domains overlap only slightly; Pancratz *et al.*, 1989).

In the case of *hb*, it is necessary to distinguish the products of maternal expression (forming an A/P gradient) from the zygotic products. A detailed developmental analysis of hb^- mutants (Lehmann & Nusslein-Volhard, 1987b) shows that the maternal hb^+ component is not in fact necessary for normal development, since oocytes lacking this product (in germ-line mosaics) can be rescued by wild-type sperm.

Nevertheless, as mentioned earlier, the hb^- mutant gap is more severe in hb/hb offspring derived from hb/hb as compared to $hb/+$ germ cells. *Rg(pbx)* is a dominant gain-of-function mutation at the *hb* locus (Bender *et al.*, 1987) which converts posterior haltere to wing in the adult fly, a phenotype also given by one class of recessive homoeotic mutations (*pbx*) in the *Ubx* domain of the bithorax complex (§5.5). The unexpected phenotype of this *hb* allele probably reflects the fact that gap-gene products directly regulate the spatial domains of expression for several homoeotic genes (including *Ubx*: White & Lehmann, 1986; Harding & Levine, 1988), as well as for pair-rule and segment-polarity genes (Ingham *et al.*, 1986; see below). Thus gap genes are indeed at the apex of the zygotic gene network which governs the number and identity of metameric units.

Zygotic hb^+ transcripts first appear in the anterior half of the syncytial blastoderm, but soon become confined to an anterior region which abuts on the Kr^+ transcript zone (see below). Slightly later, a narrow band of hb^+ transcripts appears in a posterior position corresponding to the primordia of A7/8 (Jackle *et al.*, 1986). This accords well with the hb^- mutant phenotypes listed in table 5.4, which involve the deletion of anterior regions (varying in extent depending on the presence or absence of maternal hb^+ product) plus loss of parts of A7/8. The anterior zone of hb^+ activity depends on a 2.9 kb transcript which is only expressed zygotically, using one of the two *hb* promoters. This proximal promoter is activated specifically by the maternal bcd^+ protein, which interacts with DNA sequences upstream of the cap site (Schroder *et al.*, 1988; Driever & Nusslein-Volhard, 1989). There is also a 3.2 kb RNA which is transcribed from the other *hb* promoter but which encodes the same protein product. This longer transcript is expressed both maternally in an A/P gradient, and later zygotically in two bands, one abdominal and one more central (specifically in the A7/8 primordia and just posterior to the zone of 2.9 kb hb^+ transcripts). Regulation of the distal promoter for this 3.2 kb transcript appears to be considerably more complex (Schroder *et al.*, 1988).

The distributions of hb^+, Kr^+ and kni^+ gap-gene transcripts show considerable overlaps initially, suggesting that these genes may be activated differentially by maternal A/P determinants such as the products of the bcd^+ or posterior-class genes. Subsequently, however, these overlaps are lost and the transcript domains become more sharply defined, apparently as a result of cross-regulatory influences between the gap genes (see Ingham, 1988). Thus there is a sharp boundary between the anterior hb^+ and central Kr^+ transcript domains in the cellular blastoderm. Futhermore, gap mutants alter the expression domains of other gap genes. In Kr^- mutants, the posterior boundary of hb^+ expression is shifted back; in kni^- mutants the posterior boundary of Kr^+ expression is likewise moved back, whereas its anterior boundary is moved forwards in hb^- mutants (Jackle *et al.*, 1986).

The mechanism of this cross-regulation remains obscure. The obvious explanation – namely, that each gap-gene product might act as a transcription factor, activating its own gene while repressing the gap genes active in adjacent domains – now appears oversimplistic. Thus high levels of Kr^+ protein are found only in a subset of the cell nuclei expressing Kr^+ transcripts (Gaul *et al.*, 1987) while low levels are found outside the Kr^+ transcript domain, probably reflecting post-transcriptional regulation plus diffusion of the Kr^+ protein within the syncytium. In Kr^- mutants, kni^+ transcription is diminished throughout its domain, suggesting that a low-level Kr^+-protein gradient may enhance kni^+ expression (Pancratz *et al.*, 1989). One of the two predominant Kr^+ transcripts retains a 5' intron sequence, and apparently this partially processed RNA provides the only source of new Kr^+ protein after gastrulation (Gaul *et al.*, 1987). Despite these complexities, it is clear that the DNA-binding 'zinc-finger' motif (see §5.6.2) is essential for Kr^+ protein function. Point mutations which replace one of the zinc-binding cysteines with a different amino acid abolish the biological activity of the Kr protein, resulting in a typical Kr^- phenotype (Redemann *et al.*, 1988).

(c) *Zygotic pair-rule and segment-polarity genes*

The wild-type products from these two classes of gene cooperate to define the metameric organisation of the embryonic germ band. The pair-rule genes are characteristically expressed in transient 7-stripe patterns with double-segment periodicity during the cellular blastoderm stage. The phenotypes of pair-rule mutant embryos show deletions of alternating segment-wide regions of the germ-band pattern, such that only half the normal number of segments can be discerned (Nusslein-Volhard & Wieschaus, 1980). The combined activities of the pair-rule genes generate a transient 'prepattern' on which the definitive metameric pattern is based. This definitive pattern involves the activation and persistent expression of segment-polarity genes, sometimes in 14 stripes with single-segment periodicity within the germ band. Mutations in any of the segment-polarity genes cause pattern deletions/duplications within each segment, so altering its internal polarity. But despite this apparent distinctness of function, some members of each class show characteristics of the other. Thus several pair-rule genes are expressed initially in a 7-stripe pattern, but this is rapidly modified to 14 stripes (e.g. *eve, prd*). On the other hand, some mutations of the *engrailed* segment-polarity gene give a lethal pair-rule phenotype.

Before considering some of these genes individually in greater detail, it is appropriate to outline how their spatial patterns of expression relate to the metameric organisation of the germ band (fig. 5.20). Key features linking the two are as follows:

(i)   The persistent expression of *engrailed* (en^+) gene-products in a narrow strip of cells delineating the anterior edge of each parasegment (Ingham *et al.*, 1985b); these en^+-positive cells also constitute the posterior compartment of each future segment (see §5.4.1; Kornberg *et al.*, 1985; DiNardo *et al.*, 1985).

(ii)   The persistent expression of *wingless* (wg^+) gene-products in a narrow strip of cells marking the posterior edge of each parasegment (Baker, 1987, 1988a); thus the definitive parasegment borders, and also the morphological transverse grooves, fall between adjacent en^+ and wg^+ stripes (Akam, 1987; Martinez-Arias *et al.*, 1988).

(iii) The parasegments are defined initially by two complementary pair-rule functions; even-numbered parasegments correspond to the seven stripes of *fushi tarazu* (ftz^+) expression, while odd-numbered parasegments correspond to the seven stripes of *even skipped* (eve^+) expression. In each case, the parasegment border (as defined by en^+ expression) coincides cell by cell with the anterior margin of the zone expressing ftz^+ or eve^+ products (Lawrence *et al.*, 1987).

Pair-rule genes. Several members of the pair-rule class of genes (*hairy, ftz, eve, runt* and *paired*) have been cloned and their product distributions determined by *in situ* hybridisation (RNA) or antibody staining (protein). Three of these genes (*ftz, eve, paired*) contain homeobox sequences, suggesting a DNA-binding function for their protein products (Laughon & Scott, 1984; Kuroiwa *et al.*, 1984; Macdonald *et al.*, 1986; Frigiero *et al.*, 1986); however, two others (*runt* and *hairy*) do not. Nevertheless, the $hairy^+$(h^+) ftz^+ and eve^+ proteins are all mainly nuclear in location (Carroll & Scott, 1986; Frasch *et al.*, 1987; Carroll *et al.*, 1988b). The h^+ protein contains a MYC-related DNA-binding domain (Rushlow *et al.*, 1989). At least two other pair-rule genes (*odd paired* and *odd skipped*) are known by their mutant effects on

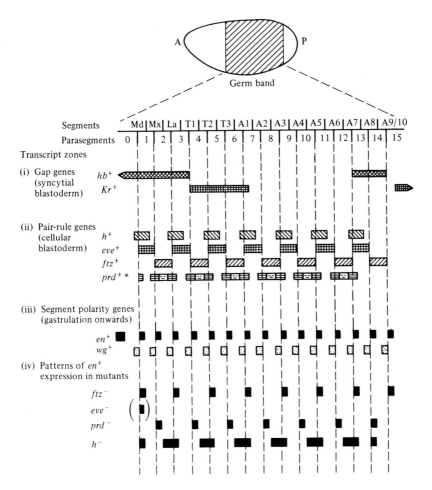

Fig. 5.20 Zygotic control of metamerisation in *Drosophila* (after Akam, 1987).
*Registration of *prd*⁺ stripes as in Ingham *et al.*, 1988).

metameric pattern and on the spatial distributions of segementation gene-products (particularly *en*⁺).

The prototypic pair-rule gene is *ftz*; its name – *fushi tarazu*, meaning 'few segments' in Japanese – refers to the phenotype of *ftz*⁻ null mutants, in which only half the normal number of segments is apparent. Detailed examination shows that the deleted regions are not alternate segments but rather parasegments, thus the anterior compartment of T1 is fused to the posterior compartment of T2, anterior T3 to posterior A1, and so on. More simply, the *ftz⁻* embryo is a chain of odd-numbered parasegments, the even-numbered ones (2,4,6, etc.) being deleted. Both in location and in extent, these missing pattern elements correspond to the seven stripes of *ftz⁺* expression in the blastoderm of wild-type embryos (Hafen *et al.*, 1984). As shown in fig. 5.20, each of the even-numbered parasegments in the germ band is defined by a stripe of *ftz⁺* expression (both RNA and protein), while odd-numbered parasegments do not express *ftz⁺*. Consistent with this, the number of *en⁺* stripes is reduced from 14 to 7 in the germ band of *ftz⁻* mutant embryos; those *en⁺* stripes which normally coincide with the anterior boundaries of *ftz⁺*-expression are specifically lost, while the alternate *en⁺* stripes are retained (fig. 5.20). Thus *ftz⁺* products are needed to activate *en⁺* expression in the even-numbered parasegments. Indeed, in the absence of *ftz⁺* function, these parasegments fail to develop properly, and the cells that would normally form them either die after germ-band extension or are reassigned to other fates (Martinez-Arias & White, 1988). In *ftz⁻* mutants, dying cells (intermingled with dividing cells) are found scattered throughout the even-numbered parasegments but also in the posterior parts of odd-numbered ones (Magrassi & Lawrence, 1988). Ectopic expression of *ftz⁺* throughout the blastoderm embryo causes a complementary (anti-*ftz*) series of pattern deletions, as shown by fusing the *ftz⁺* structural gene to a heat-shock promoter in a P-element construct, and then heat-shocking embryos which carry this construct (Struhl, 1985). The anti-*ftz* phenotype can be interpreted in various ways, but probably involves interactions of the *ftz⁺* product with other segmentation genes.

The pattern of *ftz⁺* RNA and protein expression observed in the cellular blastoderm does not emerge all at once. Initially, low levels of *ftz⁺* expression can be detected throughout the blastoderm, but this distribution becomes rapidly modulated through a 4-segment repeat to the seven *ftz⁺* stripes in alternate parasegments (Weir & Kornberg, 1985; Edgar *et al.*, 1986; Karr & Kornberg, 1989). Even this is only temporary, as shown by antibody staining studies. The *ftz⁺* protein bands soon become asymmetric, their anterior borders remaining stable and sharply demarcated while their posterior borders shift such that each stripe becomes narrower and more intense (see cover photograph); this continues through gastrulation into germ-band elongation (see Lawrence & Johnston, 1989). Even after it has faded from the epidermis, *ftz⁺* protein later reappears in a subset of cells within each of the 15 metameric units of the developing CNS (Carroll & Scott, 1985), and still later in the developing hindgut (Krause *et al.*, 1988). The 7-stripe blastoderm pattern of *ftz*⁺ expression evolves in response to spatial cues provided by (i) the zygotic gap genes and (ii) a subgroup of pair-rule genes including *hairy* (*h*), *eve* and *runt*. This is shown clearly by the altered patterns of *ftz*⁺ expression observed in *hb*⁻, *Kr*⁻, *kni*⁻, *gt*⁻, *tll*⁻, *h*⁻, *eve*⁻ or *runt*⁻ mutant embryos (Carroll & Scott, 1986; Frasch & Levine, 1987; Howard & Ingham, 1986). Some of these genes also regulate the patterns of *eve*⁺ expression in a complementary fashion, suggesting that combinations of their wild-type products which activate the *ftz*⁺ gene may repress *eve*⁺, and vice versa (Frasch & Levine, 1987). This in turn implies that the *ftz* promoter must respond positively or negatively to several different combinations of regulatory influences within the germ-band region.

The *ftz* structural gene itself is relatively small; it comprises two exons (the second with a homeobox) encoding a 1.8 kb poly(A)⁺ RNA species which is most prevalent between the early blastoderm and gastrulation stages (Kuroiwa *et al.*, 1984; Weiner et al., 1984). Upstream from the *ftz* cap site lie several *cis*-acting DNA sequences required for different aspects of *ftz*⁺ regulation, some of them quite distant from the gene they control. These sequences have been identified by fusing *ftz*-gene flanking regions to β-galactosidase genes and introducing such constructs via P elements into wild-type flies (Hiromi *et al.*, 1985; Hiromi & Gehring, 1987). Constructs containing 6.3 kbp of 5'-flanking DNA from the *ftz* locus give a spatial pattern of β-galactosidase staining similar to that for the wild-type *ftz* protein (7 blastoderm and later 15 CNS stripes). Reducing the length of *ftz* regulatory DNA down to 3 kbp results in two additional regions of expression (9 rather than 7 stripes), while a further reduction to 1.5 kbp eliminates expression specifically in the ventral nerve cord. This defines a separate 'neurogenic' control element required only for *ftz*⁺ expression in the CNS (Hiromi *et al.*, 1985). Further analysis has identified two sequence elements which cooperate to produce the striped *ftz*⁺ pattern in the blastoderm, but which are not required for CNS expression. These are: (i) the proximal 'zebra' element lying close to (within 740 bp of) the *ftz* cap site, which interacts with the products of *h*⁺, *runt*⁺ and the gap genes to generate the 7-striped pattern of *ftz*⁺ expression; and (ii) a distal enhancer element which specifically binds the *ftz*⁺ protein product (Hiromi & Gehring, 1987). This latter element itself can confer a striped pattern of expression on a basal promoter, since this can only be activated by the *ftz* enhancer in regions of the embryo where endogenous *ftz*⁺ protein is present. This implies that the *ftz*⁺ protein normally autoregulates its own gene by binding to the distal enhancer, so providing a positive feedback loop; this is required for the maintenance of the 7-striped expression pattern, though not for its initial establishment (Hiromi & Gehring, 1987).

Since the spatial domains of *ftz*+ expression do not remain constant in the early embryo (even the 7-stripe pattern is only transient), it follows that both mRNA and protein products from the *ftz* gene must turn over rapidly; their instability can be demonstrated directly after injections of transcriptional (α-amanitin) or translational (cycloheximide) inhibitors (Edgar *et al.*, 1986). However, both degradation of *ftz*+ mRNA (blocked by cycloheximide) and cellularisation of the embryo (blocked by cytochalasin) seem to play secondary roles in the *ftz*+ expression pattern, which can be established in the presence of either inhibitor, though not maintained in the presence of both (Edgar *et al.*, 1987). Post-translational modification of the *ftz*+ protein gives rise to a series of isoforms, perhaps altering the properties and/or functions of this protein (Krause *et al.*, 1988).

Among the other cloned pair-rule genes, *hairy* (*h*) and *runt* are exceptional on two counts: (i) they lack homeoboxes; (ii) together with *eve*, they regulate subsidiary pair-rule genes. Whereas the *ftz*+-expression pattern is disrupted in *h*− mutants, *h*+ expression is unaffected in *ftz*− embryos – implying that *h*+ regulates *ftz*+ but not vice versa (Howard & Ingham, 1986). Indiscriminate expression of *h*+ products can be achieved (as with *ftz*+) by heat-shocking embryos which carry a P-element construct with the *hairy* structural gene fused to a heat-shock promoter. The resulting pattern defects strongly resemble those in *ftz*− embryos, implying that *h*+ products may normally repress *ftz*+ expression (Ish-Horowicz & Pinchin, 1987). The dosage of *runt* genes (presumably reflected in the amount of *runt*+ product) seems critically important for normal development. Embryos with extra copies of the *runt* gene develop an 'anti-*runt*' mutant phenotype characterised by pattern deletions complementary to those resulting from *runt*− deficiencies (Gergen & Wieschaus, 1986).

Transcripts from both *h* and *runt* genes appear in the late syncytial blastoderm, slightly after those from the gap genes but before those from the other pair-rule genes such as *ftz*+ (Ingham, 1988). The *h*+ and *runt*+ transcript distributions overlap at first, but rapidly evolve into mutually exclusive 7-stripe patterns within the germ band (see Ingham, 1988; Gergen & Butler, 1988); this occurs while the gap-gene transcript domains are becoming sharper. There are also discrete patches of expression for both *h*+ and *runt*+ in the anterior dorsal region (Ingham *et al.*, 1985a; Ingham, 1988). The early pair-rule patterning of h^+ and $runt^+$ products seems to cue in the expression of subordinate pair-rule genes such as *ftz* and *paired*, although the details of how this is achieved remain as yet unclear (see Ingham, 1988). Both the *h* and *runt* genes are controlled via extensive regulatory sequences (>20 kbp for *h*: Howard *et al.*, 1988; see also Gergen & Butler, 1988 for *runt*). At least in the case of *h*, there are region-specific regulatory mutations which map within these 5'-flanking sequences; the mutant phenotypes conferred by these mutations affect only limited parts of the 7-stripe h^+ expression pattern (Howard *et al.*, 1988). This implies that different *h* regulatory elements interact with region-specific combinations of DNA-binding proteins, presumably the products of maternal A/P polarity and zygotic gap genes. On this basis, a prime function of *h* (and *runt*) may be to decode the complex 'prepattern' of spatial cues provided by these maternal and gap gene-products so as to generate a repeating striped pattern of expression. The nuclear h^+ protein lacks a homeodomain but has a MYC-related DNA-binding domain (Rushlow *et al.*, 1989); this same h^+ protein is also involved in adult bristle patterning, but it is not expressed in the embryonic nervous system (Carroll *et al.*, 1988b). Note that h^- mutants reduce the number of en^+ stripes within the germ band, but those which remain are much wider than usual (fig. 5.20), suggesting complex or indirect control of en^+ by h^+.

The *even skipped* (*eve*) gene resembles *ftz* in its small size (only 1.5 kbp) and possession of a homeobox sequence, albeit of a variant type (Macdonald *et al.*, 1986; Frasch *et al.*, 1987). The 1.4 kb eve^+ RNA encodes a protein of 376 amino acids, which first becomes detectable in a series of seven transverse stripes at the cellular blastoderm stage. These transient eve^+ stripes correspond to the odd-numbered parasegments, and their positions are complementary to the ftz^+ stripes (Lawrence *et al.*, 1987). Again, it is the stable anterior border of eve^+ expression in each stripe which persists and hence 'defines' the anterior border of that parasegment (Lawrence & Johnston, 1989). However, this simple pattern of eve^+ expression soon changes; the seven early eve^+ bands become narrower, and after gastrulation a second set of (weaker) eve^+ bands becomes intercalated between them, giving 14 equally spaced stripes which all disappear by the time of germ-band elongation. The anterior margins of all 14 eve^+ bands coincide with the en^+ stripes defining the anterior borders of all parasegments. Since repeating en^+ stripes are absent from eve^- embryos (fig. 5.20), eve^+ products are probably required for en^+ expression in all parasegments. The regulation of *eve* expression is complex; the initial establishment of the seven eve^+ stripes requires a series of regulatory DNA elements between −0.4 and −4.7 kbp (relative to the *eve* cap site) which interact with gap-gene products, whereas the later maintenance of *eve*+ expression involves a distal site (-5.9 to -5.2 kbp) which interacts with the products of *h*, *runt* and *eve* itself (Harding *et al.*, 1989; Goto *et al.*, 1989). This autoregulatory feature is reminiscent of *ftz* (above). Note that the establishment of a particular *eve*+ stripe depends on a subset of the proximal regulatory elements interacting with a subset of gap-gene products. Thus *Kr*+ and *hb*+ proteins bind to distinct consensus sequences and thereby interact differentially with defined *eve* promoter elements (Stanojevic *et al.*, 1989; Akam, 1989b).

Transcripts from the *paired* (*prd*) gene also show a transient 7-stripe pattern in the blastoderm (Kilcherr *et al.*, 1986), each stripe being wider than a single parasegment. This pattern changes rapidly through the disappearance of *prd*+ transcripts from the centre of each stripe, producing a more stable series of 14 narrow bands (see fig. 5.20). The *prd* gene includes M repeats (cf. *opa* in *Notch*), a variant homeobox, and the so-called *prd* sequence (encoding histidine-proline repeats). Both the *prd* repeat and the *prd*-type homeobox are features shared by the *bcd* gene (Frigiero *et al.*, 1986) and by two genes at the *gooseberry* (*gsb*) segment-polarity locus (Bopp *et al.*, 1986). Null *prd*− mutants delete alternate *en*+ stripes in the germ band, specifically those belonging to the odd-numbered parasegments (even-numbered *en*+ stripes are retained; fig. 5.20). This *en*+ pattern is complementary to that observed in *ftz*− mutants; thus *prd*+ and *eve*+ are both needed for *en*+ expression in the odd-numbered parasegments, whereas *ftz*+ and *eve*+ are required for this in the even-numbered ones (see figs. 5.20 and 5.23).

Later in development, expression of both *eve*+ and *ftz*+ proteins occurs in distinct but overlapping sets of neurones within all 15 metameric units of the ventral nerve cord. This is controlled by separate *cis*-acting DNA sequences at least in the case of *ftz* (see above). Among the many identified neurones stained by antibodies against these proteins, both RP1 and RP2 express *ftz*+, but only RP2 expresses *eve*+. In mutants lacking neuronal *ftz*+ function, RP2 shows an abnormal axon trajectory (similar to RP1) and also fails to express *eve*+. In a temperature-sensitive *eve*− mutant, RP2 is transformed into a replica of RP1 if it develops at the restrictive temperature. This implies that *eve*+ normally specifies the identity of RP2, while *ftz*+ function is required to regulate *eve*+ expression in this cell (Doe & Scott, 1988; Doe *at al.*, 1988a, b).

Segment polarity genes. The *engrailed* (*en*) segment polarity gene was first discovered through an unusual mutant allele which causes mirror-image posterior-to-anterior transformations in some adult structures such as the wing (fig. 5.21). Marked clones of homozygous *en/en* mutant cells can be induced on a heterozygous *en/+* background via X-ray-induced somatic recombination. If such clones are induced anywhere in the anterior compartment of (say) the wing disc, there is no visible pattern alternation in the adult wing, suggesting that *en*+ function is not required in anterior cells. However, if such clones are induced in the posterior compartment, the mutant *en/en* phenotype becomes apparent as a mirror-image duplication of anterior structures confined to the marked clone; (e.g. anterior-type bristles along the posterior margin of the wing where this is made up of mutant *en/en* cells; fig. 5.21). Thus *en*+ function is required specifically in posterior compartment cells. Furthermore, *en/en* clones in a posterior compartment frequently spill over the border into the anterior compartment (fig. 5.21). This further implies that the *en*+ function normally distinguishes posterior form anterior cells, such that posterior cells lacking *en*+ product can mingle freely with anterior cells (which do not express *en*+). The pattern disruption caused by *en/en* clones in posterior but not anterior compartments can be observed in the cuticle of many adult structures (including parts of the head; Lawrence & Struhl, 1982), but not in muscle or gut tissues (Lawrence & Johnston, 1984a). Thus the *en*+ function confers posterior compartment identity in the epidermis (and CNS), but not in the mesoderm or endoderm (see also §5.4.1). When *en/en* clones are induced in the posterior compartment of the wing disc, the staining properties of the mutant cells are transformed from a posterior to an anterior pattern (Brower, 1984); thus the *en*+ function is also required during larval life to distinguish posterior form anterior compartments.

Null alleles of *en* are lethal during the embryonic stages. Although *en* was orignally assigned to the pair-rule class on the basis of its mutant phenotype (Nusslein-Volhard & Wieschaus, 1980), it soon became apparent that wild-type *en*+ function is in fact required in the posterior compartment of every embryonic segment (Kornberg, 1981), as well as in the larva and adult. This proposed role was strikingly confirmed by *in situ* hybridisation of a cloned *en* probe to wild-type embryos (Kornberg *et al.*, 1985; Fjose *et al.*, 1985). From gastrulation onwards, *en*+ transcripts are detected in a narrow strip of cells marking the posterior compartment of every segment (i.e. the anterior of every parasegment) within the germ band. Staining with antibodies against the *en*+ protein reveals that these *en*+ stripes are not established all at

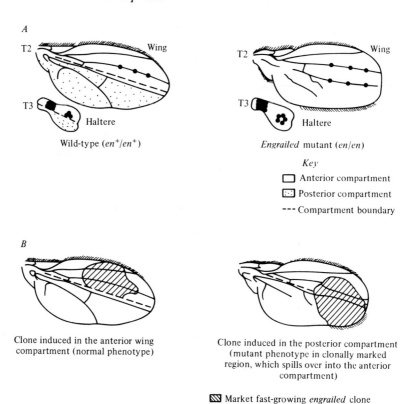

Fig. 5.21 *engrailed*[+] function in *Drosophila* (based on Garcia-Bellido *et al.*, 1979).

A Phenotype of wings and halteres in wild-type and *engrailed* mutant flies.
B Fast-growing *engrailed* clones (M^+en/M^+en) induced in the wing discs of heterozygous ($M^+en /M en^+$) flies.

once (DiNardo *et al.*, 1985). At the onset of gastrulation, the first *en*[+] stripe appears in parasegment 2, initially in the ventral region and then spreading dorsally. The stripe in parasegment 1 develops slightly later, again showing a progressive ventral-to-dorsal spread. A similar sequence is then repeated in more posterior regions, the even-numbered *en*[+] stripes always appearing a little before the neighbouring odd-numbered ones. This disparity probably reflects the different regulatory requirements for *en*[+] activation in odd- compared with even-numbered parasegments (see above). Patches of en^+ staining later arise in the anterior head region and in parts of the hindgut, although these structures are not obviously segmented.

The *en* gene itself contains a variant homeobox sequence which is split by an intron. This unusual arrangement is shared by the neighbouring *invected* (*inv*) gen, which is related in sequence and expressed in a similar spatial pattern to *en*, although its function remains unknown (Poole *et al.*, 1985; Coleman *et al.*, 1987). The presence of a homeodomain in the predicted en^+ protein suggests a role in DNA-binding (see also §5.6); this is supported by the localisation of en^+ staining in cell nuclei (DiNardo *et al.*, 1985), and confirmed more directly by the demonstration that en^+ protein binds to DNA in a sequence-specific manner (Desplan *et al.*, 1985, 1988). DNA sites which can bind the homeodomain of the en^+ protein (consensus TCAATT-AAATT) are relatively common, and are found *inter alia* clustered in the regulatory region of the *en* gene itself (Desplan *et al.*, 1988). Similar sites are bound by the ftz^+ homeodomain, and both of these protein fragments also bind to TAA repeats. The sequence specificity of en^+ binding to DNA may be modulated through association with other factors; en^+ proteins are found in large stable complexes along with a variety of other nuclear proteins (Gay *et al.*, 1988).

The *en* locus is large, since en^- mutations occur scattered across a 70 kbp region of DNA. Mutations in the proximal 50 kbp give an embryonic lethal phenotype, whereas those in the distal 20 kbp cause pattern alternations in the adult (Kuner *et al.*, 1985). Within this large region, the structural *en* gene is only 3.9 kbp long, and there are no other transcription units within 48 kbp upstream or 16 kbp downstream. These silent 'peripheral regions' include regulatory sequences essential for en^+ function (Drees *et al.*, 1987). The *en* gene produces three poly(A)$^+$ RNA species of 3.6, 2.7 and 1.4 kb. The 2.7 kb transcript is expressed and apparently provides an essential function in precellular embryos, well before the onset of the definitive pattern of en^+ expression (Karr *et al.*, 1985). The spatial distribution of en^+ transcripts then evolves from general, through transient patterns with 4- and then 2-segment periodicity, until the single-segment repeat is established during gastrulation (as described earlier). It may be that the transient double-segment pattern in some way underlies the pair-rule phenotype of null en^- mutant embryos.

The regulation of *en* transcription by other proteins will undoubtedly prove complex, though a start has been made by developing an *in vitro* cell-free transcription system using the cloned *en* gene

(Soeller *et al.*, 1988). Two protein fractions are required for accurate initiation of transcription; one of these includes two or more sequence-specific DNA-binding proteins (one bound by eight sites within 400 bp of the cap site). Another approach is to use P-element constructs with *en*-regulatory sequences fused to β-galactosidase genes, so as to study the spatial distribution of the fusion products in wild-type recipient flies. As shown in fig. 5.22, this vividly illustrates the persistent expression of *en*⁺ in posterior compartment cells, not only in the extended germ-band embryo, but also in imaginal discs during larval development (see also Brower, 1986) and even in the adult. The fusion construct used to transform the flies shown in fig. 5.22 contains only 5.7 kbp of *en* 5′-flanking sequences, but has become inserted fortuitously into the regulatory region of the endogenous *en*⁺ gene, such that expression occurs in the normal regulatory context of the *en* gene (T. Kornberg, personal communication). The insertion disrupts *en*⁺ expression in *cis*, so this fusion construct can only be maintained in heterozygotes. Note that insertion of this construct at other chromosomal locations does not give an expression pattern mimicking that of the endogenous *en* gene.

As discussed earlier, the 14 germ-band stripes of *en*⁺ expression can be used as an assay for metameric pattern. The altered distributions of *en*⁺ products in various pair-rule mutant embryos clearly implicate the wild-type products of these genes as direct or indirect regulators of the *en* gene (fig. 5.20). Thus *eve*⁺ is required for *en*⁺ expression in all parasegments, while *ftz*⁺ is needed in the even-numbered ones and *prd*⁺ in the odd-numbered ones (DiNardo & O'Farrell, 1987). Among the other pair-rule functions, *odd paired* (*opa*⁺) seems to be required along with *ftz*⁺ for *en*⁺ expression in the even-numbered parasegments (these 7 *en*⁺ stripes are deleted in *opa*⁻ as well as in *ftz*⁻ embryos). The *odd skipped* (*odd*⁺) function may play some role together with *eve*⁺ in establishing the correct position and width of each *en*⁺ stripe (DiNardo & O'Farrell, 1987). However, these pair-rule functions are all expressed transiently in the embryo and can serve only to set up the striped pat-

Fig. 5.22 Expression of an *en*-promoter/β-galactosidase fusion gene in P-element-transformed *Drosophila* embryo (*A*), larval wing disc (*B*), adult wing (*C*), and adult body (*D*).
Note: this construct contains only 5.7 kbp of *en* promoter DNA, but is fortuitously inserted into the endogenous *en* promoter region in this particular transformant strain, conferring wild-type *en*⁺ regulation of β-galactosidase expression. Photographs kindly supplied by Dr T. Kornberg, University of California, San Francisco.

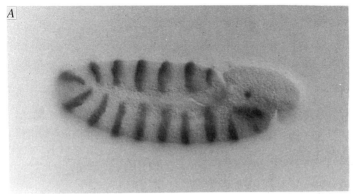

tern of *en*+ expression initially. The maintenance of this pattern through the whole of development after gastrulation (when the pair-rule products decay) must therefore depend on a separate regulatory system, which turns out to involve several segment-polarity genes (see below; DiNardo *et al.*, 1988). There are also superimposed systems of negative regulation whereby *en*+ expression is repressed (probably by known segmentation gene products) in the anterior third of the embryo and also in the interband regions between the 14 *en*+ stripes; these features are suggested by the novel patterns of *en*+ expression observed after injecting cycloheximide into early *Drosophila* embryos (Weir *et al.*, 1988).

The roles of all the segment polarity genes (of which there are at least 10) remain somewhat unclear, though there is evidence that they may be involved at more than one level in the regulatory hierarchy. Thus mutations in some of these genes – such as *wingless* (*wg*) and *dishevelled* (*dsh*) – disrupt the process of metamerisation at an early stage and alter the distribution of *en*+ products; by contrast, the *en*+ pattern is unaffected by mutations in other such genes (presumably acting later) – including *gooseberry* (*gsb*), *hedgehog* (*hh*), *fused* (*fu*) and *armadillo* (*arm*) (Perrimon & Mahowald, 1987). Several of these segment polarity genes have now been cloned, including *wg*, *arm* and *gsb*. The *gsb* locus comprises two related genes (BSH 9 and BSH 4) which share several features with the *prd* pair-rule gene, including its variant homeobox (Côté *et al.*, 1987; Baumgartner *et al.*, 1987). Both genes are expressed in 14 stripes (single-segment repeat) within the extended germ band. The stripes of BSH 9 transcripts show a pattern of V→D and A→P spreading similar to that already described for the *en*+ protein. Both *gsb*+ genes are transcribed in stripes of cells which coincide with the 14 stripes of later *prd*+ expression, and which overlap those expressing *en*+ (fig. 5.23). Whereas BSH 9 transcripts appear initially in the ectoderm and later in the mesoderm, BSH 4 transcripts seem largely confined to the neuroectoderm (Baumgartner *et al.*, 1987). There are also some regions of *gsb*+ expression both anterior and posterior to the germ band. The *armadillo* (*arm*) locus encodes two abundant 3.2 kb transcripts, both of which appear to be uniformly distributed throughout the embryonic segments (Riggelman *et al.*, 1989). Since the *arm*⁻ mutant phenotype suggests a *localised* requirement for the (single) 91 kd *arm*+ protein within each segment, this may imply local activation of the *arm*+ protein product.

The *wg* gene (Baker, 1987a, b) produces a 3 kb transcript which is first expressed in anterior and posterior regions during blastoderm

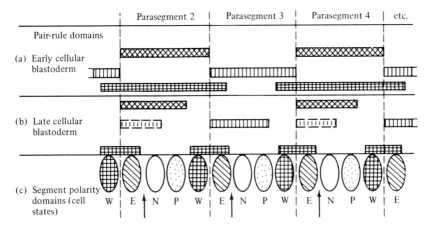

Fig. 5.23 Pair-rule and segment-polarity functions define the metameric repeat (modified from Ingham *et al.*, 1988 and Martinez-Arias *et al.*, 1988). For the sake of clarity, each parasegment is shown as only four cells wide, though this number is approximate and increases later in development.

⟶ , future segmental boundary; ▨▨▨, *ftz*⁺ domain; ▥▥▥, *prd*⁺ domain; ▯▯▯▯, primary *eve*⁺ domain; ▯▯▯, secondary *eve*⁺ domain (which arises at gastrulation); ⊕ *wg*⁺ transcript domain (W cell fate); ⊛ , *en*⁺ transcript domain (E cell state); ○ , implied domain of *nkd*⁺ activity (N cell state); ☺ , implied domain of *ptc*⁺ activity (P cell state).

Note: the transcript zones for *nkd* and *ptc* may be much broader than the N and P states shown, which indicate only that part of each parasegment whose identity depends largely on the activity of that particular gene.

cellularisation. By the extended germ band stage, *wg*⁺ transcripts are expressed mainly in epidermal tissues, forming 14 narrow stripes which mark the posteriormost cells of each parasegment (Baker, 1987a, b, 1988b). Since *en*⁺ is expressed in the anteriormost cells of each parasegment, it follows that the definitive parasegment borders fall between adjacent stripes of *wg*⁺-positive and *en*⁺-positive cells (see earlier), as do the transverse grooves formed across the germ band. Once again, there are patches of *wg*⁺ expression in both anterior and posterior regions lying outside the germ band proper (Baker, 1988b). Since adjacent patches of tissue often express *en*⁺ (see diagram in Akam, 1987), this may imply that some apparently non-segmented structures were once metameric in the ancestors of modern Diptera. The *wg*⁻ mutant can be phenocopied by injecting antisense RNA (complementary to the 3 kb *wg*⁺ transcript) into wild-type embryos

(Cabrera *et al.*, 1987a). Sequence studies reveal that the *wg* gene is related to the mouse *int-1* proto-oncogene (see § 5.6 below); both probably encode secreted or cell-surface proteins (see Bender & Peifer, 1987). Consistent with such a role – perhaps in intercellular signalling - the *wg*[+] function is not cell-autonomous; i.e. the mutant phenotype of single *wg/wg* cells can be rescued by adjacent wild-type tissue (Baker, 1988c). This suggests that the mechanism of *wg*[+] protein action must be fundamentally different from that of *en*[+], though the patterns of transcript expression for these two genes appear complementary in formal terms. Although the *wg* gene is expressed persistently (like *en*) in laval imaginal discs, neither the distribution of *wg*[+] transcripts there (markedly different between dorsal and ventral discs), nor the phenotype of a pupal-lethal *wg*[CX3] mutant, are consistent with a simple role for *wg*[+] in defining A/P pattern during later development (Baker, 1988a).

It remains to ask how these segment polarity genes cross-regulate each other so as to maintain their striped patterns of expression, which persist from the extended germ-band stage onwards (at least in the case of *en*[+]; fig. 5.22). At its inception, each parasegment is about four cells wide; the posteriormost line of cells expresses *wg*[+] while the anteriormost expresses *en*[+]. At least two further segment-polarity genes, *patched* (*ptc*) and *naked* (*nkd*), are required to define the identities of the two central cells within each parasegment, as implied by their mutant phenotypes and the novel patterns of *wg*[+] and *en*[+] expression found in *ptc*[-] or *nkd*[-] embryos (Martinez-Arias *et al.*, 1988). The *ptc*[+] protein seems to have several membrane-spanning domains (Nakano *et al.*, 1989); *ptc*[+] transcripts are at first expressed throughout the cellular blastoderm and gastrula, but then become spatially restricted to those cells of each parasegment which do not express *en*, during the extended germ-band stage. As a simplified working model, one can envisage the germ band in terms of four cell states reiterated serially, namely, ENPW/ENPW/ etc., where E and W denote the known *en*[+] and *wg*[+] domains while N and P designate the proposed *nkd*[+] and *ptc*[+] domains (Martinez-Arias *et al.*, 1988). As we have seen, the parasegment borders fall between W and E, whereas the later segment borders would arise between E and N. Notably the N domains would give rise to the larval denticle belts, which are abolished in *nkd*[-] mutants. Interactions between these four gene functions are required in order to maintain their domains of influence, as implied by the novel spatial patterns of *wg*[+] and/or *en*[+] expression observed in several segment-polarity mutants (Martinez-Arias *et al.*, 1988; DiNardo *et al.*,

1988). As we have seen, the wg^+ product is probably a secreted or cell-surface protein which could function as an intercellular signal and so influence neighbouring cells; other segment-polarity genes such as *ptc* may perhaps encode receptors mediating a response to that signal (Ingham, 1988). Since several segment-polarity genes encode products which are expressed in regions broader than those affected by the corresponding null mutants (e.g. *arm*, *ptc*), localised activation of such products may underlie their spatially restricted zones of influence.

The initial establishment of four such domains must involve their differential activation by the transient pair-rule gene activities (fig. 5.23). Both ftz^+ and eve^+ products are activators of en^+ expression (see above) but repressors of wg^+ expression. Thus in ftz^- (or eve^-) mutants, alternate zones of wg^+ expression extend throughout the missing ftz^+ (or eve^+) domains. In wild-type embryos at the late blastoderm stage, the alternating stripes of both ftz^+ and eve^+ products become narrower, decreasing from about four to three cells in width (see cover photograph). Notably, wg^+ expression is activated only in that posteriormost strip of cells which first ceases to express ftz^+ or eve^+ in each parasegment (Ingham *et al.*, 1988; fig. 5.23). The prd^+ function is also required to activate en^+ in odd-numbered parasegments, and these en^+ stripes coincide with the overlaps between prd^+ and eve^+ domains. Likewise, the opa^+ function seems to be required along with ftz^+ (and eve^+) in the even-numbered parasegments (DiNardo & O'Farrell, 1987). However, the persistent prd^+ domains also overlap the en^+ stripes in the even-numbered parasegments, where prd^+ is not required for en^+ activation (see fig. 5.23). This suggests a dependence on context which as yet remains obscure (Ingham *et al.*, 1988).

5.5 Homoeotic genes

5.5.1 Introduction

Broadly speaking, a homoeotic transformation involves the replacement of one structure (or series of structures) by another; a group of cells which would normally produce the one is redirected along an alternative path and forms an extra copy of the other instead. A homoeotic gene is one which controls such a developmental switch, its activity being required for one pathway but not for the other (the latter is therefore a default option). A hypothetical example was given earlier in §4.3, whereby recessive (null loss-of-function) and dominant (constitutive gain-of-function) mutations at the same homoeotic locus would cause opposite developmental transformations. Genes with some or all of these characteristics have cropped up repeatedly in the preceding chapter (e.g. *mec-3, lin-12* and the heterochronic mutants in *C. elegans*) and in the earlier sections of this one. Examples include the *cut*+ and *sev*+ functions in the development of external sensory organs and of R7 photoreceptors, respectively, and also some genes involved in the sex determination pathway (§ 5.2.3) or in neurogenesis (§ 5.3.2). In other cases a 'homoeotic' role appears secondary; for instance, the requirement for *bcd*+ activity (together with maternal terminal gene products) to distinguish acron from telson development, for *Kr*+ to specify Malpighian tubule rather than hindgut cells, or for *eve*+ to identify the RP2 as against RP1 neurone (§ 5.4.2). These examples show that the definition of a homoeotic gene is necessarily elastic, and also make the point that a binary choice between alternative fates may be governed by one gene product or by several acting in concert.

The *Distal-less* (*Dll*) gene encodes a homeodomain protein (see § 5.6) whose activity is required at high levels for the development of distal limb structures, but only at low levels for proximal limb development (Cohen *et al.*, 1989). Thus extreme *Dll*− mutants lack limbs completely, whereas milder mutants lack only distal limb structures. This involves the graded expression of *Dll*+ products along the proximo-distal axis of each limb, a situation which pertains in larvae as well as adults.

This section will deal with homoeotic genes affecting the identity of particular metameric units within the germ band (parasegments, at least initially). Anterior head and terminal structures outside the germ band will not be considered, although mutations in several homoeotic genes give spectacular transformations in these regions – for example the mutants *ophthalmoptera* (causing wings to sprout in the place of eyes) or *proboscipedia* (*pb*¹ causing labial palps to develop as antennae at 18 °C but as prothoracic legs at 28 °C). The *pb* locus lies in the Antennapedia complex (ANT-C, see (ii) below), and has been characterised in both molecular and genetic terms (Pultz *et al.*, 1988); however, its wild-type function is not essential during embryonic development (though it is expressed), unlike the other homoeotic genes considered below. Mutations in the *spalt* (*sal*) gene transform posterior head into anterior thoracic structures and anterior telson into posterior abdominal structures (Jurgens, 1988); *sal*+ products are expressed and required in parasegments 1/2 and 14/15. The *fork head* (*fkh*) gene is expressed and required outside the germ band in the fore- and hindgut. Both *fkh*+ and *sal*+ proteins are nuclear but lack homeodomains (Weigel *et al.*, 1989).

Among those homoeotic genes which act within the germ band, the three linked genes of the bithorax complex (BX-C) affect metameric identity in the posterior thorax and abdomen, whereas several genes clustered in the Antennapedia complex (ANT-C) perform a similar function in the anterior thorax and posterior head. After reviewing the BX-C and ANT-C (§ § 5.5.2 and 5.5.3 below) from a genetic and molecular standpoint, the early spatial regulation of both will be considered separately (§ 5.5.4 below). This last will reintroduce several themes from the preceding section (§ 5.4.2), and will show how individual parasegments become identified by expressing particular combinations of homoeotic genes during the metamerisation process (see reviews by Scott & Carroll, 1987; Akam, 1987; Ingham, 1988; Carroll *et al.*, 1988a). The homeobox is a DNA sequence of about 180 bp found in the coding regions of many *Drosophila* homoeotic and segmentation genes; it encodes a protein homeodomain of some 60 amino acids (mostly basic) which binds to DNA. Since similar sequences are found in the genomes of other metazoans, consideration of this topic is deferred until § 5.6, along with other such sequence homologies.

5.5.2 *The bithorax complex (BX-C)*

Several extensive reviews of BX-C function have appeared recently (Morata *et al.*, 1986; Duncan, 1987; Peifer *et al.*, 1987), to which the reader is referred for further details. Our present understanding of the BX-C is firmly rooted in the pioneering genetic studies of E.B. Lewis, so it is appropriate to begin with an outline of his model (see Lewis, 1978) and the supporting evidence provided by various BX-C mutant phenotypes. More recent modifications of this model will then be discussed in the light of extensive molecular studies of BX-C function.

In essence, Lewis's model proposes that all segments posterior to T2 are controlled by subfunctions of the BX-C, with successively more posterior segments requiring the activity of more and more BX-C sites. Thus T2 would be a 'ground state' in which the entire BX-C is inactive; element 1 of the BX-C would be active in T3 to distinguish it from T2, element 2 in A1 to distinguish it from T3, and so on back to A8 where the entire BX-C would be active. Such a model would predict that null mutations of element 1 should convert T3 into a replica of T2, while similar mutations affecting element 2 would transform A1 into T3, and so on. The phenotypes of several BX-C mutants conform admirably to these predictions. Deletion of the entire BX-C is lethal, but these BX-C⁻ embryos develop far enough for each segment to be identified on the basis of its cuticular pattern (note that the number and size of segments remain essentially unchanged). Whereas the head, T1 and T2 segments appear normal, all segments from T3 to A8 inclusive show a T2-like phenotype, as judged by their denticle belts and triradiate Keilin's organs (fig. 5.24). One can only imagine the multiwinged and multilegged creature that would result if such a BX-C$^-$ embryo could develop to adulthood! So far, the nearest approach to this is the justly famous four-winged fly shown in fig. 5.25, where segment T3 has been transformed into a replica of T2, with an extra pair of wings replacing the halteres normally found on T3.

At this point the complications begin. The fly shown in fig. 5.25 results from a triple mutant combination of separate deficiencies in the *anterobithorax* (*abx*), *bithorax* (*bx*) and *postbithorax* (*pbx*) subfunctions of the BX-C. Looking at these individually, *bx*$^-$ mutants show a conversion of anterior metathorax (aT3) into anterior mesothorax (aT2), such that the anterior but not posterior part of each haltere becomes converted into wing. Similarly, *pbx*$^-$ mutants transform pT3 into pT2, with only the posterior part of each haltere changed to wing. Confusingly, *abx*$^-$ mutants are rather similar in phenotype to *bx*$^-$ mutants, although a slightly more anterior set of structures is affected; however, there are also weaker *abx*$^-$ mutant effects similar to those caused by *pbx*$^-$ deficiencies (Peifer & Bender, 1986). All three subfunctions are defined by groups of mutations which map to separate sites within the left half of the BX-C (see later). The *bithoraxoid* (*bxd*) class of mutations defines a fourth subfunction; in *bxd*$^-$ mutants aA1 is converted towards aT3, such that metathoracic-type legs may appear on A1 (Lewis, 1978). Note that aA1 does not revert to the aT2 condition in *bxd*$^-$ mutants, since *abx*$^+$/*bx*$^+$/*pbx*$^+$ sites are still functional. This implies that the BX-C subfunctions are additive or at least superimposed in successively more posterior regions. All of the mutant phenotypes described so far involve recessive loss-of-function mutations. There are also dominant mutations in the BX-C which cause complementary gain-of-function phenotypes. Perhaps the clearest example is *Contrabithorax* (*Cbx*^{D}), which has the converse effect to *pbx*$^-$ deficiencies, causing pT2 to develop like pT3 (posterior wing converted to haltere). This phenotype can be explained by the *pbx*$^+$ element becoming active inappropriately in pT2 (normally it is not active anterior to pT3; see below).

To summarise so far, there are clear requirements for *abx*$^+$/*bx*$^+$ function in aT3, for *pbx*$^+$ in pT3, and for *bxd*$^+$ in aA1. If separate structural genes performing similar subfunctions were to define each of the abdominal compartments, then nearly 20 such BX-C genes would be needed in order to confer unique compartmental identities from T3 back to A8. A more plausible view – and one supported by the molecular evidence (see below) – is that the BX-C comprises only a few structural genes, each of which is active across a broad domain but subject to spatial regulation by multiple *cis*-acting sites. These would affect the amount or distribution of a given BX-C gene-product within smaller A/P subdivisions such as compartments. Mutations in any one such regulatory site would thus affect only a limited spatial region. By contrast, null mutations in one of the structural genes would inactivate all of the regulatory sites affecting the expression of that gene. This is indeed the case, since *Ultrabithorax* (Ubx^-) mutations eliminate the abx^+/bx^+, pbx^+ and bxd^+ subfunctions together; the '*Ubx* domain' may be said to comprise all of these regulatory sites plus the Ubx^+ structural gene whose expression they regulate. Likewise, several *infra-abdominal* (iab^+) subfunctions affecting the anterior part of the abdomen are abolished in null *abdominal-A* ($abd-A^-$) mutants, while more posterior iab^+ subfunctions are eliminated in *Abdominal-B* ($Abd-B^-$) mutants (Karch *et al.*, 1985; Sanchez-Herrero *et al.*, 1985). Moreover, homozygotes for triple point mutations in these three structural genes (Ubx^-, $abd-A^-$, $Abd-B^-$) show the same embryonic-lethal phenotype as that caused by deleting the entire BX-C (fig. 5.24), implying that all BX-C functions are executed by these three structural genes.

This view of BX-C function is supported by genetic complementation tests. Whereas abx^-/bx^- mutations can partially complement pbx^-/bxd^- mutations, neither class can complement Ubx^- mutations (Lewis, 1978). Similarly, the *iab-2*, *-3*, *-4* and sometimes *iab-5* subfunctions cannot complement $abd-A^-$ mutations, while the *iab-5*, *-6*, *-7* and *-8/9* subfunctions cannot complement $Abd-B^-$ mutations (Karch *et al.*, 1985). This implies that abx^+, bx^+, pbx^+ and bxd^+ are all regulators of the Ubx^+ coding unit, since none of these subfunctions can have any effect in the absence of Ubx^+ products. Likewise, $iab-2^+$, -3^+, and -4^+ would regulate the $abd-A^+$ gene, whereas $iab-6^+$, -7^+ and $-8/9^+$ would regulate the $Abd-B^+$ gene ($iab-5^+$ may influence both). These three 'executive' BX-C genes also contain one homeobox each (see §5.6), located in the 3' exons of *Ubx*, *abd-A* and *Abd-B*, respectively. Null mutations in each of these three structural genes define three broad domains of influence in the posterior thorax and abdomen of the fly, namely Ubx^+ from pT2 (see below) to aA1 inclusive (ps 5 and 6), $abd-A^+$ from pA1 to aA4 (ps 7 to 9), and $Abd-B^+$ from pA4 to aA9 (ps 10 to 14). Within these domains, the regulatory subfunctions define smaller spatial units such as compartments or parasegments (see below).

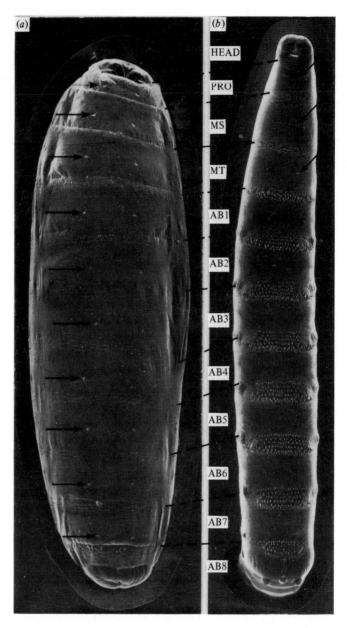

Fig. 5.24 Photographs of BX-C⁻ (*A*) and wild-type (*B*) *Drosophila* embryos, reprinted by permission from Prof. E. B. Lewis (California Institute of Technology, Pasadena) and from *Nature* **176**, 565–70. Copyright © 1978, Macmillan Journals Ltd. Note the presence of Keilin's organs (arrowed) and narrow denticle bands (typical of thorax) around most of the thoracic and abdominal segments in

Fig. 5.25 Four-winged adult *Drosophila* showing complete conversion of segment T3 into T2 (i.e. carrying a second pair of wings in place of halteres); this mutant is deficient in the wild-type functions of the *bx*, *abx* and *pbx* sites within the bithorax complex (see text). Photograph kindly supplied by Prof. E. B. Lewis (California Institute of Technology, Pasadena). Unpublished, from Biology Annual Report 1983 Cal. Tech.; permission granted for use.

Fig. 5.24 (*cont.*)

(*A*), as compared to the absence of Keilin's organs and wider denticle bands around the abdominal (AB) segments in (*B*). Thus most of the abdominal segments have been transformed towards a thoracic (T2-like) phenotype in the absence of BX-C gene functions (*A*). PRO, prothorax; MS, mesothorax; MT, metathorax.

Although Lewis (1978) originally envisaged additive expression of BX-C genes in successively more posterior segments, the situation *in vivo* is complicated by cross-regulatory interactions between the three structural gene products (e.g. *Ubx*⁺ expression is down-regulated by *abd-A*⁺ and *Abd-B*⁺ in the more posterior regions where their domains overlap; see later).

The spatial domains of most or all of these BX-C activities appear to be parasegmental rather than segmental. As pointed out by Lawrence & Morata (1983), the phenotype of BX-C⁻ embryos is not quite that predicted for a chain of T2 segments (fig. 5.24); rather, each segment from T2 back to A8 has a hybrid identity, resembling the anterior compartment of T2 (aT2) plus the posterior compartment of T1 (pT1). In view of the parasegment (ps) model proposed later (Martinez-Arias & Lawrence, 1985), we may reinterpret the BX-C⁻ phenotype more simply; i.e. ps 0 to 4 are normal, whereas ps 5 to 14 inclusive are converted into replicas of ps4 (which comprises pT1 + aT2). More posterior structures (derived from ps 15 and the telson) do not show any conversion towards thorax and appear to be independent of BX-C control (Sato & Denell, 1986). Consistent with this view, Morata & Kerridge (1981) have described a further *Ubx* subfunction required to distinguish pT2 from pT1 (i.e. in its absence, pT2 is converted into pT1). This *postprothorax* (*ppx*⁺) element appears to act independently during embryogenesis, but is later subsumed by the *abx*⁺/*bx*⁺ subfunctions (Sanches-Herrero & Morata, 1984; Casanova *et al.*, 1985b). Other evidence confirms that the *Ubx*⁺ domain extends from pT2 to aA1 inclusive (Hayes *et al.*, 1984; Struhl, 1984a), i.e. ps 5 and 6. In the absence of *Ubx*⁺ activity, pT2 and pT3 both develop like pT1, while aT3 and aA1 develop like aT2; more simply, ps 5 and 6 are transformed into replicas of ps 4. The earlier confusion over BX-C domains stems from the relative paucity of posterior cuticular markers in the larva, such that segmental identities were assigned mainly on the basis of anterior features such as denticle belts (Hayes *et al.*, 1984). The parasegmental domains of BX-C (and ANT-C) gene action are most clearly illustrated by their respective patterns of expression in the embryonic germ band (see §5.5.4 below) though subsequent development modifies and refines these initial domains to a considerable extent. The parasegmental model also suggests that the *ppx*⁺/*abx*⁺/*bx*⁺ subfunctions of the *Ubx*⁺ domain should act together in ps 5, whereas *pbx*⁺ and *bxd*⁺ should act in ps 6. This interpretation is supported by the genetic organisation of the BX-C locus, since *abx*⁻ and *bx*⁻ mutations map in separate clusters within the *Ubx* transcription unit, whereas *pbx⁻* and *bxd⁻* mutations map beyond its 5′ end (see fig. 5.26).

The entire BX-C has now been cloned as a series of overlapping DNA fragments, 195 kbp representing the left-hand end of the complex (mainly the *Ubx* domain; Bender *et al.*, 1983), and a further 215 kbp from the right-hand end (the *abd-A* and *Abd-B* domains; Karch *et al.*, 1985). A large number of BX-C mutations have been mapped within this 400 kbp region of DNA, some of which are shown in fig. 5.26. Most of the X-ray-induced mutations result from significant deletions and/or rearrangements of DNA, while many spontaneous mutations involve insertions of transposable DNA elements (particularly the 7.3 kbp 'gypsy' element). Only a tiny proportion of this huge length of DNA encodes the protein products of its three structural genes (*Ubx, abd-A, Abd-B*), although much larger regions are represented in their primary transcripts. One interesting feature of the BX-C (fig. 5.26) is that its regulatory subfunctions are arranged on the chromosome in the same linear order as the parasegments in which they are first required. The effects of mutations in each regulatory site are seen predominantly in a single parasegment, but also to a lesser extent in more posterior parasegments. Thus as successive regulatory subfunctions become superimposed on the pattern of BX-C gene expression in more posterior parasegments, so the influence of anterior

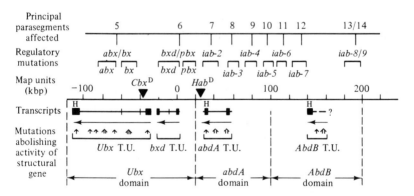

Fig. 5.26 The bithorax complex (BX–C) (adapted from Peifer *et al.*, 1987). T.U., transcription unit; H, location of homeobox; ■ , exon; ——, intron; ↑ ↟ etc., approximate sites of mutations abolishing activity of structural gene; ⌐¬ , DNA region within which mutations defining a regulatory site are clustered; ▼ , insertion of DNA causing a dominant gain-of-function mutation; 3′◄——5′, direction of transcription (as shown).

subfunctions becomes less apparent but does not wholly disappear (Lewis, 1978).

Despite this linear arrangement of regulatory sites, chromosomal continuity of the entire BX-C is not obligatory; it remains functional even if split by breakpoints which separate one (or two) of its functional domains onto a different chromosome (Struhl, 1984a; Tiong *et al.*, 1987). This confirms that the three domains are operationally independent as regards *cis*-acting influences, though each is regulated in part by *trans*-acting products from the others (see (e) below).

The *Ubx* domain is by far the best characterised in molecular terms, hence the *abd-A* and *Abd-B* domains will not be further considered except as regulators of Ubx^+ expression [see (e) below]. Apparent point mutations which eliminate Ubx^+ activity have been mapped to either end of a 70 kbp genomic region, suggesting that they may inactivate the 5' or 3' exons of a very long transcript (Akam *et al.*, 1984; fig 5.26). Some of the regulatory sites/mutations affecting Ubx^+ expression map within this long transcription unit (*abx*, *bx*), whereas others map beyond its 5' end (*pbx*, *bxd*,). Despite its large size (70 kbp), the *Ubx* transcription unit contains only four exons, two being extremely short (the internal 'microexons', each only 51 bp in length). There are two major size classes among the spliced Ubx^+ RNAs (3.2 and 4.3 kb), differing principally in their use of alternative polyadenylation sites (O'Connor *et al.*, 1988). The larger 5' and 3' exons are represented in all Ubx^+ RNA species, but at least five variant Ubx^+ products can be generated by differential splicing of the internal microexons. Further variation is provided by two separate splice donor sites located 27 bp apart at the 3' end of the first exon (O'Connor *et al.*, 1988). These various splicing options lead to a family of Ubx^+ proteins with constant N-terminal (247 amino acids) and C-terminal (99 amino acids) regions separated by three optional elements of 9, 17 and 17 amino acids (Kornfeld *et al.*, 1989). Alternative splicing patterns and the choice of polyadenylation site are both regulated in a stage- and tissue-specific manner. For instance, a nonsense codon in one of the micro-exons blocks the reading frame of all *Ubx* RNAs containing it and hence inactivates their protein products, whereas those Ubx^+ RNAs which lack this microexon are unaffected (Wienzierl *et al.*, 1987). This particular mutation disrupts the epidermal pattern but has no apparent effect in the adult nervous system (where Ubx^+ is also required; see (g) below), suggesting that Ubx^+ RNAs containing this microexon are essential for Ubx^+ function in the epidermis but not in the nervous system. Embryonic Ubx^+ RNAs are initiated from an upstream promoter and all contain a long (*c.* 1 kb) leader sequence which is not represented in the *Ubx*[+] proteins; however, this leader does include two methionine codons, one of which is followed by a short open reading frame that could encode a small 69-amino-acid protein (Saari & Bienz, 1987). The function of this (if any) remains to be determined.

In *situ* hybridisation has been used to examine the spatial distribution of *Ubx*[+] transcripts during the course of development (Akam, 1983; Akam & Martinez-Arias, 1985), while antibody staining methods have been used to localise *Ubx*[+] proteins in both wild-type and mutant embryos (White & Wilcox, 1984, 1985a, b; Cabrera *et al.*, 1985; Beachy *et al.*, 1985; White & Akam, 1985; Struhl & White, 1985; Carroll *et al.*, 1988a). The principal findings relating to the spatial regulation of BX-C genes (mainly *Ubx*) are summarised under (a)–(g) below; (h) deals briefly with some recent information on *Ubx* promoter function, which should be considered in the light of the complex spatial regulation of this gene.

(a) Ubx[+] *transcript patterns in wild-type embryos*
Initially, *Ubx*[+] transcripts are detectable in a broad central zone of the late syncytial blastoderm. By the cellular blastoderm stage, *Ubx*[+] RNAs are found principally in ps 6, but after gastrulation they are expressed prominently in reiterated parts of ps 6–12 inclusive, plus lower amounts in ps 5 and 13 (Akam & Martinez-Arias, 1985).

(b) Ubx[+] *protein patterns in wild-type embryos*
After germ-band shortening, *Ubx*[+] proteins are detected at high concentrations in parts of ps 6, at lower levels in some cells of ps 7–12, and in still lower amounts in a few cells of ps5 and ps13 (White & Wilcox, 1984, 1985a; Beachy *et al.*, 1985). Note that *Ubx*[+] protein expression is not uniform in all cells of a parasegment [see also (e) below]. Anti-*Ubx* staining is largely confined to cell nuclei, consistent with the inferred DNA-binding role of the homeodomain encoded by the 3' exon of the *Ubx* gene.

(c) *Specification of the domains in which* Ubx[+] *is expressed*
The initial parasegmental pattern of *Ubx*[+] expression (above) is established by combinations of maternal A/P, gap and pair-rule functions (see § 5.4.2). This is clearly implied by the altered distributions of *Ubx*[+] products observed in maternal A/P, gap and pair-rule mutant embryos (Ingham *et al.*, 1986; Ingham & Martinez-Arias, 1986; Duncan, 1986). However, the *Ubx*[+] pattern soon becomes refined within each parasegment through a variety of further regulatory influences, includ-

ing other homoeotic and some segment-polarity gene products. Because similar considerations apply to the early domains of expression for the other BX-C (and ANT-C) genes, further consideration of this topic is deferred until §5.5.4 below.

(d) *Influence of regulatory subfunctions on* Ubx+ *expression*

In homozygous *abx−*, *bx−*, *pbx−* or *bxd−* embryos, the distribution of *Ubx+* proteins is altered in a characteristic way (White & Wilcox, 1985b). In particular, *bxd−* mutations strongly reduce the intensity of *Ubx+* staining in ps 6 (largely comprising aA1) so that it comes to resemble ps5 (largely aT3). Similarly, the dominant *Cbx*D mutation causes intense *Ubx+* staining in ps 5 as well as in ps 6, again consistent with the observed morphological changes (White & Wilcox, 1985b; Beachy *et al.*, 1985; Casanova *et al.*, 1985b; Cabrera *et al.*, 1985; White & Akam, 1985). Because the *abx−* and *bx−* groups of mutations map within the *Ubx* transcription unit, it was originally thought that one or both might represent alternative *Ubx* exons; however, they do not correspond in position to the two internal microexons, but rather lie within long intronic regions, suggesting a *cis*-acting regulatory function. The *pbx−* and *bxd−* groups of mutations map 5′ to the *Ubx* gene and are spread across a large region of DNA, at least part of which is transcribed (the *bxd* transcription unit; see below), with the *pbx−* mutations lying more distally. The original *pbx*1 mutation arose simultaneously with *Cbx*1 as a result of X-irradiation, and the two were later separated by recombination. Molecular analysis of these two mutations (which give largely complementary phenotypes) shows that *pbx*1 involves a DNA deletion of 17 kbp from a region lying >40 kbp upstream from the *Ubx* cap site, whereas *Cbx*1 involves the insertion of this same DNA segment in reverse orientation at a new position within the *Ubx* transcription unit. This causes the inappropriate expression in ps 5 of *Ubx+* products at concentrations normally confined to ps 6, hence the *Cbx*1 mutant phenotype. Conversely, these elevated amounts of *Ubx+* products cannot be expressed even in ps 6 if the *pbx+* regulatory site is deleted, as in *pbx*1 (Bender *et al.*, 1983). The situation with *bxd* is more complex; *bxd−* mutations map across a 40 kbp region lying between *pbx* and the *Ubx* gene, but the most severe *bxd−* phenotypes result from breakpoints close to the 5′ end of *Ubx*. Much of this *bxd* region is transcribed to give a 26 kb primary transcript, from which several processed RNAs of 1.1 to 1.3 kb are derived. There are suggestions of an inverse relationship between *Ubx+* and *bxd+* transcript levels, the latter being higher in ps 7–12 where the former is lower, and vice versa in ps 6. Although *bxd*[+] function is essential to establish the wild-type *Ubx*[+] expression pattern, none of the *bxd*[+] RNAs synthesised during early development appears to have significant protein-coding potential (Lipshitz *et al.*, 1987). By contrast, during late larval, pupal and adult stages the *bxd* region produces a late 0.8 kb transcript with good protein-coding potential, derived from a single exon which is spliced out as an intron from all the embryonic *bxd*[+] RNAs. It remains to be seen whether a protein is translated from this late *bxd*[+] RNA, and if so what function it might perform; however, the regulatory influence of *bxd*[+] on *Ubx*[+] expression during embryogenesis is unlikely to be mediated via any *bxd* [+] protein product. Recent data (Sanchez-Herrero & Akam, 1989) suggests that overlapping parts of the *iab* regulatory region are similarly transcribed in restricted spatial domains of the abdomen; the significance of these transcripts, like those from the *bxd* locus, remains to be determined.

(e) *Influences of* abd-A[+] *and* Abd-B[+] *expression on* Ubx[+] *pattern* *Ubx*[+] protein levels are markedly down-regulated in those cells of ps 7–13 which also express *abd-A*[+], and additionally by *Abd-B*[+] expression in ps 13 (later also in ps 10–12) (Struhl & White , 1985; White & Wilcox, 1985b). Cell by cell examination of antibody staining for *Ubx*[+] and *abd-A*[+] proteins reveals a mosaic pattern of expression, with some cells expressing one, some the other, some both and some neither; moreover, this pattern becomes modulated in more posterior parasegments by the action of successive regulatory elements within the BX-C (see review by Peifer *et al.*, 1987). The wild-type *abd-A* gene is initially expressed in ps 7–13, overlapping with *Ubx*[+] expression in ps 7–12 and with early *Abd-B*[+] expression in ps 13; higher levels of *abd-A*[+] expression are found throughout this domain after gastrulation. In *abd-A*[−] mutants, ps 7–12 tend towards the ps 6 condition, though this effect is more complete anteriorly (in ps 7 - 9). The C1 deletion removes both the 3' exon of *abd-A* and the 5' exon of *Ubx*, so fusing the promoter and 5' exon of *abd-A* onto the 3' exons of *Ubx* (both breakpoints occur within introns). This hybrid gene is functional and expressed from ps 5 to 14 (the combined domains of both *Ubx*[+] and *abd-A*[+] expression). The spatial distribution of its products in ps 5 and 6 suggests that the *abx*[+]/*bx*[+] control element is active, though here regulating the *abd-A* rather than the *Ubx* promoter (Casanova *et al.*, 1988; Rowe & Akam, 1988).

Early expression of *Abd-B*[+] transcripts is limited to ps 13–15, but later this zone extends anteriorly into ps 10–12 as well. *Abd-B*[−] mutants change ps 10–13 towards ps 9, but also cause abnormal development of ps 14. The *Abd-B* gene appears to encode two distinct functions, the m (morphogenetic) element active in ps 10–13 and the r (regulatory) element required in ps 14 (Casanova & White, 1987). These functions probably involve distinct *Abd-B+* products, since some mutations abolish the m function only, some the r function only, and some both (these last are null *Abd-B−* alleles). Several distinct transcripts are derived from the *Abd-B* locus (Kuziora & McGinnis, 1988a; Sanchez-Herrero & Crosby, 1988) using two different promoters governing the m and r functions, respectively. One group of transcripts is expressed maximally in ps13 and anteriorly as far as ps 10 (presumably encoding the m function), while another is expressed only in ps14 and perhaps ps15 (likely to specify the r function). An *Abd-B+* transcript expressed in ps14 encodes a small protein containing a homeodomain; this may be a *trans*-acting regulator which confers the r function (DeLorenzi *et al.*, 1988). Several regulatory mutations in the right half of the BX-C show semidominant effects in the head region (e.g. the appearance of posterior abdominal tergites and/or genitalia), but also act recessively in the posterior abdomen (causing losses among these structures). Most of these *tumorous head* (*tuh*) mutations map to the regulatory regions of *Abd-B*, though one maps near *abd-A* (Kuhn & Packert, 1988). Plausibly, these mutations may cause inappropriate *Abd-B+* (or *abd-A+*) expression in the head (dominant effect), but might also diminish its function in posterior regions (recessive effect). Another dominant regulatory mutation in the *Abd-B* domain is *Transabdominal* (*Tab*); in *Tab*/+ heterozygotes, part of the T2 notum is changed into a sexually dimorphic region of abdominal cuticle (Celniker & Lewis, 1987).

(f) *Trans-regulatory functions needed to maintain* Ubx+ *pattern*

The *Polycomb* (*Pc*) gene was originally proposed as a negative regulator of BX-C function, since in *Pc−* mutants most of the embryonic segments resemble A8, where all BX-C subfunctions should be active according to the Lewis (1978) model. In fact, *Ubx+* staining is at a low and fairly uniform level throughout in *Pc−* mutant embryos (Wedeen *et al.*, 1986). This reflects the similarly derepressed expression of *abd-A+* and *Abd-B+* products, both of which down-regulate *Ubx+* expression [see (e) above]. Thus the spatial controls which normally restrict the domains of homoeotic gene expression appear to be relaxed in the absence of *Pc+* activity. Clones of *Pc−* cells in the adult epidermis show ectopic expression of several BX-C and ANT-C genes and also of *engrailed* (Busturia & Morata, 1988). However, many other genes

(Jurgens, 1985) appear to act in a similar way to *Pc*, including *Polycomblike* (*Pcl;* Duncan, 1982), *extra sex combs* (*esc*; Struhl, 1981b, 1983), *polyhomoeotic* (*ph*; Dura *et al.*, 1987; Dura & Ingham, 1988) and *super sex combs* (*sxc*; Ingham, 1984). These *Pc*-group genes do not specifically regulate *Ubx*⁺ or even BX-C function, since mutations in any of them cause widespread derepression of multiple homoeotic genes, including those in the distal BX-C and ANT-C. The *Pc*⁺ protein product can be immunolocalised on *Drosophila* polytene chromosomes in salivary-gland nuclei, and is found at some 60 discrete chromosomal sites; these include the BX-C and ANT-C as well as several other genes in the *Pc* group, consistent with a transcriptional repressor function (Zink & Paro, 1989).

In *esc*⁻ mutant embryos, the pattern of *Ubx*⁺ expression is at first normal in ps 6–12; later, *Ubx*⁺ transcripts are expressed ectopically throughout the germ band, in high amounts during its extension but at much lower levels during its shortening (Struhl & Akam, 1985). This final state again reflects generalised *abd-A*⁺/*Abd-B*⁺ as well as *Ubx*⁺ expression (cf. in *Pc*⁻ mutants). By contrast, the *trithorax* (*trx*) gene apparently encodes an opposite (positive) regulatory function, since in *trx*⁻ mutants many adult segments tend towards a thoracic (T2-like) condition. However, the absence of *trx*⁺ product cannot simply inactivate the BX-C, since embryonic segmentation is little disrupted in *trx*⁻ mutants. Moreover, in *trx*⁻/*esc*⁻ double-mutant homozygotes, both the adult and embryonic segment patterns are relatively normal (Ingham, 1983). Taken together, all this suggests that the *trans*-regulatory gene products function together to maintain but not to establish the spatial domains of BX-C (and ANT-C) gene action. An indirect and cooperative role is implied by the fairly normal body pattern produced in the absence of both negative (*esc*⁺) and positive (*trx*⁺) elements in this network. The molecular basis of this regulation remains to be clarified. Unlike their homoeotic gene targets, most or all of these *trans*-regulatory genes are expressed maternally in the oocyte as well as zygotically in the embryo (Haynie, 1983; Lawrence *et al.*, 1983; Ingham, 1984; Breen & Duncan, 1986).

(g) *Pattern regulation by the BX-C in other germ layers*

Ubx⁺ expression is not uniform in all germ layers of the embryo. As we have seen, one of the internal microexons is essential for epidermal *Ubx*⁺ function but apparently dispensable in the CNS (Wienzierl *et al.*, 1987). In the epidermis, CNS and somatic mesoderm, *Ubx*⁺ expression extends from ps 5 to 13, whereas in the visceral mesoderm it is confined to a single metamere (ps 7; Bienz *et al.*, 1988); *Ubx*⁺ is not expressed at all in the endoderm (Lawrence, 1985). In the larval nervous system and imaginal discs, antibody staining detects high concentrations of *Ubx*⁺ proteins in the posterior regions of ps 6 and to a lesser extent ps 5 (Brower, 1987), though the boundaries of expression are not strictly parasegmental. Although the BX-C genes are active in the mesoderm and CNS, this does not prove that they determine pattern autonomously in these tissues, as they do in the epidermis. In the case of the ventral nerve cord, there is a marked distinction between thoracic ganglia (large and paired) and abdominal ganglia (small and fused). In *bxd*⁻ mutants (converting aA1 towards aT3) there is a fourth pair of thoracic-type ganglia in the A1 position. By contrast, in dominant *Hyperabdominal* (*Hab*ᴰ) mutants which convert T3 and A1 towards A2, there are only two such paired ganglia (T1 and T2), the T3 position being occupied by an additional set of fused abdominal-type ganglia. When using weak mutant alleles of either class, the ventral nerve cord is sometimes altered in the absence of corresponding cuticular changes (and vice versa). This implies that thoracic/abdominal CNS pattern is controlled by autonomous expression of the BX-C genes, and is not imposed by the overlying epidermis (Teugels & Ghysen, 1983). Similarly, patterns of neuron branching characteristic of T2 become duplicated in T3 as a result of BX-C mutations converting T3 to T2 (Thomas & Wyman, 1984). These findings have been confirmed with the aid of an antibody which is specific for thoracic as opposed to abdominal features of the larval ganglia (Ghysen *et al.*, 1985); note also that the domains of BX-C action in the CNS appear to be parasegmental, as in the epidermis (Teugels & Ghysen, 1985).

The situation in the somatic mesoderm is more complex. The metameric patterning of the musculature can be investigated by clonal marking in much the same way as epidermal patterns (§ 5.4.1), using a cell-autonomous marker expressed in internal tissues. One such marker system involves the induction of clones homozygous for a temperature-sensitive mutation in the succinate dehydrogenase gene, whose product (or lack of it) can be readily detected by histochemical staining (Lawrence, 1981). Adult muscles form precise parasegmental sets underlying segmental units of ectoderm (Lawrence, 1982, 1985). Early investigations implied that muscle pattern might be dictated by the overlying epidermis (Lawrence & Brower, 1982). Later, autonomous BX-C action in the mesoderm was inferred from gynandromorph mosaics (see § 5.2.3) in which the female cells are wild-type while the male cells carry a BX-C homoeotic mutation. Two such mutations were used, one converting A4 and the other A6/7 towards the A5 condition; a sex-specific A5 muscle found in males but not females provides a convenient marker (Lawrence & Johnston, 1984b). Mosaics with mutant male internal tissues produce extra copies of the A5 marker muscle in the appropriate locations (A4 or A6/7), even if the abdomen is entirely overlain by wild-type female epidermis. However, more recent studies using this approach suggest that muscle pattern depends on homoeotic gene expression neither in the overlying epidermis nor even in the mesoderm itself, but rather in the underlying CNS whose neurons innervate the muscles (Lawrence & Johnston, 1986).

(h) *The* Ubx *promoter*

Sequence comparisons between the *Ubx* leader and promoter regions from several dipterans (*Musca* and *Drosophila* species) reveal a number of conserved elements which may be targets for regulatory interactions (Wilde & Akam, 1987). The molecular mechanisms underlying the control of *Ubx*+ function are now being studied through *in vitro* transcription systems, and also via fusions between *Ubx* regulatory sequences and β-galactosidase genes, which can be introduced via P elements into wild-type flies. Both approaches are yielding useful insights, although the task ahead is formidable [see (a)–(g) above]. *In vitro* transcription, using cell-free extracts from staged embryos, mimics the temporal pattern of *Ubx*+ expression found *in vivo*, and also identifies *cis*-acting regulatory sequences close to the cap site (between −300 and +100). Several of the proteins present in active extracts show sequence-specific binding to the *Ubx* promoter; one, for instance, binds to multiple GAGA motifs (Biggin & Tijan, 1988). Among these protein factors is the *zeste*+ gene product, which is known to be required for transvection in the *Ubx* and some other loci (e.g, *white*). Transvection describes the ability of *cis*-acting sequences near a mutant structural gene to cross-regulate a wild-type copy of that gene on a different chromosome, provided the two sites are closely paired. The *zeste*+ protein binds to multiple sites close to the 5' end of the *Ubx* gene and is required for its transcriptional activation; no such activation can be elicited by *zeste*+ protein if the binding sites are deleted. Similar experiments with the *white* locus reveal that the *zeste*-binding elements can be moved closer to or further away from the cap site without affecting their ability to activate transcription in the presence of *zeste*+ protein (Biggin *et al.*, 1988). Thus the function of *zeste*+ product in transvection appears similar to that of an enhancer-binding protein.

Ubx/β-galactosidase gene fusions containing about 4 kbp of *Ubx* promoter DNA are expressed in wild-type embryos in reiterated patterns which extend through much of the germ band (Bienz *et al.*, 1988). It now appears that some 25 kbp of DNA from the 5'-flanking region of the *Ubx* gene (extending into the *bxd* site) are necessary in order to confer an approximately normal pattern of spatial regulation on such fusion constructs. However, sequences essential for correct *Ubx*+ expression in ps 7 of the visceral mesoderm (VM) all lie within 4 kbp 5' to the cap site, including at least one element located between −1.7 and −3.1 kbp. Although the epidermal domain of *Ubx*+ expression overlaps with that of *Antp*+ anteriorly (see below) and with that of *abd-A*+ posteriorly [(e) above], this is not the case in the VM, where these three genes are active in mutually exclusive domains (ps 6 for *Antp*+, ps 7 for *Ubx*+, and ps 8 back for *abd-A*+). Notably, in *abd-A*− mutants the VM domain of *Ubx*+ expression extends posteriorly from ps 7, filling the whole of the VM domain in which *abd-A*+ is normally active (Bienz & Tremml, 1988; Tremml & Bienz, 1989a); *Ubx*+ products also autoregulate their own synthesis within the ps 7 domain. Thus *abd-A*+ expression from ps 8 back may preclude *Ubx*+ expression in the posterior regions of VM. It remains to be seen whether some elaboration of this autoregulation/mutual exclusion mechanism might stabilise or maintain the domains of *Ubx*+ expression in epidermis, CNS or somatic mesoderm. A purified *Ubx*+ protein binds tightly to DNA sites near its own promoter (consistent with inferred autoregulation), and also near the P1 promoter of the *Antennapedia* gene (Beach *et al.*, 1988). Although the *eve*+ protein represses the *Ubx* promoter *in vitro* (Biggin & Tijan, 1989), *eve*+ function is essential for expression of *Antp*+, *Ubx*+ and *abd-A*+ proteins in the VM (Tremml & Bienz, 1989b).

5.5.3 The Antennapedia complex (ANT-C)

The Antennapedia complex (ANT-C) comprises a second group of linked homoeotic genes, this time involved in anterior development. The three genes *Antennapedia* (*Antp*), *Sex combs reduced* (*Scr*) and *Deformed* (*Dfd*) control metameric identities in the anterior thorax and posterior head, much as the three BX-C genes do in more posterior body regions. Moreover, all three of these genes contain a homeobox sequence, as do their BX-C counterparts. In other respects, however, the molecular organisation of the ANT-C differs somewhat from that of the BX-C (fig. 5.27). Most notably, the ANT-C includes several other genes which do not function in the specification of anterior segments but play other roles in pattern formation, such as the *ftz* pair-rule gene,

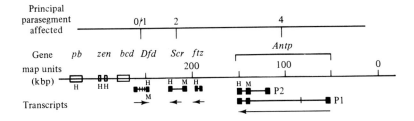

Fig. 5.27 The Antennapedia complex (ANT-C) modifed from
Gehring & Hiromi, 1986). ■■■, exon; ——, intron; 3′ ◄— 5′,
direction of transcription; H, location of homeobox; M, location of
M repeats; ☐, approximate locations of further genes within the
ANT-C; P1 and P2, distal and proximal promoters for *Antp* gene.
Notes: several additional transcription units of unknown function
have been omitted from this map; X lies between *Antp* and *ftz* near
170 kbp on the DNA scale, while Y2, Y3 and Z lie between 240
and 270 kbp in the *Dfd* region (see Gehring & Hiromi, 1986). The
newly identified homeobox-containing *lab* gene lies near *pb* at the
left-hand end of the cluster.

the *bcd* maternal A/P gene (see §5.4.2) and the *zen* pair of zygotic D/V
genes (see §5.3.1); homeoboxes are also present in these genes. The
ANT-C also contains transcription units of unknown function (see
Gehring & Hiromi, 1986), plus at least two further homoeotic genes,
proboscipedia (*pb*; Pultz *et al.*, 1988) and *labial* (*lab*; Mlodzik *et al.*, 1988)
which act mainly in the anterior head, though *lab*⁺ expression also
occurs in the posterior midgut (see later).

Among the major homoeotic functions, the eponymous *Antp* gene
has been characterised in greatest detail. The original *Antp* homoeotic
mutations are mainly dominant gain-of-function alleles (hereafter designated *Antp*ᴰ) which cause second legs to develop in the place of
antennae. Null *Antp*⁻ mutations produce the converse effect, with
antennae appearing in the place of second legs. Since this is an embryonic lethal phenotype, it can be studied in adults only by inducing
mutant clones locally and comparing their cuticular features with
those of adjacent wild-type areas (Struhl, 1981c). This null mutant pattern implies a specific function for the *Antp*⁺ gene in the second
thoracic segment, or more probably ps 4. This is confirmed by inducing triple mutant clones which simultaneously lack the *Antp*⁺, *Scr*⁺ and
Ubx⁺ functions. As we have seen, *Ubx*⁻ mutations convert ps 5 and 6
towards ps 4, whereas *Scr*⁻ mutations *inter alia* transform ps 3 towards
ps 4; effectively, therefore, all three thoracic segments resemble T2

(mainly ps 4) in *Ubx⁻/Scr⁻* mutants. In the additional absence of *Antp⁺* function, all three pairs of legs are converted into antennae (Struhl, 1982), consistent with a controlling role for *Antp⁺* in ps 4.

The distribution of wild-type *Antp⁺* products has been studied by *in situ* hybridisation (Levine *et al.*, 1983; Martinez-Arias, 1986) and by antibody staining (Wirz *et al.*, 1986; Carroll *et al.*, 1986a, 1988a). *Antp⁺* transcripts accumulate in the ectoderm of ps 4 and 5, but only in ps 5 of the mesoderm (Martinez-Arias, 1986); they are also detectable in much of the embryonic CNS (though mainly in ps 4 and 5), as well as in the proximal parts of all six pairs of thoracic imaginal discs (Levine *et al.*, 1983). The distribution of *Antp⁺* proteins broadly confirms this picture; they are nuclear in location, as predicted from the homeo-domain encoded by the 3' exon of the *Antp* gene. The altered distributions of *Antp⁺* staining shown in other homoeotic mutants suggest the following spatial controls for *Antp⁺* expression in the epidermis:

(i)   Positive autoregulation by the *Antp⁺* product itself.
(ii) Negative control in ps 5 and more posterior regions by the *Ubx⁺* and other BX-C gene products (cells in ps 5 that express *Ubx⁺* do not express *Antp⁺*, and vice versa (Carroll *et al.*, 1986a, 1988a).
(iii) Generalised negative control by *trans*-regulatory genes such as *Pc⁺* (see above); in *Pc⁻* mutants there is low-level *Antp⁺* staining throughout much of the embryo (Carroll *et al.*, 1986a), implying indiscriminate *Antp⁺* expression which is down-regulated by the BX-C functions (cf. for *Ubx⁺*).
(iv) Various segmentation gene products (see §5.5.4).

Needless to say, there remain further complexities awaiting explanation (see e.g. Wirz *et al.*, 1986). Overall, *Antp⁺* appears to be the primary determinant of pattern in ps 4, but probably acts in concert with other homoeotic functions such as *Ubx⁺* in ps 5 (Martinez-Arias, 1986). *Antp⁺* may also function in conjunction with *Scr⁺* in ps 3 (§5.5.4 below). BX-C deletions cause high-level *Antp⁺* expression to extend back as far as A7 (ps 12) in the embryonic nervous system; however, *Scr⁻* mutations do not result in a similar anterior shift of *Antp⁺* expression in the CNS (Wirz *et al.*, 1986). Thus the mode of *Antp⁺* regulation in ps 3 may be different from that in ps 5 and more posterior parasegments.

The molecular organisation of the *Antp* gene is also complex. As shown in fig. 5.27, it is over 100 kbp long (larger even than *Ubx*) and includes two promoters some 70 kbp apart, a total of eight exons and two alternative polyadenylation sites (Stroeher *et al.*, 1986; Schneuwly *et al.*, 1986; Laughon *et al.*, 1986). There are four major poly(A)$^+$ transcripts, two starting from the proximal P2 promoter in front of exon 3 and two from the distal P1 promoter in front of exon 1 (see also Garber *et al.*, 1983). All *Antp*$^+$ mRNAs have long (1–2 kb) 5′ leader and 3′ trailer sequences which are untranslated. Which of two 5′ leaders (1.5 or 1.7 kb) is present depends on the promoter used, whereas the 3′ trailer present reflects a choice between the two polyadenylation sites (1.4 kbp apart). There is evidence that both choices are developmentally regulated; for instance, there is a marked preference for the second polyadenylation site in neural tissue (Stroeher *et al.*, 1986; Laughon *et al.*, 1986). Four slightly different *Antp*$^+$ proteins are derived from alternatively spliced transcripts (Bermingham & Scott, 1988). Differential regulation of the two *Antp* promoters is implied by their spatially distinct patterns of transcript accumulation in a subset of wild-type imaginal discs (Jorgensen & Garber, 1987; see § 5.5.4 below). Despite the huge (70 kbp) separation between the two *Antp* promoters, both drive very similar protein-coding regions, which start in exon 5 and encode a 42.8 kd protein rich in glutamine and proline. Exons represented in this protein include both M repeats (cf. in the *paired* and *Notch* genes) and a 3′ homeobox located in exon 8 (Stroeher *et al.*, 1986; Schneuwly *et al.*, 1986). This latter encodes a protein homeodomain which can bind in isolation to specific DNA sequences, including TAA repeats found in the *Antp* gene region itself (Muller *et al.*, 1988a; Mihara & Kaiser, 1988).

It remains to explain the dominant *Antp*^{D} phenotype which converts antenna to second leg. Since *Antp*$^+$ function is needed for normal metameric identity in the thoracic region (particularly ps 4), it is plausible to suggest that *Antp*$^+$ activity is specifically *not* required in the anterior head. Inappropriate expression of *Antp*$^+$ in this region would therefore result in the acquisition of thoracic characteristics, such as replacement of antennae by second legs. This interpretation has been directly confirmed by linking an *Antp*$^+$ cDNA to a heat-shock promoter and introducing this via a P-element construct into wild-type flies. When individuals carrying this construct are heat-shocked during development, the *Antp*$^+$ protein becomes expressed indiscriminately throughout the body; this causes the anticipated transformation of antennae to second legs, and also converts part of the dorsal head into dorsal mesothorax (Schneuwly *et al.*, 1987a). Studies of several *Antp*^{D} dominant mutations suggest that the *Antp*$^+$ protein-coding region is transcribed inappropriately in the anterior head (Jorgensen & Garber, 1987). Many of these *Antp*^{D} mutations result from chromosomal inversions involving the *Antp* gene and a second gene called *rfd*, which is normally expressed in the head and is transcribed in the opposite direction to *Antp*. These inversions result in a reciprocal exchange of promoters and first exons between the two genes, such that the *Antp*+ protein-coding unit (exons 5–8) is now transcribed from the *rfd* promoter along with one or more of the 5' *rfd* exons. Like *rfd*+ normally, this fused *rfd/Antp* gene is expressed in the head, where ectopic *Antp*+ protein production causes the characteristic antenna-to-leg transformation (Frischer *et al.*, 1986; Schneuwly *et al.*, 1987b).

Null mutations in the *Sex combs reduced* (*Scr*) gene show a partial conversion of T1 towards T2 (more precisely, ps 3 to ps 4) and of the labial segment towards the maxillary state (ps 2 towards ps 1; Struhl, 1983). This suggests that the *Scr*+ function influences pattern in ps 2 and 3. The *Scr* gene comprises two exons separated by a large intron; the major 3.9 kb transcript includes both M repeats and a homeobox sequence in the 3' exon (Kuroiwa *et al.*, 1985). In gastrula embryos, *Scr*+ transcripts accumulate mainly in the posterior head and anterior thoracic regions of the germ band (roughly ps 2/3); later in development the suboesophagal and prothoracic ganglia of the ventral CNS show strong hybridisation with *Scr* probes (Kuroiwa *et al.*, 1985; see also Mahaffey & Kaufman, 1987). Antibodies agains the *Scr*+ protein generallly confirm this picture; prominent *Scr*+ expression starts in ps 2 and then extends into ps 3 in the germ band (Riley *et al.*, 1987). *Scr*+ products are also expressed in the mesoderm and CNS of the ps 2/3 region (Mahaffey & Kaufman, 1987). During larval development, the *Scr*+ protein accumulates widely in the prothoracic leg disc, but only in certain cells of the humeral, labial and antennal discs; it is also expressed by adepithelial cells in all thoracic leg discs (Glicksman & Brower, 1988). These complex patterns of *Scr*+ expression are modulated by (i) the products of other homoeotic genes such as *Antp*+ (*Scr*+ is expressed ectopically in *Antp*− mutants; Riley *et al.*, 1987); (ii) *trans*-regulatory functions such as *Pc*+ (as before); and (iii) a variety of segmentation gene products (mutations in *ftz*, *hb*, *Kr*, or *gt* all alter the spatial distribution of *Scr*+ protein; Riley *et al.*, 1987; see § 5.5.4).

The mutant phenotype caused by deficiencies at the *Deformed* (*Dfd*) locus is more difficult to define; dorsally, posterior head structures are transformed towards thorax, while ventrally structures are deleted in *Dfd*− mutants (Merrill *et al.*, 1987). The wild-type *Dfd* gene comprises five exons distributed over an 11 kbp region of DNA; the major 2.8 kb RNA species encodes a 63.5 kd protein which again contains a 3' homeobox (Regulski *et al.*, 1987). In normal germ-band embryos, *Dfd*+ transcripts are confined to a single anterior band which includes parts of the future mandibular and maxillary segments (Chadwick & McGinnis, 1987), corresponding mainly to ps 0 and 1 (Martinez-Arias *et al.*, 1987). *Dfd+* is also expressed in parts of the eye/antennal disc during larval development. The initial parasegmental domains of both *Scr+* (ps 2 + 3) and *Dfd+* (ps 0 + 1) transcripts become modified to a segmental pattern during formation of the gnathal appendages. Indeed, there is a marked dorsal/ventral dispartiy in the boundaries of both *Dfd+* and *Scr+* expression; these are segmental dorsally (and later throughout the epidermis) but parasegmental ventrally (and later in the ventral nerve cord; Mahaffey *et al.*, 1989). Expression of *Dfd+* products is regulated by pair-rule segmentation genes, as shown by the altered *Dfd+* protein distributions observed in mutants for eight of the nine pair-rule genes (Jack *et al.*, 1988). The *Dfd+* protein also autoactivates expression of its own gene, as shown by using P-element constructs containing a *Dfd* cDNA linked to a heat-shock promoter (giving indiscriminate *Dfd+* expression when heat-shocked; Kuziora & McGininis, 1988b). This autoregulatory feature may account for the stability of the determined state controlled by the *Dfd* selector gene. Substitution of the *Ubx+* homeodomain into the *Dfd+* protein changes the specificity of this latter, causing activation of *Antp+* rather than autoactivation of *Dfd+* expression (Kuziora & McGinnis, 1989).

A further homeobox gene, designated *F90-2*, is expressed in a zone anterior to that of *Dfd+* in the ectoderm, but also in the posterior midgut (endoderm). This gene maps close to *Dfd* in the ANT-C, and its homeobox is split unusually be an intron (Hoey *et al.*, 1986); M repeats are also present in the 5′ part of the protein-coding region. All of these features correspond to those described for the recently isolated *labial* gene (Mlodzik *et al.*, 1988). Patterns of expression for the *labial+*, *Dfd+*, *Scr+* and *proboscipedia+* proteins have been described in the embryonic head region; while the first three are expressed in non-overlapping domains, this is not true for *pb+*, whose function is inessential in the embryo (Mahaffey *et al.*, 1989). Both *ftz* and *zen* genes have been reviewed earlier (§ §5.4.2 and 5.3.1); their relationship to the main homoeotic genes of the ANT-C remains unclear, although the similarity of the *ftz* and *Antp* homeobox sequences is intriguing. Homoeotic genes in both the ANT-C and BX-C show joint and mutual spatial regulation in the germ band embryo, so defining parasegmental identities; this topic is discussed briefly in the next subsection.

5.5.4 Spatial regulation of ANT-C and BX-C genes in the embryo

In the flour beetle *Tribolium*, several homoeotic mutations affecting both anterior and posterior development map close together in a single gene complex (HOM-C; Beeman, 1987). Mutations in six HOM-C complementation groups give phenotypes which are reminiscent of both ANT-C and BX-C mutants in *Drosophila*; moreover, the map order of these HOM-C genes is the same as the linear sequence of segments in which they act (see also Beeman *et al.*, 1989). This suggests that the separate *Drosophila* ANT-C and BX-C clusters might have arisen through the splitting of a single ancestral gene-complex (as in *Tribolium*), much as the globin genes have become split into separate α- and β-type clusters in higher but not lower vertebrates (see §3.2). The possible role of homoeotic genes in arthropod evolution is a fascinating topic for speculation, though rather outside the scope of the present text. However, we may note in passing that many of the ANT-C and BX-C functions serve to repress the production of supernumerary pattern elements such as wings and legs. Such functions would be needed only to modify the anterior and posterior ends in a primitive myriapod-like ancestor, where a large number of central segments are virtually identical in pattern. However, many insects do not lay down primordia for all of the future body segments during early embryogenesis, but rather bud off more and more segments posteriorly as development proceeds (e.g. in locust). It will be interesting to discover how segmentation and homoeotic gene functions are utilised during this latter pattern of development.

The existence of a single *Tribolium* homoeotic gene-complex emphasises the need in *Drosophila* to consider the interwoven functions of ANT-C and BX-C genes together, despite their separation in DNA terms. Several features of the preceding discussion (§§5.5.2 and 5.5.3 above) suggest how ANT-C and BX-C gene activities might be deployed initially in the embryonic germ band. This becomes more apparent if we list the major homoeotic genes alongside the parasegments in which their activities are most essential, i.e where they have a primary pattern-determining role, at least in the epidermis:

Dfd	in parasegment 1	(ANT-C);
Scr	in parasegment 2	(ANT-C);
Antp	in parasegment 4	(ANT-C);
Ubx	in parasegment 6	(BX-C);
Abd-B	in parasegment 14	(BX-C);
cad	in parasegment 15	(unlinked).

Note the exclusion (for the moment) of *abd-A* from this list and also the inclusion of *cad*. This last is a homeobox-containing gene which lies outside the BX-C and ANT-C; it is expressed both maternally (see §5.4.2) and zygotically. Although maternal *cad*$^+$ products form a posterior-to-anterior gradient in the early embryo (Mlodzik *et al.*, 1985; Macdonald & Struhl, 1986), zygotic *cad*$^+$ expression is confined to a single posterior band in ps 15, where it appears to be the main pattern-determining element (see also Mlodzik & Gehring, 1987). The beginning of the parasegmental sequence (ps 0/1) is defined by the *Dfd*$^+$ domain and the end (ps 15) by the *cad*$^+$ domain. In between, the major homoeotic genes are expressed predominantly in alternate (even-numbered) parasegments, namely *Scr*$^+$ in ps 2, *Antp*$^+$ in ps 4, *Ubx*$^+$ in ps 6 and *Abd-B*$^+$ in ps 14. The intervening odd-numbered parasegments express lower amounts of the two homoeotic gene-products characteristic of their neighbours, i.e. *Scr*$^+$/*Antp*$^+$ in ps 3 and *Antp*$^+$/*Ubx*$^+$ in ps 5. In ps 7, *abd-A*$^+$ activity joins that of *Ubx*$^+$, but thereafter the pattern breaks down. There is no single parasegment in which *abd-A*$^+$ influence can be said to predominate; rather, both *Ubx*$^+$ (anteriorly) and/or *Abd-B*$^+$ (posteriorly) act in conjunction with *abd-A*$^+$ to define ps 7–13. Notably, these abdominal parasegments all develop along rather similar paths until quite late, whereas other parasegments develop divergently after the germ-band stage. However, transcripts of the *iab* regulatory sites *are* expressed within spatially restricted abdominal domains; type I transcripts (extending as far as *iab-3/-4*) in ps 8–15, type II transcripts (extending only as far as *iab-5/-6*) in ps 11–15, and type III transcripts (*iab-7* only) in ps 13–15 (Sanchez-Herrero & Akam, 1989). The overall pattern of homoeotic gene expression in the germ band embryo is summarised in fig. 5.28 (based on Akam, 1987). Transcripts from both promoters of the *Antp* gene are initiated at the cellular blastoderm stage, but because of the enormous size of those transcribed from the distal (P1) promoter they are not completed until after gastrulation. Only *Antp*$^+$ P2 transcripts are expressed in ps 3 and in ps 13–15, whereas both P1 and P2 promoters are used in ps 4–12. The pattern of *Abd-B*$^+$ expression changes later in development as its m transcripts become detectable in ps 10–12; however, during germ-band extension this gene is only expressed in ps 13 (m) and 14–15 (r). The main homoeotic genes (apart from the end members *Dfd* and *cad*) are expressed in rather broad and overlapping domains at this stage (fig. 5.28*C*); however, intense transcription of each gene is at first confined to a single parasegment in the late cellular blastoderm (fig. 5.28*A*), namely *Scr*$^+$ in ps 2, *Antp*$^+$ in ps 4, *Ubx*$^+$ in ps 6 and *Abd-B*$^+$ in ps 14.

What regulatory features might underlie this early spatial pattern of expression? At this stage, the germ band is divided by gap-gene prod-

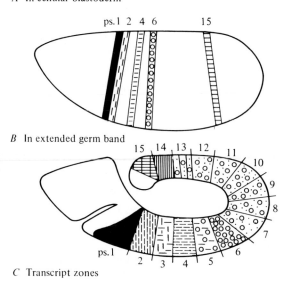

A In cellular blastoderm

ps.1 2 4 6 15

B In extended germ band

15 14 13 12 11 10 9 8 7

ps.1 2 3 4 5 6

C Transcript zones

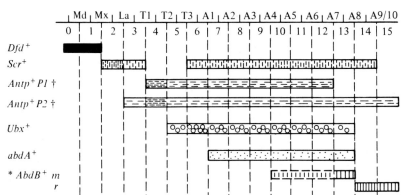

	Md	Mx	La	T1	T2	T3	A1	A2	A3	A4	A5	A6	A7	A8	A9/10
	0	1	2	3	4	5	6	7	8	9	10	11	12	13	14	15

Dfd^+

Scr^+

$Antp^+ P1$ †

$Antp^+ P2$ †

Ubx^+

$abdA^+$

* $AbdB^+$ m

r

cad^+

Fig. 5.28 Early domains of homoeotic gene expression (modified from Akam, 1987). *A* In cellular blastoderm. *B* In extended germ band. *C* Transcript zones. ■, Dfd^+; ▨, Scr^+; ▨, $Antp^+$; ▨, Ubx^+; ▨, *abd-A*$^+$; ▥, *Abd-B*$^+$; ▤, cad^+.

Notes: zones of $Antp^+$ and Scr^+ expression in posterior regions are not shown in part *b*; see part *C* for details. Density of cross-hatching or speckling indicates relative intensity of expression in parts *B* and *C*.

*Domains of expression from the two *Abd-B* promoters (m and r) are shown separately; note that the m domain is at first confined to ps 13, but later extends anteriorly into ps 10–12 (dashed).

†P1 and P2 refer to the alternative distal and proximal *Antp* promoters.

ucts into broad but discrete domains, and further subdivided by pair-rule products into repeating metameric units (see §5.4.2). Since the primary domains of homoeotic gene activity are in even-numbered parasegments (apart from *Dfd* and *cad*), one might anticipate a prominent regulatory role for *ftz*+. In the abdominal regions of wild-type embryos (jointly controlled by *Ubx*+, *abd-A*+ and *Abd-B*+), there are transient peaks of *Ubx*+ transcripts in ps 8, 10 and 12, as well as most prominently in ps 6 – all suggestive of pair-rule patterning (Ingham & Martinez-Arias, 1986). There is widespread disruption of the spatial patterns of homoeotic gene expression in *ftz*⁻ and some other pair-rule mutants. Whereas the distributions of *Scr*+, *Antp*+ and *Ubx*+ transcripts are fundamentally altered in *ftz*⁻ embryos, they are much less affected in *opa*⁻ embryos, which also show deletions of the even-numbered parasegments. This suggests that *ftz*+ has a dual function in setting the spatial limits of homoeotic gene expression as well as in metamerisation, whereas *opa*+ acts only in the latter process (Ingham & Martinez-Arias, 1986). A similar conclusion was reached by Duncan (1986), based on the mutant phenotypes of dominant *ftz*ᴰ alleles which cause pattern deletions out of phase with those in *ftz*⁻ mutants.

The products of maternal A/P polarity genes and zygotic gap genes also control the initial domains of homoeotic gene expression. For example, in *Kr*⁻ gap mutants, the *Ubx*+ domain is altered in line with the changes observed in pair-rule (*ftz*+, *h*+) and segment polarity (*en*+) expression patterns (Ingham *et al.*, 1986). Similarly, Harding & Levine (1988) have monitored the effects of various gap-gene deficiencies (in *hb*⁻, *Kr*⁻ and *kni*⁻ embryos) on the patterns of *Antp*+ and *Abd-B*+ expression. *Abd-B*+ is negatively regulated by *kni*+ activity, but may be positively controlled by a more posterior gap gene such as *tll*. The case of *Antp*+ is more complex, since the two promoters seem to be differentially regulated. The P1 promoter depends absolutely on *Kr*+ function (*Antp*+ is expressed within the *Kr*+ domain), whereas the P2 promoter responds differentially to *bcd*+, *hb*+ and posterior class products (but these last may act indirectly via removal of maternal *hb*+ products; see §5.4.2a, Irish *et al.*, 1989a,b; Ingham, 1988). Similarly, the initial *Ubx*+ domain depends on activation by posterior-class and repression by *hb*+ products. Unique combinations of maternal and gap-gene products define the initial domains of *Ubx*+ and *Antp*+ transcription, along with pair-rule products (particularly *ftz*+) which determine the parasegmental phasing of their peak expression (Irish *et al.*, 1989b). This pattern becomes further modified by interactions with the products of other homoeotic genes (Carroll *et al.*, 1988a). Such interactions are mediated through the DNA-binding and activation domains of these regulatory proteins, presumably acting as transcription factors to modulate the pattern of homoeotic gene expression. This can be explored by contransfecting cultured *Drosophila* cells with two constructs, one containing a target promoter fused to a reporter gene while the other encodes a candidate regulatory protein. Such studies reveal that *ftz*[+] protein activates the *Ubx* promoter and that *Antp*[+] protein auto-activates its own P1 promoter (Winslow *et al.*, 1989). Likewise, the *Ubx*[+] protein binds near and activates its own promoter but represses the *Antp* P1 promoter (Krasnow *et al.*, 1989).

5.6 The homeobox and other conserved sequences

5.6.1 The homeobox

The homeobox was first characterised as a short region (180 bp) of sequence homology present in several *Drosophila* homoeotic (*Ubx* and *Antp*) and segmentation (*ftz*) genes (McGinnis *et al.*, 1984a; Scott & Weiner, 1984). Soon afterwards, closely related sequences were discovered in the genomes of other higher metazoans (notably in vertebrates; McGinnis *et al.*, 1984b). Many of the vertebrate homeoboxes are as much as 90% homologous to the *Antp* prototype, suggesting close conservation during the course of metazoan evolution. In *Drosophila*, *Antp*-like homeobox sequences are found mostly in the BX-C and ANT-C gene clusters, with one copy each in the *Ubx*, *abd-A* and *Abd-B* genes of the former (Regulski *et al.*, 1985), and in the *Antp*, *Scr*, *Dfd*, *ftz*, *pb*, *lab*, and z1/z2 *zen* genes of the latter (Gehring & Hiromi, 1986). However, in *Drosophila* there are several variant types of homeobox which may be only 40–60% homologous to the *Antp* prototype; these include the *prd* class (also found in *gsb*, *bcd* etc), the *en/inv* class, the *cad* class and the *eve* class. An even more divergent homeobox has been isolated from the H2.0 gene, which is expressed in a tissue- (rather than region-) specific manner in the visceral mesoderm (Barad *et al.*, 1988). In each case, the homeobox sequence is exonic (though split by an intron in *en/inv* and differently in *lab*); it is generally located towards the 3′ end of the gene and is translated into a 'homeodomain' protein sequence which is even more tightly conserved than the homeobox.

The homeodomain comprises about 60 amino acids, many of which are basic (cf. in histones). Distantly related sequences (*c.* 25% identity) are found in the yeast MAT α2 and MAT a1 proteins which are involved in DNA-binding during switches of mating type (Shepherd *et al.*, 1984; see also Hicks, 1987 for a brief review of mating-type switching); still more remotely related sequences occur in several procaryotic DNA-binding proteins (Laughon & Scott, 1984). The common feature in all of these proteins is a 'helix-turn-helix' motif in which two α helices are separated by a β turn (see review by Struhl, K., 1989). These relationships suggest a DNA-binding role for both insect and vertebrate proteins containing homeodomains, and so far all such proteins have proved to be nuclear in location. The *en*⁺ protein shows sequence-specific binding to DNA (Desplan *et al.*, 1985), as do the mouse *Hox-1.5* protein (which binds upstream of its own gene; Fainsod *et al.*, 1986) and the mouse *Hox-1.3* protein (Odenwald *et al.*, 1989). The homeodomains from the *en*⁺ and *ftz*⁺ proteins bind to similar target sequences (Desplan *et al.*, 1988), including TAA repeats which are also recognised by the *Antp*⁺ homeodomain (Muller *et al.*, 1988a; Mihara & Kaiser, 1988). The *eve*⁺ protein binds to two distinct types of sequence, one GC-rich and found upstream of its own gene, the other AT-rich and found near the 5′ end of the *en* gene (Hoey & Levine, 1988); moreover, both *prd*⁺ and *zen*⁺ proteins compete for the *eve*-gene binding sites, whereas only *zen*⁺ competes with *eve*⁺ protein for the *en*-gene binding sites. Such overlaps in binding specificity could result in competition for the same target sites, which might perhaps underlie the mutually exclusive expression patterns often seen for homeodomain proteins. But in many cases such proteins clearly bind to distinct target sites, and certain regions within the homeodomain have been shown to confer this specificity. For instance, single amino-acid changes at the C-terminal end of the recognition helix in the *prd*⁺ homeodomain can alter its binding specificity to that of the *ftz*⁺ or *bcd*⁺ homeodomain (Treisman *et al.*, 1989). In general, homeodomain proteins represent one particular class of DNA-binding transcription factors (see also end of § 5.5.4). This leaves open the question of whether such proteins govern key developmental decisions in other animal groups, as is generally the case in *Drosophila*.

The phylogenetic distribution of homeoboxes is intriguing; sequences closely related to the *Antp* prototype are found in higher but not lower eucaryotes – for instance in vertebrates, arthropods, echinoderms and molluscs, but not in nematodes, slime moulds or yeast (Holland & Hogan, 1986). However, less tightly conserved homeobox sequences have since been discovered both in the nematode *C. elegans* (in the *mec-3*, *unc-86* and *mab-5* genes; Way & Chalfie, 1988; Finney *et al.*, 1988; Costa, M. *et al.*, 1988) and even in yeast (the *PHO 2* regulatory gene; Burghlin, 1988). Thus the homeobox itself appears to be an ancient eucaryotic (not even metazoan) device. Even if its inferred switch function is universal, we should not expect this to be deployed in the same way across such a diverse range of organisms. This is particularly relevant when we consider the prominence of homeobox genes in controlling metamerisation in *Drosophila* (e.g. *bcd*, *ftz*, *prd*, *eve*, *en*, plus the ANT-C and BX-C homoeotic genes). Vertebrates show evidence of metameric organisation (see below), particularly in the embryonic somites, hind-brain, vertebrae, spinal ganglia and ribs. Could homeobox genes fulfil similar controlling roles in both vertebrate and arthropod metamerisation (see e.g. Struhl, 1984b)? The answer is, intriguingly, both yes and no. On the one hand, there is evidence that mouse homeobox genes are expressed in overlapping but region-specific patterns during embryogenesis (see below and fig. 5.30), rather reminiscent of homoeotic gene expression in the *Drosophila* germ band (fig. 5.28). On the other hand, there is no obvious evolutionary link between arthropod and vertebrate metamerisation. Rather, these processes appear to have evolved independently within the two great subdivisions of triploblastic animals. Arthropods and annelids both belong to the protostome group which also includes the non-segmented molluscs, whereas vertebrates belong to the deuterostome group which also includes the non-segmented echinoderms (see Hogan *et al.*, 1985). There is no strong evidence that the last common ancestor of both groups was itself segmented. Present-day molluscs and echinoderms contain several well-conserved homeobox sequences (Holland & Hogan, 1986); moreover, at least one such homeobox gene is expressed during early sea urchin development (Dolecki *et al.*, 1986). Clearly, genes such as these cannot function in metamerisation, though they could well be involved in other developmental switches (cf. *mec-3* or *unc-86* in *C. elegans*; H2.0 or *cut* in *Drosophila*). Indeed, there is considerable evidence to suggest that homeobox gene products often participate in modular switch mechanisms which can be utilised in many different regulatory contexts (see also later). Another plausible function for homeobox gene products might be the provision of positional information (required in all animals. whether segemented or not), as suggested by the position-specific expression of the *mab-5* homeobox gene in *C. elegans* (non-segmented; Costa, M. *et al.*, 1988). This view gains credence from the recent discovery of striking parallels between the major clusters of homeobox genes in mouse and in *Drosophila*, in terms of their organisation and expression patterns (see fig. 5.29; Graham *et al.*, 1989). These features suggest a common ancestral cluster of homeobox genes, even though the last common ancestor of protostomes and deuterostomes may have been unsegmented.

Before surveying the spatial patterns of expression for several mouse homeobox genes, it is worth describing briefly how metameric structures arise in mouse and chick embryos (for further details, see Hogan *et al.*, 1985). After formation of the primitive streak, ingressing mesodermal cells begin to condense into somite blocks on either side of the midline. These are formed in an antero-posterior (rostro-caudal) sequence as Hensen's node regresses posteriorly. There is evidence for some kind of prepattern in the mesoderm before it becomes overtly organised into somites, visible as whorls of cells terms somitomeres. In the head region, seven pairs of anterior somitomeres do not become organised into proper somites, whereas the remainder condense in rostro-caudal sequence into 65 pairs of somites. Later in development, each somite contributes cells to the dermal layer of the skin (dermatome), to the skeletal structures of the vertebral column and ribs (sclerotome), and to the musculature (myotome). Evidence of this basic metameric organisation becomes obscured in the skin, but persists throughout life in the sclerotome-derived structures (axial skeleton) and to some extent in those derived from the myotome (obvious in the repeating muscle blocks of fish, but also, for example, in the mammalian intercostal muscles). The vertebrae are not formed one from each somite; rather, the posterior part of one somite contributes the anterior half of a vertebra, while the anterior part of the next somite contributes the posterior half of that same vertebra (see e.g. Bagnall *et al.*, 1988). This resegmentation process is rather reminiscent of the transition from parasegments to segments in *Drosophila*, though it is difficult to see how the two processes can be related (see above). This implies that anterior (*a*) half-somites are in some way different from posterior (*p*) half segments (cf. compartments in *Drosophila*), a point confirmed by grafting experiments in chick embryos. Although cells taken from *a* (or *p*) half-somites can mix freely with other *a* (or *p*) cells at the graft site, *p* cells will not mix with *a* cells nor vice versa; wherever graft and recipient cells from unlike halves abut, extra segment borders are formed (Stern & Keynes, 1987).

The vertebrate CNS is also organised in a basically metameric pattern; this becomes apparent early on as a series of transient swellings termed neuromeres, which are repeated at somite-length intervals in the hindbrain and, in lower vertebrates, the spinal cord (Hanneman *et al.*, 1988). Later evidence of this patterning is provided by the series of cranial and spinal nerves plus associated ganglia (e.g. the dorsal root ganglia along the vertebral column). This metameric patterning of the

CNS may be imposed by the underlying mesoderm, as shown by rotating sections of somitic mesoderm or of neural tube in early chick embryos; whereas the former procedure disrupts the segmental patterning of the CNS, the latter does not (instead the neural structures become reorganised in accordance with the undisturbed mesoderm; Keynes & Stern, 1984). However, the segmental character of the hindbrain neuromeres (termed rhombomeres) appears to be fundamental. Mouse rhombomeres show an interesting deployment of homeobox genes (see later; Lewis, 1989), while chick rhombomeres are defined by cell lineage restrictions (Fraser *et al.*, 1990; cf. compartments, Lawrence, 1990).

Vertebrate homeobox-containing genes have been cloned from mouse (>30), zebrafish, human, chick and *Xenopus*; the summary below will concentrate mainly on mouse (reviewed by Holland & Hogan, 1988a; see also Graham *et al.*, 1989). There are four large clusters of homeobox genes in the mouse genome designated hox 1, hox 2, hox 3 and hox 5 on chromosomes 6, 11, 15 and 2, respectively (Hart *et al.*, 1985, 1987; Rabin *et al.*, 1986; Duboule *et al.*, 1986; Bucan *et al.*, 1986; Lonai *et al.*, 1987; Schughart *et al.*, 1988; Sharpe *et al.*, 1988; Graham *et al.*, 1989). Because of earlier confusion as to chromosomal location, the hox 3 cluster now includes *Hox-6.1/-6.2* as well as *Hox-3.1/-3.2/-3.3*, while the hox 5 cluster includes *Hox-4.1* as well as *Hox-5.1/ -5.2/-5.3/-5.5/-5.6* (Graham *et al.*, 1989; Dollé *et al.*, 1989). Members of each cluster are designated by an additional number, which denotes the order of discovery rather than the chromosomal arrangement of these genes within each cluster (hence *Hox-1.1* to *Hox-1.7* in the hox 1 cluster, and so on; Rubin *et al.*, 1987). The human genome contains several cognate clusters of homeobox genes on homologous chromosomes, namely, *HOX* 1 on chromosome 11, *HOX* 2 on chromosome 17 and *HOX* 3 on chromosome 12 (corresponding to the mouse hox 1, 2 and 3 clusters, respectively; Hauser *et al.*, 1985; Rabin *et al.*, 1986). Conservation of such large chromosomal domains may indicate an important role for the homeobox genes they contain, though this should not be assumed *a priori*. The chromosomal organisation of the *Hox* genes is strikingly similar within the four hox clusters (albeit with some gaps), and also resembles that in the *Drosophila* BX-C/ANT-C clusters. Corresponding members of different hox clusters (e.g. *Hox-1.4/-2.6/-5.1* or *Hox-1.7/-2.5/-3.2*; fig. 5.29) are not only related in sequence but also show similar expression domains (Gaunt *et al.*, 1989; Bogorad et al., 1989). More surprisingly, several specific *Hox* genes are clearly related in sequence to a corresponding member of the BX-C/ANT-C group,

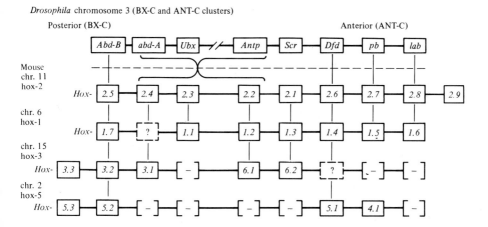

Fig. 5.29 Organisation of homeobox gene clusters in mouse as compared to *Drosophila* (after Graham *et al.*, 1989). Approximate gene maps are shown for the four main mouse clusters and the *Drosophila* BX–C/ANT- C clusters. Vertical alignments are based on sequence relationships between homeoboxes (see Graham *et al.*, 1989. *Hox-5.5* and *Hox-5.6* lie to the left of *Hox-5.3*.

suggesting a common ancestral cluster of similar overall organisation. For instance, the protein specified by the mouse *Hox-2.6* gene (together with its human, *Xenopus* and zebrafish counterparts) shows significant sequence similarities to the protein encoded by the *Drosophila Dfd* gene, even extending to regions outside the homeodomain (Graham *et al.*, 1988). A mouse homologue of the *Drosophila caudal* gene has also been identified; this *Cdx-1* gene is expressed in the epithelial cells lining the intestine (i.e. restricted to endoderm), starting at about 14 days of foetal development and continuing into adulthood (Duprey *et al.*, 1988). Sequence relationships among the homeodomains suggest the alignments shown in fig. 5.29 between the four major hox clusters in mouse and *Drosophila* BX-C/ANT-C clusters; in all cases the gene arrangement shown corresponds to the chromosomal order, and in at least two cases (BX-C/ANT-C and hox 2) this is mirrored in the A/P expression domains for individual genes within the cluster.

The conserved chromosomal arrangement of these genes suggests that duplications of the prototype cluster have occurred in mouse and other vertebrates; in *Drosophila* the cluster has become split, but as mentioned earlier both BX-C-like and ANT-C-like genes are found in a single cluster (the HOM-C) in the beetle *Tribolium* (Beeman *et al.*, 1989). Moreover, this arrangement appears to be mirrored in the domains of expression for each gene along the A/P axis. In the case of hox 2, genes at the left-hand end of the cluster (*Hox-2.5* or *-2.4*) are expressed more posteriorly than genes at the right-hand end (*Hox-2.8* or *-2.7*; see below and Graham *et al.*, 1989). Indeed, the rostral boundary of expression for each gene in the hox 2 cluster (except *Hox-2.9*; see later) correlates with its chromosomal position; the same may be true for some members of the other hox clusters (Graham *et al.*, 1989).

Two further mouse homeoboxes are strikingly similar to the *en* prototype in *Drosophila*. Moreover, the region of homology between these genes and *en* extends considerably beyond the homeodomain (Joyner *et al.*, 1985). These two mouse genes are unlinked (unlike the *Drosophila en/inv* pair), with *En-1* located on chromosome 1 and *En-2* on chromosome 5 (Joyner & Martin, 1987).

All of the above genes are expressed during embryogenesis, usually starting at some point after the 7.5-day (presomite gastrula) stage. Most show maximal expression at around the 12.5-day stage illustrated in fig. 5.30, though some are expressed persistently in certain adult tissues (e.g. *Hox-1.4* in testis; Wolgemuth *et al.*, 1986, 1987; Rubin *et al.*, 1986). A variety of homeobox gene transcripts are expressed sequentially when embryonal carcinoma cells (cancerous but undifferentiated stem cells) are induced to differentiate into endoderm in culture (Colberg-Poley *et al.*, 1985a,b; Breier *et al.*, 1986). This appears to be a feature of induction by retinoic acid (RA) specifically rather than of differentiation *per se* (Deschamps *et al.*, 1987; Mavilio *et al.*, 1988). However, RA is a known vertebrate morphogen (see Slack, 1987) which acts through a receptor protein related to the steroid receptors (Giguère *et al.*, 1987; Petkovitch *et al.*, 1987). It will be interesting to determine whether the pattern alterations caused by RA during limb regeneration are mediated via induction of homeobox genes (see below). The induction process may sometimes involve post-transcriptional RNA stabilisation (as for *Hox-1.1*; Dony & Gruss, 1988), and/or an increase in the rate of transcription (as for *Hox-1.3*; Murphy *et al.*, 1988).

Figure 5.30 presents a summary (based on Holland & Hogan, 1988a and Graham *et al.*, 1989) of recent data on the spatial distribution of mouse homeobox gene transcripts. For simplicity, discussion is confined to the segmented axial structures of the somite chain and CNS at the 12.5-day stage of development. An overview of expression patterns in other tissues of segmental origin is given in Holland & Hogan (1988a), along with a summary of data on earlier stages of development (7.5–7.75 and 8–8.5 days *post coitum*). Note that transcript expression often tends to fade out posteriorly, such that the anterior limits of expression are usually better defined than the posterior limits.

(i) *En-1* is expressed along virtually the entire length of the neural tube, whereas *En-2* transcripts are confined to a narrow band in the metencephalon of the brain (Joyner & Martin, 1987; Davis *et al.*, 1988; Davidson *et al.*, 1988). Monoclonal antibodies against conserved parts of the *Drosophila en+/inv+* proteins detect related *en* proteins in animals ranging from annelids and arthropods to vertebrates (Patel *et al.*, 1989). Whereas arthropods show segmentally reiterated patterns of *en* expression, vertebrates show an anterior expression domain in the CNS, corresponding to the *En-2* transcript zone in mouse (e.g. in *Xenopus*: Brivanlou & Harland, 1989).

(ii) The anterior boundary of *Hox-1.5* expression in the CNS lies just anterior to the otic vesicle (Fainsod *et al.*, 1987; Gaunt, 1987); this domain may perhaps abut on that of *En-2*. *Hox-1.5* is also expressed throughout much of the mesodermal chain of prevertebrae.

(iii) A number of homeobox genes show similar anterior boundaries of expression within the myelencephalon, posterior to the otic vesicle in the CNS; these include *Hox-1.2, -1.3, -1.4, -2.1, -2.6* and *-6.1* (Dony & Gruss, 1987; Krumlauf *et al.*, 1987; Toth *et al.*, 1987; Holland & Hogan, 1988b; Sharpe *et al.*, 1987; Graham *et al.*, 1988, 1989). However, these boundaries are not all identical; the rostral (anterior) limits of expression for several hox 2 genes fall within this general region (Graham *et al.*, 1989). Notably, the most anterior boundaries are those given by *Hox-2.8* and *Hox-2.7*, followed successively by those for *Hox-2.6, -2.1, -2.2, -2.3* and *-2.4*. The order of their rostral expression limits within the CNS is precisely the same as their order on the chromosome (see fig. 5.29). Corresponding members of other clusters often show similar anterior limits of expression, e.g. in the case of the *Dfd*-related *Hox-2.6, -1.4* and *-5.1* genes (Graham *et al.*, 1989). The anterior boundaries of expression for hox 2 genes within the myelencephalon correspond to no obvious morphological features at the 12.5-day stage of development. Earlier, however, this hindbrain region is derived from a series of eight rhombomeres, and the various hox 2 expression limits may be related to these units. In any case, the evidence suggests that the spatial organisation of the hindbrain is complex, involving the differential expression of hox 2 and other genes. Within the 9.5-day mouse hindbrain, the anterior limit of *Hox-2.1* expression is at the junction between rhombomere 8 (r8) and the spinal cord; other anterior expression limits are at the r6/r7 junction for *Hox-2.6*, the r4/r5 junction for *Hox-2.7*, and the r2/r3 junction for *Hox-2.8* (Wilkinson *et al.*, 1989b; note the 'double-segment' spacing). The *Hox-2.9* gene is expressed at

A In central nervous system. *Brain structures: Pr,* prosencephalon; Ms. mesencephalon; Mt. metencephalon; My, myelencephalon; *V-X,* cranial nerves.

Positions along the spinal cord are given by the corresponding dorsal root ganglia (DRG). C1-8, cervical DRG; T1-13, thoracic DRG; L1-6, lumbar DRG.

*Note that many anterior boundaries within this region of the hindbrain are non-identical, but do not correlate with obvious morphological features at the 12.5 day stage of development.

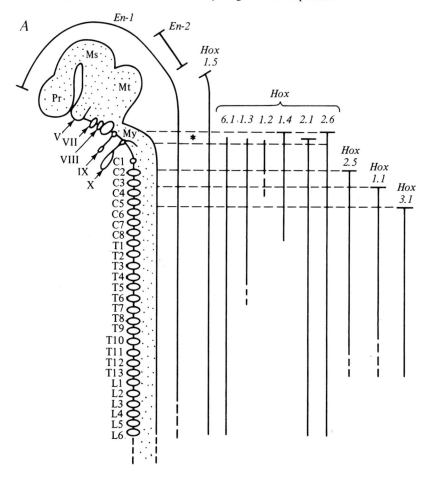

Fig. 5.30 Homeobox gene expression in 12.5 day mouse embryo CNS and axial mesoderm (after Holland & Hogan, 1988a, modified according to Graham *et al.,* 1989).

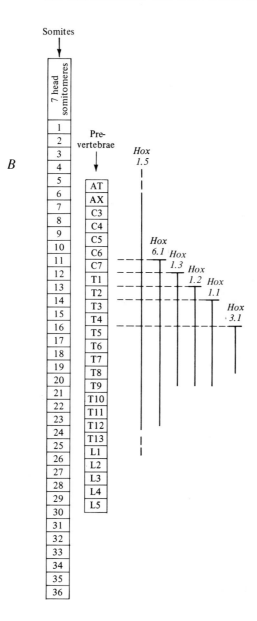

B In axial mesoderm. AT, atlas; AX, axis; C3–7, cervical vertebrae;
T1–13, thoracic vertebrae; L1–6, lumbar vertebrae. The preceding
sequence of somites (including 7 head somitomeres) is also shown;
note that the anterior and posterior halves of each somite contribute
to different vertebrae (see text).

this stage only in a single rhombomere, r4, while the *Krox-20* finger-protein gene (see later) is expressed in the adjacent r3 and r5 rhombomeres (Murphy *et al.*, 1989). Overall, r3, r4 and r5 at least express unique combinations of genes. In the mesoderm, *Hox-6.1* expression begins between the sixth and seventh cervical prevertebrae, whereas *Hox-1.3* expression begins between the seventh cervical and first thoracic, and *Hox-1.2* expression between the first and second thoracic prevertebrae. There may be different posterior limits of expression for genes showing similar anterior limits (fig. 5.30).

(iv) The *Hox-2.5* gene shows an anterior boundary of CNS expression at about the first dorsal root ganglion (DRG) in the spinal cord; as implied by the chromosomal organisation of the hox 2 locus, this *Hox-2.5* limit lies posterior to that of *Hox-2.4* expression (Graham *et al.*, 1989). *Hox-1.1* expression in the CNS begins at about the third DRG (Fienberg *et al.*, 1987), but in the mesoderm, *Hox-1.1* expression begins between the second and third thoracic prevertebrae.

(v) The most posteriorly expressed homeobox gene yet described is *Hox 3.1*, whose expression in the CNS starts at the fifth DRG and in the mesoderm between the fourth and sixth thoracic prevertebrae (Utset *et al.*, 1987; Le Mouellic *et al.*, 1988; Breier *et al*, 1988).

It must be stressed that the patterns of expression shown in fig. 5.30 are not static, but rather change dynamically as development proceeds. Indeed, the *En-2* domain is remarkable not only for its narrowness but also for its stability! Some genes are expressed early on (soon after gastrulation starts) in broad antero-posterior (A/P) zones, e.g. *Hox-1.5* in the posterior presomitic mesoderm and ectoderm (Gaunt, 1987), and *Hox-3.1* even more posteriorly in the extra-embryonic allantois (a mesodermal tissue that eventually contributes to the extra-embryonic blood vessels). These early domains of expression may perhaps provide A/P positional information to which later-expressed homeobox genes respond. Yet further clustered mouse genes contain homeoboxes related to that in the *Drosophila* muscle-specific gene *Msh*. One of these mouse genes, designated *Hox-7.1*, is expressed in a variety of embryonic tissues, including the neural crest, limb buds, and later the interdigital mesenchyme of all four limbs (i.e. regions of cells programmed to die in order to separate the digits; Hill *et al.*, 1989).

From this summary, it is clear that there are several similarities between the patterns of expression of mouse *Hox/En* genes in 12.5-day embryos and those of the *Drosophila* homoeotic genes in germ-band embryos. Thus the mouse genes are expressed in broad and overlapping domains, but often with distinctive anterior (or posterior) limits. Although there is a general correspondance between the domains of expression in the CNS and in the prevertebral chain for any given gene, there are also some differences (fig. 5.30). At least the hox 2 cluster appears to be organised in the antero-posterior order of expression of its component genes (Graham *et al.*, 1989; see above), as is the case for the *Drosophila* BX-C/ANT-C clusters. It remains to be seen whether this generalisation holds true for the other homeobox gene clusters identified in mouse and in other vertebrates (including *Xenopus*, human, chick and zebrafish), or indeed in other insects. The genetic organisation of the single HOM-C gene cluster in *Tribolium* suggests that this may indeed be the case (see Akam, 1989a). Taken together, these data imply that both insect and vertebrate clusters of homeobox genes may have evolved from a common ancestral gene cluster, possibly concerned with the provision of positional information (Akam, 1989a). If this is true, then one might expect features of the same clustered organisation to be apparent for homeobox genes in present-day non-segmented animals. All of these features suggest that the mouse homeobox genes may play a role in regionalising the embryo, though they do not assign specific segmental identities (except possibly in the hindbrain?). But even in *Drosophila*, many parasegments are defined by combinations of homoeotic genes.

We should remember that the *Drosophila* ANT-C contains several homeobox genes with diverse patterning functions (e.g. *Antp, ftz, zen*); there is no *a priori* reason why the large array of mouse homeobox genes (see above) should not encompass equally diverse functions. Another point which should be stressed is that these localised expression patterns do not in themselves prove whether homeobox genes control regionalisation during early mouse development. Proof that this is indeed the case will require genetic evidence, emerging either from homoeotic-type mutations (whether natural or induced by directed mutagenesis *in vivo*), or else from experimental manipulations such as the ectopic expression of homeobox structural genes under the control of foreign promoters in transgenic mouse embryos. The latter approach will probably give more definitive information in the long run, although the former provides some interesting clues. There are few mouse mutants whose phenotypes can be described as homoeotic (which would involve, say, the replacement of lumbar by cervical vertebrae, or vice versa). Both *pudgy (pu/pu)* and *rachiterata (rh/rh)* homozygotes come close to this, showing axial skeletal abnormalities such as fusion or loss of vertebrae in particular regions (see Hogan *et al.*, 1985). Other developmental mutants in mouse include: (i)

Hypodactyly (*Hd*), which maps near *Hox-1.5* (Mock *et al.*, 1987); (ii) *tail-short* (*ts*), which lies close to the hox 2 cluster (Muncke *et al.*, 1986); and (iii) *Dominant hemimelia* (*Dh*), which maps near the *En-1* gene (Hill *et al.*, 1987; Joyner & Martin, 1987). However, in no case has it been demonstrated that the mutant phenotype results from a mutation in the homeobox gene itself (rather than a closely linked gene). A transgenic mouse carrying a *Hox-1.4* gene-construct in all cells expresses this gene with appropriate tissue specificity, but does so at abormally high levels in the embryonic gut, resulting in an aberrant pattern of colon development (megacolon phenotype) which is heritable (Wolgemuth *et al.*, 1989). Similarly, ectopic expression of the *Hox-1.1* gene (under the control of a chick β-actin promoter) in transgenic mice results in severe craniofacial defects which are lethal shortly after birth (Balling *et al.*, 1989). This pattern of defects is similar to that found in homozygous *far* mutants, and to that induced by retinoic acid treatment during gestation; all three may be linked to abnormalities in the cranial neural crest cells. These transgenic results strongly suggest a patterning role for mouse *Hox* genes.

Turning briefly to *Xenopus*, early studies discovered homeobox genes expressed during postgastrular embryonic development (Carrasco *et al.*, 1984), or during oogenesis (Muller *et al.*, 1984). The *XlHbox 6* gene is expressed in posterior regions of the neural tissue induced during gastrulation by the underlying mesoderm; this gene is transcribed when ectoderm is placed in contact with mesoderm in culture, but not when either tissue is cultured alone (Sharpe *et al.*, 1987). Notably, *XlHbox 6* expression is more readily inducible in dorsal than ventral ectoderm, suggesting some predisposition in the former for neural development. The *Xhox-36* gene is transcribed in posterior regions of both mesoderm and ectoderm, suggesting a regional rather than tissue specificity (Condie & Harland, 1987). The *Xhox-1* region contains two homeobox genes, one transcribed maternally during oogenesis and again zygotically from midgastrulation to the swimming tadpole stage (Harvey *et al.*, 1986). This *Xhox-1A* gene is expressed mainly in somitic mesoderm during postgastrular development; injection of excess synthetic *Xhox-1A* mRNA into fertilised eggs causes a disruption of somitogenesis (fusion and chaotic arrangement of somites), implying a patterning role for this gene during somite formation (Harvey & Melton, 1988).

The zygotically expressed *XlHbox 1* gene has two promoters, generating long- and short-form proteins (the former with an extra 82 N-terminal amino acids: Cho *et al.*, 1988). The *XlHbox 1* products are expressed within a limited A/P domain in both neural and mesodermal tissues (de Robertis *et al.*, 1989). Injection into zygotes of an antibody against the long-form protein transforms the developing anterior spinal cord (which normally expresses the long-form protein) into hindbrain structures (Wright, C.V.E. *et al.*, 1989). Similar, though not identical, abnormalities are caused by injection of short-form mRNA, perhaps implying antagonistic functions for the short- and long-form proteins. The *XlHbox 1* gene is closely related to human, mouse (*Hox-6.1*) and newt (*NvHbox 1*) homologues; at least the *Xenopus* and newt versions play some role in limb morphogenesis. Specifically, *XlHbox 1* protein forms an A/P gradient in forelimb but not hindlimb mesoderm (an early molecular difference between arm and leg), and is also expressed in the inner ectodermal layer of both fore- and hindlimbs (Oliver *et al.*, 1988a, b). Earlier in development (before the appearance of limb buds), *XlHbox 1* expression is detectable in the lateral plate mesoderm of the forelimb but not hindlimb field, i.e. that region shown by experimental manipulation to be capable of forming an arm. A complementary gradient of *Hox-5.2* protein (encoded by the *Xenopus* homologue of *Hox-5.2*) exists in both fore- and hindlimb buds; anti-*Hox-5.2* staining is most intense in the postero-distal mesoderm, as opposed to the antero-proximal maximum for *XlHbox 1* protein (Oliver *et al.*, 1989). The newt *NvHbox 1* homologue is expressed in both limb and tail tissues (using exons 1 + 2 and 2 only, respectively) in adult as well as embryonic stages. Since its *XlHbox 1* counterpart is expressed in embryos but not adults, this may relate to the fact that limb regeneration is possible in adult newts but not in adult *Xenopus* (Savard *et al.*, 1988). Retinoic acid (RA) is a limb morphogen present in a posterior-to-anterior gradient in the chick limb bud. In newts, only RA treatment can cause proximalisation of the blastema formed after distal amputation of a limb, such that an RA-treated regenerate forms more proximal limb structures than those lost, which is never otherwise the case. The links between homeobox genes and the RA morphogen in limb development and regeneration remain unclear (see Brockes, 1989); *Hox-1.6* expression may be directly induced by RA in embryonal carcinoma cultures, but there is little apparent change in *NvHbox 1* expression following RA treatment of regenerating newt limbs.

Two other *Xenopus* genes with variant homeoboxes are: (i) *XlHbox 8* – whose expression is restricted to a narrow band of endoderm (persistently in the pancreas, but also in the embryonic duodenum; Wright *et al.*, 1988); and (ii) *Xhox-3*, whose homeobox is related to that of the *Drosophila eve* gene. Expression of *Xhox-3* is initially graded along the A/P axis of the gastrula/neurula mesoderm (highest at the posterior end), but later becomes bimodal, with transcripts concentrated in the tadpole tail bud and in the anterior nervous system (see Ruiz i Atalba & Melton, 1989a, b). Injection of synthetic *Xhox-3* mRNA into early embryos at successively more anterior sites causes a graded loss of anterior structures, implying a role for *Xhox-3* in axial patterning. Expression of yet another homeobox gene, *Mix-1*, is rapidly induced by MIF (see § 2.8), but mainly in future endoderm cells (Rosa, 1989).

Overall, the ease with which early *Xenopus* embryos can be manipulated has helped to provide experimental support for the view that homeobox genes are involved in key patterning events during early development, such as neural induction, limb morphogenesis, axial pattern and somitogenesis. In other respects, the picture in *Xenopus* is similar to that in mouse, although the confused nomenclature of *Xenopus* homeobox genes makes comparisons difficult. In human embryos, stage- and tissue-specific patterns of expression have been described for several homeobox- containing genes (Mavilio *et al.*, 1986; Simeone *et al.*, 1986). Apart from the large HOX 1 and HOX 2 clusters mentioned earlier (Rabin *et al.*, 1986), there are also three homeoboxes linked close together in the HOX 3 cluster on chromosome 12. All three homeoboxes apparently belong to the same transcription unit, whose primary transcript is spliced alternatively to give several mature mRNAs. These messengers contain a common 5' non-coding exon followed by different protein-coding regions that include one out of the three homeodomains (Simeone *et al.*, 1988).

An extension of the homeodomain homology is found in the POU domain, so called because of its occurrence in the mammalian transcription factors Pit-1 (also known as GHF-1) and Oct-1/Oct-2 (also know as NF-A1 and NF-A2), as well as in the *C. elegans unc-86* lineage gene (see also § § 2.2.2 and 4.3; reviewed in Robertson, 1988a, Herr *et al.*, 1988 and Levine & Hoey, 1988). The POU domain is in fact a bipartite DNA-binding structure; the 60 amino-acid homeodomain (towards the C-terminus) and 75 amino-acid POU-specific region (towards the N-terminus) are joined together by a flexible linker (Sturm & Herr, 1988). However, the homeodomain may be largely responsible for DNA binding (Garcia-Blanco *et al.*, 1989). Whereas *unc-86+* function is required for the appearance of certain types of neurones (in its absence these cells repeat the division of their mothers: see §4.3), both Oct-2 and Pit-1 are tissue-specific transcription factors (in pituitary and lymphoid cells, respectively), while Oct-1 is ubiquitous. It is likely that these four proteins act in fundamentally similar ways, and that the apparent contrasts mainly reflect differences in experimental approach between the nematode and vertebrate systems. Expression of a cDNA encoding Oct-2 in Hela cells is sufficient to activate several B-cell-specific promoters (never normally active in these cells; Muller *et al.*, 1988b). However, addition of excess Oct-1 protein can activate transcription from such promoters *in vitro* (LeBowitz *et al.*, 1988), demonstrating that both factors interact with the same target octamer sequences, though not necessarily with the same efficiency. This latter feature may be influenced by sequences flanking the octamer, by interactions with other proteins, or by non-DNA-binding parts of the factor itself (see Robertson, 1988a, Levine & Hoey, 1988, Herr *et al.*, 1988, and references therein). Note that a number of enhancer-binding proteins, as well as Oct-2, are needed to activate the Ig genes fully in B cells (see §2.2.2). In the case of the rat pituitary gland, the Pit-1 transcription factor is apparently utilised in two related but distinct cell types (Nelson *et al.*, 1988), namely the lactotrophs expressing the prolactin (P) gene (Ingraham *et al.*, 1988) and the somatotrophs expressing the growth hormone (GH) gene (Bodner *et al.*, 1988). In the latter case, this transcription factor (variously called Pit-1 or GHF-1) appears to be the primary determinant of cell identity. Multiple *cis*-acting binding sites for Pit-1/GHF-1 are found in the 5′-flanking regions of both GH and P genes. Expression of Pit-1 cDNA in Hela cells leads to the activation of fusion genes containing regulatory sequences from either the GH or P genes (Ingraham *et al.*, 1988), suggesting that Pit-1 expression may suffice to confer a pituitary phenotype. Since pituitary cells can only express either the GH *or* the PH gene *in vivo* (never both together in the same cell), further cell-specific factors must also be involved. Recent evidence is somewhat contradictory as to whether Pit-1/GHF-1 activates both GH and P gene promoters or only the former (compare Castrillo *et al.*, 1989, with Mangalam *et al.*, 1989). In any case it is clear that multiple control elements (both a distal enhancer and proximal promoter) are required to give a normal lactotroph-specific pattern of P gene expression (Crenshaw III *et al.*, 1989). If pituitary-derived GH-expressing cells are fused with non-expressing cells, GH expression is extinguished, apparently through the repression of GHF-1/Pit-1 production (McCormick *et al.*, 1988). Several further octamer-binding proteins are present during mouse embryogenesis, one of which (Oct-4) is expressed in primordial germ cells of both sexes, and later in oocytes but not in testis or sperm (Scholer *et al.*, 1989). A large family of POU-domain genes is expressed specifically during mammalian brain development (He *et al.*, 1989).

5.6.2 Other homologies between Drosophila *and vertebrate genes*

A number of developmentally important genes in *Drosophila* share sequence homologies with vertebrate genes of diverse function. Several examples have been mentioned earlier in this chapter, including: *dorsal*, related to the *rel* group of vertebrate proto-oncogenes (Steward, 1987); *snake* and *easter*, which encode serine proteases (deLotto & Spierer, 1986; Chasan & Anderson, 1989); *Toll* and *sevenless*, whose predicted protein products have features typical of integral membrane proteins (Hashimoto *et al.*, 1988; Basler & Hafen, 1988); *Notch* and *Delta*, whose products are also membrane proteins, but with tandem EGF-like repeats in the extracellular domain (Wharton *et al.*, 1985b; Vassin *et al.*, 1987); and *decapentaplegic*, one of whose products shows homology to vertebrate TGF-β (Padgett *et al.*, 1987). The fact that these examples have proved so illuminating underlines a curious irony in *Drosophila;* despite the enormous sophistication of genetics and molecular biology in this organism, we often have no clear picture as to how the protein products from important genes interact in a given developmental pathway. In vertebrates, many cascades of protein interactions (e.g. in blood-clotting) have been elucidated in great detail without reference to the genes involved. However, the benefits are two-way between *Drosophila* and the vertebrates; the discovery of conserved sequences such as the homeobox (see above) has facilitated the cloning of genes which may be important in vertebrate development despite the lack of sophisticated genetics. The ability to misregulate the expression of such genes (via fusion constructs introduced into transgenic mice) can circumvent the need for mutants involving them, and should soon allow a clearer answer to the question of whether these genes control key aspects of early vertebrate development.

A few further examples of sequence conservation deserve more detailed comment. The *Drosophila wingless* (*wg*) segment polarity gene is related to the mouse proto-oncogene *int-1* (Cabrera *et al.*, 1987a; Rijsewijk *et al.*, 1987; see also Bender & Peifer, 1987). This mouse gene apparently becomes active (inappropriately) in mammary tissue as a result of viral integration by the mouse mammary tumour virus (MMTV) genome, leading to mammary cancers. The *int-1* gene is normally expressed in spatially restricted regions of the embryonic neural tube (Wilkinson *et al.*, 1987) and also in postmeiotic male germ cells in the adult (Shackleford & Varmus, 1987). Note that the neural tube expression of *int-1* is largely continuous antero-posteriorly and shows no sign of segmental repetition. This is in sharp contrast to its *Drosophila wg* counterpart, which is expressed only in the posteriormost cells of every parasegment (Baker, 1987, 1988b; see §4.2.2 above). There is thus no obvious correlation between the expression patterns of these two homologues, although both genes apparently encode similar secreted or cell-surface proteins. Injection of mouse *int-1* mRNA into *Xenopus* eggs results in widespread overexpression of the corresponding *int-1* protein during early embryonic development. Strikingly, this often leads to a bifurcation of the anterior neural plate and duplication of the underlying axial mesoderm (McMahon & Moon, 1989). This effect is a direct result of *int-1* action, since it is abolished by substitution of a single conserved cysteine in the *int-1* protein.

Pax-1 is a mouse gene isolated on the basis of its homology to the *Drosophila paired* box (using a *gsb* probe; Deutsch *et al.*, 1988). A 3.1 kb *Pax-1* transcript is expressed in a segmentally reiterated pattern in the axial mesoderm of mouse embryos; *Pax-1* probes hybridise first to the perichordal condensations around the notochord and then later to the intervertebral discs (fig. 5.31). Each vertebra is derived from parts of two adjacent somites, hence each intervertebral disc originates from the centre of a somite; the reiterated zones of *Pax-1* expression are thus in register throughout with the original somite sequence (Deutsch *et al.*, 1988). Homozygotes for a point mutation in the *paired* box of *Pax-1* display the *undulated* mutant phenotype, in which skeletal abnormalities are displayed along the entire vertebral column (Balling *et al.*, 1988). Several further genes with *paired*-box homology (e.g. *Pax-2*) are present in the mouse genome, and these may be expressed in different segmentally reiterated patterns. The evidence cited above strongly implies, but does not as yet prove, a key role for these genes in the vertebrate segmentation process. A novel homeobox gene related to *paired* is also expressed (along with several previously described homeobox genes) in mammalian haematopoietic cells (Kongsuwan *et al.*, 1988).

Finally, mention must be made of the so-called 'zinc-finger' proteins, whose prototype is the *Xenopus* transcription factor TFIIIA (see Enver, 1985). These proteins contain a repeated motif in which two cysteines are separated by a 'loop' of 12 amino acids from two histidines; these four amino acids between them bind coordinately to a zinc atom (see Evans & Hollenberg, 1988, for details). Such zinc-finger structures bind to DNA, and all proteins in this class appear to be nuclear. The zinc-finger motif is repeated several times in the predicted protein sequences from both *Kruppel* and *hunchback* gap genes (Rosenberg *et al.*, 1986; Tautz *et al.*, 1987). Other *Drosophila* genes which contain *Kr*-related zinc-finger sequences include the zygotic D/V gene

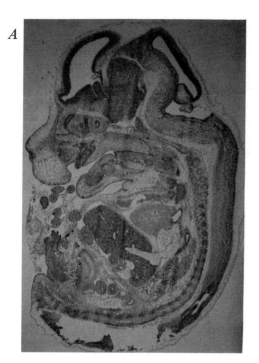

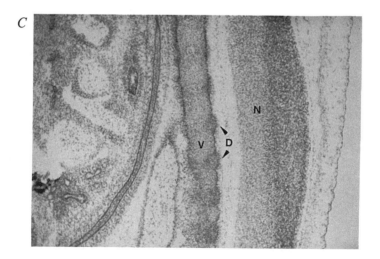

Fig. 5.31 *In situ* hybridisation of *Pax-1* to 14-day mouse embryos.
Photographs kindly provided by Dr G. Dressler, and reprinted by
permission from the copyright holders, Cell Press. From U.
Deutsch, G. R. Dressler & P. Gruss (1988), *Cell*, **53**, 617–25.
Parts *A* and *B*, mid-sagittal section of whole embryo;
magnification ×10.
Parts *C* and *D*, sagittal section through developing spinal column,

B

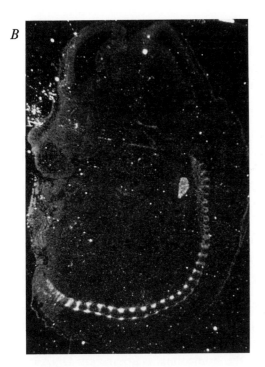

D

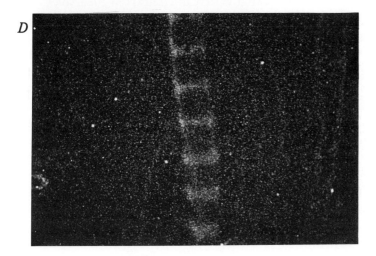

showing the neural tube (N), centrum of vertebrae (V) and
invertebral discs (D); magnification ×30.

Parts *A* and *C*, stained sections; Parts *B* and *D*, dark field
images of *A* and *C* respectively. Positive hybridisation to *Pax-1*
transcripts shows up as bright speckling against dark background.
Note strong expression of the *Pax-1* gene in the intervertebral
discs.

snail (Boulay *et al.,* 1987). In these and other cases, the predicted protein product shows invariant positioning of the crucial cys and his residues as well as a conserved stretch of seven amino acids (the HC link) in the loop region (Schuh *et al.,* 1986). Similar sequences are found in vertebrate and other eucaryotic genomes (Chowdhury *et al.,* 1987). Indeed, if the definition of sequence conservation is relaxed so as to include proteins with non-repetitive clusters of cysteine residues (which probably also bind zinc atoms coordinately), then a variety of known eucaryotic transcription factors can be accommodated within the zinc-finger class (including yeast GAL4 and the vertebrate steroid receptor superfamily; see Evans & Hollenberg, 1988). As mentioned previously (§3.3.1), the N-terminal zinc-finger in vertebrate steroid receptors is a primary determinant of their sequence specificity in binding to DNA sites flanking their target genes (see Green *et al.,* 1988; Severne *et al.,* 1988). Such variant zinc-finger motifs are present in the *knirps* gap-gene product (Nauber *et al.,* 1988), and also in the predicted protein from the *knirps-related* gene (Oro *et al.,* 1988). Yet another variant finger motif is found in the erythroid-specific transcription factor GF-1/Eryfl (Tsai *et al.,* 1989; Evans & Felsenfeld, 1989; see §3.2.3). Thus the zinc-finger, like the homeobox, is an ancient DNA-binding device which is utilised in a variety of contexts in eucaryotes.

The vertebrate zinc-finger class includes the *ZFX* and *ZFY* genes on the X and Y chromosomes (see Burgoyne, 1989; Hodgkin, 1988b), and the mouse *Krox-20* gene whose transcription is rapidly induced during the G_0/G_1 transition following mitotic stimulation of quiescent cells in culture (Chavrier *et al.,* 1988). In mouse, the *mKr2* gene is expressed during embryogenesis in all major structures of the developing nervous system, as well as in certain central and peripheral neurones in adults (Chowdhury *et al.,* 1988). The *Krox-20* gene is expressed more selectively in the developing mouse nervous system, specifically in two alternate rhombomeres (3 and 5) of the early hindbrain, and also later in neural crest cells (Wilkinson *et al.,* 1989a). Yet another zinc-finger protein plays a key role in mating-type switching in yeast, a process which also involves the homeobox-related MAT a1 and MAT α2 products. Briefly, switches of mating type depend upon the excision and reinsertion of a DNA 'cassette', requiring *HO* endonuclease activity (for a brief review, see Hicks, 1987). The *HO* gene is normally repressed by the *SIN 3* gene product, but this can be relieved by a positively acting transcription factor encoded by the *SWI 5* gene. This *SWI 5* factor is in fact a zinc-finger protein (Stillman *et al.,* 1988). Here we see a variant of what may be a widespread switching device in eucaryotes, whereby zinc-finger proteins regulate the transcription of homeobox genes. In the yeast mating-type example, this is complicated by interpolation of the *HO* function and cassette mechanism. In *Drosophila*, a group of zinc-finger-containing gap-gene products (*hb, Kr, kni*) regulate a series of homeobox-containing pair-rule (*ftz, eve, prd*) and homoeotic (ANT-C and BX-C) genes. Needless to say, the situation is not entirely simple, since the primary influence of gap-gene products on pair-rule patterning is exerted via the *runt* and *hairy* genes which do not contain homeoboxes (Ingham, 1988). Moreover, at least some zygotic gap genes (e.g. *hunchback*) are themselves regulated by the maternal homeodomain-containing *bcd*[+] protein. Nevertheless, these examples serve to illustrate how variations on a simple regulatory module can be built up into complex regulatory hierarchies. Over the next few years, the unravelling of these hierarchies, not only in *Drosophila* and *C. elegans* but also in vertebrates, will doubtless ensure that any future editions of this book are even thicker than the present one! There is also the tantalising hope that common themes will emerge even more strongly, suggesting fundamental genetic constraints on how animal development is regulated. However, as we have seen in the case of homeobox, such common themes may either reveal a basic unity between apparently unrelated processes or else conceal a bewildering diversity of functions beneath a simple DNA motif. Surveying the immense variety of present-day animals, we must be prepared for both of these possibilities to be true.

Notes added in proof

Chapter 1

§1.4

3′ cleavage of pre-mRNAs requires at least four separable proteins, but none of these contains as essential RNA component; in addition, the non-specific poly(A) polymerase is needed to cleave several types of pre-mRNA (Takagaki *et al.*, 1989).

Chapter 2

§2.2.2

Schatz *et al.*, (1989) have isolated the the RAG-1 gene, probably encoding the V(D)J recombinase.

§2.4.2b

Blanco *et al.*, (1989) describe two forms of TFIIIA protein, one of which (39 kd) declines during oogenesis while the other (42 kd) increases; both forms bind with comparable affinity to both types of 5S gene. The 39 kd form activates transcription from both gene-types, whereas the 42 kd form activates somatic-type but represses oocyte-type 5S-gene transcription. This mechanism too can explain the switch from oocyte + somatic to somatic-only 5S gene expression.

§2.7.2

Davidson (1990) gives a recent review of cell-fate specification in various types of embryo.

§ 2.8

During mesoderm formation in *Xenopus*, the MyoD homologue is expressed only in the developing muscle (myotome), whereas the *Xtwi* gene (homologue of the *Drosphila twist* gene; § 5.3.1) is expressed in a complementary pattern in notochord and lateral plate (Hopwood *et al.*, 1989a, b; Woodland, 1989).

Chapter 3

§ 3.1

Robertson (1990) reviews some of the complications involving MyoD and myogenesis.

§ 3.2.2

Transgenic mice carrying both γ- and β-globin genes (human) linked to the LAR display a normal foetal (γ) to adult (β) switch, suggesting that the LAR activates one gene or the other (but not both) in a mutually exclusive fashion (Enger *et al.*, 1990).

§ 3.2.3

The chicken β-globin stage-specific selector element interacts with the adult erythroid-specific protein NF-E4 (Gallarda *et al.*, 1989).

Chapter 4

§ 4.3.2

The EMS, C and D somatic founder-cells are generated by three stem-cell-like divisions from the germline precursors. P_1, P_2 and P_3, respectively. The stem-cellike fates of these latter three cells are specified by the *cib-1*[+] function. Maternal-effect lethal mutations of the *cib-1* gene cause these cells to skip one division and to adopt the fate that normally characterises the somatic daughter (i.e. EMS but not P_2 formed from a mutant P_1 cell, etc.; Schnabel & Schnabel, 1990).

§ § 4.4.3 and 4.6

Four heterochronic genes regulate the temporal aspects of the dauer larva formation. The stage at which this process is initiated depends upon a decrease in *lin-14*[+] activity resulting from negative regulation by the *lin-4*[+] function. Dauer morphogenesis further requires that *lin-28*[+] inhibits *lin-29*[+] activity in the early larva (Liu & Ambros, 1989).

§4.7

mec-3 fusion-gene products are expressed in three cell types; the touch recep-tors, the PVD neurones which respond to harsh touch, and the FLP cells which ultrastructurally resemble touch receptors. *mec-3*[+] activity is also required for cell identity/function in at least the first two cases. In a *mec-17* mutant, *mec-3*[+] expression is not maintained in the touch receptors, but is unaffected in the PVD and FLP cells (Way & Chalfie, 1989).

Chapter 5

§5.2.3

The *Sxl*[+] protein binds to a splice site in the *tra*[+] primary transcript (Inoue *et al.*, 1990).

§5.4.2a

Bic-D encodes a cytoskeletal protein similar to C-terminal parts of the myosin heavy chain; this may help to transport or localise the *nos*[+] signal (Wharton & Struhl, 1989; Suter *et al.*, 1989).

§5.4.2b/c

Carroll (1990) reviews the regulation of pair-rule stripes by maternal and gap-gene products.

§§5.5.2e and 5.5.4

Down regulation of homoeotic gene expression by other homoeotic gene-products may not be functionally significant (see Gonzales-Reyes *et al.*, 1990).

§5.6.1

Members of the murine hox 5 gene-cluster are transcribed in overlapping domains within developing limb buds. The most 5′ members (*Hox-5.6/-5.5*) are expressed later and in tightly limited postero-distal zones. More 3′ members (*Hox-5.3/-5.2*) are expressed earlier and in broader zones extending antero-proximally across the limb bud; lastly, *Hox-5.1* products are also detectable outside the limb field (Dollé *et al.*, 1989a; Lewis & Martin, 1989). This superim-posed expression of successive hox 5 genes may be related to the gradient of RA morphogen within the limb bud, which itself may depend on the graded expression (maximal distally) of transcripts encoding the cellular RN-binding protein (note that RA-receptor RNAs do not show graded expression in the limb bud; Dollé *et al.*, 1989b).

References

Abel, E. & Maniatis, T. (1989). Action of leucine zippers. *Nature (Lond.)* **341**, 24–5.

Adelson, D.L. & Humphreys, T. (1988). Sea urchin morphogenesis and cell-hyalin adhesion are pertubed by a monoclonal antibody specific for hyalin. *Development*, **104**, 391–402.

Akam, M.E. (1983) The localisation of *Ultrabithorax* transcripts in *Drosophila* tissue segements. *EMBO J.* **2**, 2075–84.

Akam, M. (1986). Developmental genes – mediators of cell communication? *Nature (Lond.)* **319**, 447–8.

Akam, M. (1987). The molecular basis of metameric pattern in the *Drosophila* embryo. *Development*, **101**, 1–22.

Akam, M. (1989a). *Hox* and HOM: homologous gene clusters in insects and vertebrates. *Cell*, **57**, 347–9.

Akam, M. (1989b). Making stripes inelegantly. *Nature (Lond.)* **341**, 282–3.

Akam, M.E. & Martinez-Ariaz, A. (1985). The distribution of *Ultrabithorax* transcripts in *Drosophila* embryos. *EMBO J.* **4** 1689–1700.

Akam, M., Moore, H. & Cox, A. (1984). *Ultrabithorax* mutations map to distant sites within the bithorax complex of *Drosophila*. *Nature (Lond.)* **309**, 635–8.

Albert, P.S. & Riddle, D.L. (1988). Mutants of *Caenorhabditis elegans* that form dauer-like larvae. *Devl. Biol.* **126**, 270–93.

Alberts, B., Bray, D. Lewis, J., Raff, M., Roberts, K. & Watson, J.D. (1983). *Molecular Biology of the Cell*. Garland Publishing, New York & London.

Albertson, D.G. (1985). Mapping muscle protein genes by *in situ* hybridisation using biotin-labelled probes. *EMBO J.* **4**, 2493–8.

Alexandraki, D. & Ruderman, J.V. (1985). Expression of α- and β-tubulin genes during development of sea urchin embryos. *Devl. Biol.* **109**, 436–45.

Alonso, M.C. & Carbrera, C.V. (1988). The *achaete-scute* gene complex of *Drosophila melangaster* comprises four homologous genes. *EMBO J.* **7**, 2585–91.

Ambros, V. & Horvitz, H.R. (1984). Heterochronic mutants of the nematode *Caenorhabditis elegans*. *Science*, **226**, 409–16.

Ambros, V. & Horvitz, H.R. (1987). The *lin-14* locus of *Caenorhabditis elegans* controls the time of expression of specific postembryonic developmental events. *Genes & Dev.* **1**, 398–414.

Ambrosio, L., Mahowald, A.P. & Perrimon, N. (1989). Requirement of the *Drosophila raf* homologue for *torso* function. *Nature (Lond.)* **342**, 288–90.

Amrein, H., Gorman, M. & Nothiger, R. (1988). The sex-determining gene *tra-2* of *Drosophila* encodes a putative RNA-binding protein. *Cell*, **55**, 1025–35.

Anagnou, N.P., Karlsson, S., Moulton, A.D., Keller, G. & Nienhuis, A.W. (1986). Promoter sequences required for function of the human γ globin gene in erythroid cells. *EMBO J.* **5**, 121–6.

Anderson, D.M. & Smith, L.D. (1978). Patterns of synthesis and accumulation of heterogeneous RNA in lampbrush stage oocytes of *Xenopus laevis* (Daudin). *Devl. Biol.* **67**, 274–85.

Anderson, D.M., Richter, J.D., Chamberlin, M.E., Price, D.H., Britten, R.J., Smith, L.D. & Davidson, E.H. (1982). Sequence organisation of the poly (A)+RNA synthesised and accumulated in lampbrush chromosome stage *Xenopus laevis* oocytes. *J. Mol. Biol.* **155**, 281–309.

Anderson, K.V. (1987). Dorsal–ventral embryonic pattern genes of *Drosophila*. *Trends Genet.* **3**, 91–7.

Anderson, K.V. & Nusslein-Volhard, C. (1984). Information for the dorsal-ventral pattern of the *Drosophila* embryo is stored as maternal mRNA. *Nature (Lond.)* **311**, 223–6.

Anderson, K.V., Jurgens, G. & Nusslein-Volhard, C. (1985a). Establishment of dorsal-ventral polarity in the *Drosophila* embryo: genetic studies on the role of the *Toll* gene product. *Cell*, **42**, 779–89.

Anderson, K.V., Bokla, L. & Nusslein-Volhard, C. (1985b). Establishment of dorsal–ventral polarity in the *Drosophila* embryo: the induction of polarity by the *Toll* gene product. *Cell*, **42**, 791–89.

Andrews, M.T. & Brown, D.D. (1987). Transient activation of ooctye 5S RNA genes in *Xenopus* embryos by raising the level of the *trans*-acting factor TFIIIA. *Cell*, **51**, 445–53.

Angelier, N. & Lacroix, J.C. (1975). Complexes de transcription d'origines nucléolaire et chromosomique d'ovocytes de *Pleurodeles waltlii* et *P. poireti* (Amphibiens, Urodeles). *Chromosoma*, **51**, 323–35.

Angerer, L.M. & Angerer, R.C. (1981). Detection of poly (A)+ RNA in sea urchin eggs and embryos by quantitative *in situ* hybridisation. *Nucleic Acids Res.* **9**, 2819–21.

Angerer, L.M., De Leon, D.V., Angerer, R.C., Showman, R.M., Wells, D.E. & Raff, R.A. (1984). Delayed accumulation of maternal histone mRNA during sea urchin oogenesis. *Devl. Biol.* **101**, 477–84.

Angerer, L., DeLeon, D., Cox, K., Maxson, R., Kedes, L., Kaumeyer, J., Weinberg, E. & Angerer, R. (1985). Simultaneous expression of early and late histone messenger RNAs in individual cells during development of the sea urchin embryo. *Devl. Biol.* **112**, 157–66.

Angerer, L.M., Chambers, S.A., Yang, Q., Verkatesan, M., Angerer, R.C. & Simpson, R.T. (1988). Expression of a collagen gene in mesenchyme lineages of the *Strongylocentrotus purpuratus* embryo. *Genes & Dev.* **2**, 239–46.

Angerer, R.C. & Davidson, E.H. (1984). Molecular indices of cell lineage specification in sea urchin embryos. *Science*, **226**, 1153–60.

Antoniou, M., de Boer, E., Habets, G. & Grosveld, F. (1988). The human β globin gene contains multiple regulatory regions: identification of one promoter and two downstream enhancers. *EMBO J.* **7**, 377–84.

Ao, A., Monk, M., Lovell-Badge, R. & Melton, D.W. (1988). Expression of injected HPRT minigene DNA in mouse embryos and its inhibition by antisense DNA. *Development*, **104**, 465–71.

Arion, D., Meijer, L., Brizuela, L. & Beach, D. (1988). *cdc2* is a component of the M phase-specific histone H1 kinase: evidence for identity with MPF. *Cell*, **55**, 371–8.

Arora, K., Rodrigues, V., Joshi, S. Shanbhag, S. & Siddiqi, O. (1987). A gene affecting the specificity of the chemosensory neurons of *Drosophila*. *Nature (Lond.)* **330**, 62–5.

Ashburner, M. (1972). Patterns of puffing activity in the salivary gland chromosomes of *Drosophila*: VI Induction by ecdysone in salivary glands of *D. melanogaster* cultured *in vitro*. *Chromosoma*, **38**, 255–81.

Ashburner, M., Chihara, C., Meltzer, P. & Richards, G. (1973). Temporal control of puffing activity in polytene chromosomes. *Cold Spring Harbor Symp. Quant. Biol.* **38**, 655–62.

Atchison, M.L. & Perry, R.P. (1986). Tandem kappa immunoglobulin promoters are equally active in the presence of the kappa enhancer: implications for models of enhancer function. *Cell*, **46**, 253–62.

Atchison, M.L. & Perry, R.P. (1987). The role of the κ enhancer and its binding factor NF-κB in the development regulation of κ gene transciption. *Cell*, **48**, 121–8.

Augereau, P. & Chambon, P. (1986). The mouse immunoglobulin heavy-chain enhancer effect on transcription *in vitro* and binding of proteins present in Hela and lymphoid B cell extracts. *EMBO J.* **5**, 1791–7.

Austin, J. & Kimble, J. (1987). *glp-1* is required in the germ line for regulation of the decision between mitosis and meiosis in *C. elegans. Cell*, **51**, 589–99.

Austin, J. & Kimble, J. (1989). Transcript analysis of *glp-1* and *lin-12*, homologous genes required for cell interactions during development in *C. elegans. Cell*, **58**, 565–71.

Babiss, L.E., Bennett, A., Friedman, J.M. & Darnell, J.E. (1986). DNase I hypersensitive sites in the 5′ flanking region of the rat serum albumin gene : correlation between chromatin structure and transcriptional activity. *Proc. Natl. Acad. Sci. (USA)* **83**, 6504–8.

Bagnall, K.M., Higgins, S.J. & Sanders, E.J. (1988). The contribution made by a single somite to the vertebral column: experimental evidence in support of resegmentation using the chick-quail chimaera model. *Development*, **103**, 69–85.

Baker, B.S. (1989). Sex in flies: the splice of life. *Nature (Lond.)* **340**, 521–4.

Baker, B.S. & Wolfner, M.F. (1988). A molecular analysis of *doublesex*, a bifunctional gene that controls both male and female sexual differentiation in *Drosophila melanogaster*. *Genes & Dev.* **2**, 477–89.

Baker, N.E. (1987). Molecular cloning of sequences from *wingless*, a segment polarity gene in *Drosophila*: the spatial distribution of a transcript in embryos. *EMBO J.* **6**, 1765–73.

Baker, N.E. (1988a). Transcription of the segment polarity gene *wingless* in the imaginal discs of *Drosophila* and the phenotype of a pupal-lethal *wg* mutation. *Development*, **102**, 489–97.

Baker, N.E. (1988b). Localisation of transcripts from the *wingless* gene in whole *Drosophila* embryos. *Development*, **103**, 289–98.

Baker, N.E. (1988c). Embryonic and imaginal requirements for *wingless*, a segment polarity gene in *Drosophila*. *Devl. Biol.* **125**, 96–108.

Baker, S.M. & Platt, T. (1986). Pol I transcription: which comes first, the end or the beginning? *Cell*, **47**, 839–40.

Balak, K., Jacobson, M., Sunshine, J. & Rutishauser, U. (1987). Neural cell adhesion molecule expression in *Xenopus* embryos. *Dev. Biol.* **119**, 540–50.

Baldwin, T.J. & Burden, S.J. (1989). Muscle-specific gene expression controlled by a regulatory element lacking a MyoD1-binding site. *Nature (Lond.)* **341**, 716–9.

Balling, R., Deutsch, U. & Gruss, P. (1988). *undulated*, a mutation affecting the development of the mouse skeleton, has a point mutation in the paired box of *Pax-1*. *Cell*, **55**, 531–5.

Balling, R., Mutter, G., Gruss, P. & Kessel, M. (1989). Craniofacial abnormalities induced by ectopic expression of the homeobox gene *Hox-1.1* in transgenic mice. *Cell*. **58**, 337–47.

Banerjee, V., Renfranz, P.J., Pollock, J.A. & Benzer, S. (1987a). Molecular characterisation and expression of *sevenless*, a gene involved in neuronal pattern formation in the *Drosophila* eye. *Cell*, **49**, 281–91.

Banerjee, V., Renfranz, P.J., Hinton, D.R., Rabin, B.A. & Benzer, S. (1987b). The *sevenless$^+$* protein is expressed apically in cell membranes of developing *Drosophila* retina; it is not restricted to R7. *Cell*, **51**, 151–8.

Banerji, J. Olson, L. & Schaffner, W. (1983). A lymphocyte-specific cellular enhancer is located downstream of the joining region in immunoglobulin heavy chain genes. *Cell*, **33**, 729–40.

Baniamad, A., Muller, M, Steiner, C. & Renkawitz, R. (1987). Activity of two different silencer elements of the chicken lysozyme gene can be compensated by enhancer elements. *EMBO J.* **6**, 2297–303.

Barad, M., Jack, T., Chadwick, R. & McGinnis, W. (1988). A novel tissue-specific *Drosophila* homeobox gene. *EMBO J.* **7**, 2151–61.

Baralle, F.E., Shoulders, C.C., & Proudfoot, N.J. (1980). The primary structure of the human ε globin gene. *Cell*, **21**, 621–6.

Barberis, A., Superti-Furga, G. & Busslinger, M. (1987). Mutually exclusive interaction of the CCAAT-binding factor and of a displacement protein with overlapping sequences of a histone gene promoter. *Cell*, **50**, 347–59.

Bark, C. Weller, P. Zabielski, J., Janson, L. & Petterson, V. (1987). A distant enhancer element is required for polymerase III transcription of a U6 RNA gene. *Nature (Lond.)* **328**, 356–8.

Basler, K. & Hafen, E. (1988). Control of photoreceptor cell fate by the *sevenless* protein requires a functional tyrosine kinase domain. *Cell*, **54**, 299–311.

Bauerle, P.A. & Baltimore, D. (1988a). Activation of DNA-binding activity in an apparently cytoplasmic precursor of the NF-κB transcription factor. *Cell*, **53**, 211–7.

Bauerle, P.A. & Baltimore, D. (1988b). IκB: a specific inhibitor of the NF-κB transcription factor. *Science*, **242**, 540–5.

Baumgartner, S., Bopp, D., Burri, M. & Noll, M. (1987). Structure of two genes at the *gooseberry* locus related to the *paired* gene and their spatial expression during *Drosophila* embryogenesis. *Genes & Dev.* **1**, 1247–67.

Beach, L.R. & Palmiter, R.D. (1981). Amplification of the metallothionein I gene in cadmium-resistant mouse cells. *Proc. Natl. Acad. Sci. (USA)* **78**, 2110–4.

Beachy, P.A., Helfland, S.L. & Hogness, D.S. (1985). Segmental distribution of bithorax complex proteins during *Drosophila* development. *Nature (Lond.)* **313**, 545–9.

Beachy, P.A., Krasnow, M.A., Gavis, E.R. & Hogness, D.S. (1988). An *Ultrabithorax* protein binds sequences near its own and the *Antennapedia* P1 promoters. *Cell*, **55**, 1069–81.

Beall, C.J. & Hirsh, J. (1987). Regulation of the *Drosophila* dopa decarboxylase gene in neuronal and glial cells. *Genes & Dev.* **1**, 510–20.

Beckendorf, S.K. & Kaftos, F.C. (1976). Differentiation in the salivary glands of *Drosophila melanogaster*: characterisation of the glue proteins and their developmental appearance. *Cell*, **9**, 365–73.

Beeman, R.W. (1987). A homoeotic gene cluster in the red flour beetle. *Nature (Lond.)* **327**, 247–9.

Beeman, R.W., Stuart, J.J., Haas, M.S. & Denell, R.E. (1989). Genetic analysis of the homoeotic gene complex (HOM-C) in the beetle *Tribolium castaneum*. *Devl. Biol.* **133**, 196–209.

Beerman, W. (1964). Control of differentiation at the chromosomal level. *J. Exp. Zool.* **157**, 49–62.

Behringer, R.R., Hammer, R.E., Brinster, R.L., Palmiter, R.D. & Townes, T.M. (1987). Two 3′ sequences direct adult erythroid-specific expression of human β globin genes in transgenic mice. *Proc. Natl. Acad. Sci. (USA)* **84**, 7056–60.

Belerth, T., Burri, M., Thoma, G., Bopp, D., Richstein, S. Frigiero, G., Noll, M. & Nusslein-Volhard, C. (1988). The role of localisation of *bicoid* RNA in organising the anterior pattern of the *Drosophila* embryo. *EMBO J.* **7**, 1749–56.

Bell, L.R., Maine, E.M., Schedl, P. & Cline, T.W. (1988). *Sex-lethal*, a *Drosophila* sex determination switch gene, exhibits sex-specific RNA splicing and sequence similarity to RNA-binding proteins. *Cell*, **55**, 1037–46.

Bellard, M., Dretzen, G., Bellard, F., Kaye, J.S., Pratt-Kaye, S. & Chambon, P. (1986). Hormonally induced alterations of chromatin structure in the polyadenylation and transcription termination regions of the chicken ovalbumin gene. *EMBO J.* **5**, 567–74.

Belote, J.M., Handler, A.M., Wolfner, M.F., Livak, K.J. & Baker, B.S. (1985a). Sex-specific regulation of yolk protein gene expression in *Drosophila*. *Cell*, **40**, 339–48.

Belote, J.M., McKeown, M.B., Andrew, D.J., Scott, T.N., Wolfner, M.F. & Baker, B.S. (1985b). Control of sexual differentiation in *Drosophila melanogaster*. *Cold Spring Harbor Symp. Quant. Biol.* **50**, 605–10.

Bender, W. (1985). Homoeotic gene products as growth factors. *Cell*, **43**, 559-60.

Bender, W. & Peifer, M. (1987). Oncogenes take wing. *Cell*, **50**, 519–20.

Bender, W. Akam, M., Karch, F., Beachy, P.A., Peifer, M., Spierer, P., Lewis, E.B. & Hogness, D.S. (1983). Molecular genetics of the *bithorax* complex in *Drosophila melanogaster*. *Science*, **221**, 23–9.

Bender, M., Turner, F.R. & Kaufman, T.C. (1987). A developmental genetic analysis of the gene *Regulator of postbithorax* in *Drosophila melanogaster*. *Devl. Biol.* **119**, 418–32.

Benson, S., Sucov, H., Stevens, L., Davidson, E. & Wilt, F. (1987). A lineage-specific gene encoding a major matrix protein of the sea urchin embryo spicule: I authentication of the cloned gene and its developmental expression. *Devl. Biol.* **120**, 499–506.

Berg, J.M. (1989). DNA binding specificity of steroid receptors. *Cell*, **57**, 1065–8.

Bergman, Y., Rice, D., Grosschedl, R. & Baltimore, D. (1984). Two regulatory elements for immunoglobulin κ light chain gene expression. *Proc. Natl. Acad. Sci. (USA)* **81**, 7041–5.

Bermingham, J.R. & Scott, M.P. (1988). Developmentally regulated alternative splicing of transcripts from the *Drosophila* homeotic gene *Antennapedia* can produce four different proteins. *EMBO J.* **7**, 3211–22.

Bernard, O., Hozumi, N. & Tonegawa, S. (1978). Sequence of mouse immunoglobulin light chain genes before and after somatic changes. *Cell*, **15**, 1133–44.

Beyer, A.L. & Osheim, Y.N. (1988). Splice site selection, rate of splicing, and alternative splicing on nascent transcripts. *Genes & Dev.*, **2**, 754–65.

Bieker, J.J., Martin, P.L. & Roeder, R.G. (1985). Formation of a rate-limiting intermediate in 5S gene transcription. *Cell*, **40**, 119–27.

Bienz, M. & Tremml, G. (1988). Domain of *Ultrabithorax* expression in *Drosophila* visceral mesoderm from autoregulation and exclusion. *Nature (Lond.)* **333**, 576–9.

Bienz, M., Saari, G., Tremml, G., Muller, J., Zust, B. & Lawrence, P.A. (1988). Differential regulation of *Ultrabithorax* in two germ layers of *Drosophila*. *Cell*, **53**, 567–76.

Bier, K. (1963). Synthese, interzellulärer Transport, und Abbua von Ribonukleinsaure in Ovar der Stubenfliege *Musca domestica*. *J. Cell Biol.* **16**, 436–40.

Biggin, M.D. & Tijan, R. (1988). Transcription factors that activate the *Ultrabithorax* promoter in developmentally staged extracts. *Cell*, **53**, 699–711.

Biggin, M.D. & Tijan, R. (1989). A purified *Drosophila* homeodomain protein represses transcription *in vitro*. *Cell*, **58**, 433–40.

Biggin, M.D., Bickel, S., Benson, M. Pirotta, V. & Tijan, R. (1988). *zeste* encodes a sequence-specific transcription that activates the *Ultrabithorax* promoter *in vitro*. *Cell*, **53**, 713–22.

Bindereif, A. & Green, M.R. (1987). An ordered pathway of snRNP binding during mammalian pre-mRNA splicing complex assembly. *EMBO J.* **6**, 2415–24.

Bird, A.P. (1984). DNA methylation – how important in gene control? *Nature (Lond.)* **307**, 503–4.

Bird, A.P. (1986). CpG-rich islands and the function of DNA methylation. *Nature (Lond.)* **321**, 209–13.

Birnstiel, M.L., Busslinger, M. & Strub, K. (1985). Transcription termination and 3′ processing: the end is in site! *Cell*, **41**, 349–56.

Bishop, J.O., Pemberton, R. & Baglioni, C. (1972). Reiteration frequency of haemoglobin genes in the duck. *Nature New Biol.* **235**, 231–4.

Blau, H.M. (1988). Hierarchies of regulatory genes may specify mammalian development. *Cell*, **53**, 673–4.

Blochlinger, K., Boder, R., Jack, J., Jan, L.Y. & Jan, Y.N. (1988). Primary structure and expression of a product from *cut*, a locus involved in specifying sensory organ identity in *Drosophila*. *Nature (Lond.)* **333**, 629–33.

Bodine, D.M. & Ley, T.J. (1987). An enhancer element lies 3' to the human $^A\gamma$ globin gene. *EMBO J.* **6**, 2997–3004.

Bodner, M., Castrillo, J.-L., Theill, L.E., Deerinck, T., Ellisman, M. & Karin, M. (1988). The pituitary-specific transcription factor GHF-1 is a homeobox-containing protein. *Cell*, **55**, 505–18.

Boehm, K.D., Hood, R.L. & Ilan, J. (1988). Induction of vitellogenin in primary monolayer cultures of cockerel hepatocytes. *Proc. Natl. Acad. Sci. (USA)* **85**, 3450–4.

Bogenhagen, D.F., Sakonju, S. & Brown, D.D. (1980). A control region in the centre of the 5S gene directs specific initiation of transcription: II The 3' border of the region. *Cell*, **19**, 27–35.

Bogorad, L.D., Utset, M.F., Awgulewitsch, A.W., Miki, T., Hart, C.P. & Ruddle, F.H. (1989). The developmental expression of a new murine homeobox gene: *Hox-2.5*. *Devl. Biol.* **133**, 537–49.

Bolten, S.L., Powell-Abel, P., Fischoff, D.H. & Waterson, R.H. (1984). The *sup-7(5)X* gene of *Caenorhabditis elegans* encodes a transfer RNA-Trp UAG amber suppressor. *Proc. Natl. Acad. Sci. (USA)* **81**, 6784–8.

Bonner, J.J. & Pardue, M.-L. (1977a). Polytene chromosone puffing and *in situ* hybridisation measure different aspects of RNA metabolism. *Cell*, **12**, 227–34.

Bonner, J.J. & Pardue, M.-L. (1977b). Ecdysone-stimulated RNA synthesis in salivary glands of *Drosophila melanogaster*: assay by *in situ* hybridisation. *Cell*, **12**, 219–25.

Bopp, D., Buri, M., Baumgartner, S., Frigiero, G. & Noll, M. (1986). Conservation of a large protein domain in the segmentation gene *paired* and in functionally related genes of *Drosophila*. *Cell*, **47**, 1033–40.

Boswell, R.E. & Mahowald, A.P. (1985). *tudor* – a gene required for the assembly of the germ plasm in *Drosophila melanogaster*. *Cell*, **43**, 97–104.

Boulay, J.L., Dennefeld, C. & Alberga, A. (1987). The *Drosophila* developmental gene *snail* encodes a protein with nucleic acid binding fingers. *Nature (Lond.)* **330**, 395–7.

Bourouis, M. & Richards, G. (1985). Remote regulatory sequences of the *Drosophila* glue gene *sgs-3* as revealed by P-element transformation. *Cell*, **40**, 349–57.

Boveri, T. (1910). Über die Teilung centrifugierte Eier von *Ascaris megalocephala*. *Wilhelm Roux' Arch. Entwicklungsmech. Org.* **30**, 101–25.

Boycott, A.E., Diver, C., Garstrang, S.L. & Turner, F.M. (1930). The inheritance of sinistrality in *Limnaea peregra* (Mollusca, Pulmonata). *Phil. Trans. Roy. Soc. Lond. B*, **219**, 51–131.

Bozzoni, I., Fragapane, P., Annesi, F., Pierandrei-Amaldi, P., Amaldi, F. & Beccari E, (1984). Expression of two *Xenopus laevis* ribosomal protein genes in injected frog oocytes: a specific splicing block interferes with the L1 RNA maturation. *J. Mol. Biol.* **180**, 987–1005.

Brandhorst, B.P. (1980). Simultaneous synthesis, translation and storage of mRNA including histone mRNA in sea urchin eggs. *Devl. Biol.* **79**, 139–49.

Brandhorst, B.P. & Newrock, K.M. (1981). Posttranscriptional regulation of protein synthesis in *Ilyanassa* embryos and isolated polar lobes. *Devl. Biol.* **83**, 250–4.

Brandis, J.W. & Raff, R.A. (1978). Translation of oogenetic mRNA in sea urchin eggs and early embryos: demonstration of change in translational efficiency following fertilization. *Devl. Biol.* **67**, 99–113.

Breathnach, R., Benoist, C., O'Hare, K., Gannon, F. & Chambon, P. (1978). Ovalbumin gene: evidence for a leader sequence in mRNA and DNA sequences at the exon-intron boundaries. *Proc. Natl. Acad. Sci. (USA)* **75**, 4853–7.

Breen, T.R. & Duncan, I. (1986). Maternal expression of genes that regulate the bithorax complex of *Drosophila melanogaster. Devl. Biol.* **118**, 442–56.

Breier, P., Bucan, M., Francke, U., Colberg-Poley, A.M. & Gruss, P. (1986). Sequential expression of murine homeobox genes during F9 EC cell differentiation. *EMBO J.* **5**, 2209–15.

Breier, G., Dressler, G.R. & Gruss, P. (1988). Primary structure and developmental expression pattern of *Hox 3.1*, a member of the murine *Hox 3* homeobox gene cluster. *EMBO J.* **7**, 1329–36.

Breitbart, R.E., Nguyen, H.T., Medford, R.M., Destrée, A.T., Mahdavi, V., & Nadal-Ginard. B. (1985). Intricate combinatorial patterns of exon splicing generate multiple regulated troponin T isoforms from a single gene. *Cell*, **41**, 67–82.

Brenner, S. (1974). The genetics of *Caenorhabditis elegans. Genetics*, **77**, 71–94.

Briggs, R. & King, T.J. (1957). Changes in the nuclei of differentiating endoderm cells as revealed by nuclear transplantation. *J. Morphol.* **100**, 269–312.

Briggs. R., Green, E.V. & King, T.J. (1951). An investigation of the capacity for cleavage and differentation in *Rana pipiens* eggs lacking 'functional' chromosomes. *J. Exp. Zool.* **116**, 455–94.

Brivanlou, A.H. & Harland, R.M. (1989). Expression of an *engrailed*-related protein is induced in the anterior neural ectoderm of early *Xenopus* embryos. *Development*, **106**, 611–7.

Brock, M.L. & Shapiro, D.J. (1983). Estrogen stabilises vitellogenin mRNA against cytoplasmic degradation. *Cell*, **34**, 207–14.

Brockes, J.P. (1989). Retinoids, homeobox genes and limb morphogenesis. *Neuron*, **2**, 1285–94.

Brower, D.L. (1984). Posterior-to-anterior transformation in *engrailed* wing imaginal discs of *Drosophila. Nature (Lond.)* **310**, 496–7.

Brower, D.L. (1985). The sequential compartmentalisation of *Drosophila* segments revisited. *Call*, **41**, 361–4.

Brower, D.L. (1986). *engrailed* gene expression in *Drosophila* imaginal discs. *EMBO J.* **5**, 2649–56.

Brower, D.L. (1987). *Ultrabithorax* gene expression in *Drosophila* imaginal discs and larval nervous system. *Development*, **101**, 83–92.

Brower, D.L., Smith, R.J. & Wilcox, M. (1982). Cell shapes on the surface of the *Drosophila* wing imaginal disc. *J. Embryol. Exp. Morphol.* **67**, 137–51.

Brower, D.L., Wilcox, M., Piovant, M., Smith, R.J. & Reger, L.A. (1984). Related cell-surface antigens expressed with positional specificity in *Drosophila* imaginal discs. *Proc. Natl. Acad. Sci. (USA)* **81**, 7485–9.

Brower, D.L., Piovant, M. & Reger, L.A. (1985). Developmental analysis of *Drosophila* position-specific antigens. *Devl. Biol.* **108**, 120–30.

Brown, D.D. (1984). The role of stable complexes that repress and activate eucaryotic genes. *Cell*, **37**, 359–65.

Brown, D.D. & Schlissel, M.S. (1985). A positive transcription factor controls the differential expression of two 5S RNA genes. *Cell*, **42**, 759–67.

Bryant, P.J. (1971). Regeneration and duplication following operations *in situ* on the imaginal discs of *Drosophila melanogaster*. *Devl. Biol.* **26**, 606–15.

Bryant, P.J. (1988). Localised cell death caused by mutations in a *Drosophila* gene coding for a transforming growth factor-β homologue. *Devl. Biol.* **128**, 386–95.

Bucan, M., Yang-Feng, T., Colberg-Poley, A.M., Wolgemuth, D.J., Guenet, J-L., Franke, U. & Lehrach, H. (1986). Genetic and cytogenetic localisation of the homeobox containing genes on mouse chromosome 6 and human chromosome 7. *EMBO J.* **5**, 2899–2905.

Burch, J.B.E. & Weintraub, H. (1983). Temporal order of chromatin structural changes associated with activation of the major chicken vitellogenin gene. *Cell*, **33**, 65–76.

Burghlin, T.H. (1988). The yeast regulatory gene PHO 2 encodes a homeobox. *Cell*, **53**, 339–40.

Burns, A.T.H., Deeley, R.G., Gordon, J.I., Udell, D.S., Mullinix, K.P. & Goldberger, R.F. (1978). Primary induction of vitellogenin mRNA in the rooster by 17β oestradiol. *Proc. Natl. Acad. Sci. (USA).* **75**, 1815–9.

Burrows, P.D., Beck-Engeser, G.B. & Wabl, M.R. (1983). Immunoglobulin heavy chain class switching in a pre-B cell line is accompanied by DNA rearrangement. *Nature (Lond.)* **306**, 246–6.

Burtis, K.C. & Baker, B.S. (1989). *Drosophila doublesex* gene controls somatic sexual differentiation by producing alternatively spliced mRNAs encoding related sex-specific polypeptides. *Cell* **56**, 997–1010.

Busslinger, M., Hurst, J. & Flavell, R.A. (1983). DNA methylation and the regulation of globin gene expression. *Cell*, **34**, 197–206.

Busturia, A. & Morata, G. (1988). Ectopic expression of homeotic genes caused by the elimination of the *Polycomb* gene in *Drosophila* imaginal epidermis. *Development*, **104**, 713–20.

Cabrera, C.V., Botas, J. & Garcia-Bellido, A. (1985). Distribution of *Ultrabithorax* proteins in mutants of *Drosophila* bithorax complex and its transregulatory genes. *Nature (Lond.)* **318**, 569–71.

Cabrera, C.V., Alonso, M.C., Johnston, P., Phillips, R.G. & Lawrence, P.A. (1987a). Phenocopies induced with antisense RNA identify the *wingless* gene. *Cell*, **50**, 659–63.

Cabrera, C.V., Martinez-Arias, A. & Bate, M. (1987b). The expression of three members of the *achaete-scute* complex correlates with neuroblast segregation in *Drosophila*. *Cell*, **50**, 425–33.

Caldwell, D.C. & Emerson, C.P. (1985). The role of cap trimethylation in the transitional activation of stored maternal histone mRNA in sea urchin embryos. *Cell*, **42**, 691–700.

Callan, H.G. (1963). The nature of lampbrush chromosomes. *Int. Rev. Cytol.* **15**, 1–34.

Calzone, F.J., Lee, J.J., Lee, N., Britten, R.J. & Davidson, E.J. (1988a). A long nontranslatable poly(A) RNA stored in the egg of the sea urchin *Strongylocentrotus purpuratus*. *Genes & Dev.* **2**, 305–18.

Calzone, F.J., Thézé, N., Thiebaud, P., Hill, R.L., Britten, R.J. & Davidson, E.H. (1988b). Developmental appearance of factors that bind specifically to *cis*-regulatory sequences of a gene expressed in the sea urchin embryo. *Genes & Dev.* **2**, 1074–88.

Cameron, R.A., Hough-Evans, B.R., Britten, R.J. & Davidson, E.H. (1987). Lineage and fate of each blastomere of the eight-cell sea urchin embyro. *Genes & Dev.* **1**, 75–84.

Campbell, A.D., Long, M.W. & Wicha, M.S. (1987). Haemonectin, a bone marrow adhesion protein specific for cells of granulocyte lineage. *Nature (Lond.)* **329**, 744–6.

Campos-Ortega, J.A. (1983). Topological specificity of phenotype expression of neurogenic mutations in *Drosophila*. *Roux'Arch. Devl. Biol.* **192**, 317–26.

Campos-Ortega, J.A. & Hartenstein, V. (1985). *The embryonic development of* Drosophila melanogaster. Springer-Verlag, Berlin.

Carrasco, A.E. & Malacinski, G.M. (1987). Localisation of *Xenopus* homeobox gene transcripts during embryogenesis and in the adult nervous system. *Devl. Biol.* **121**, 69–81.

Carrasco, A.E., McGinnis, W., Gehring, W.J. & de Robertis, E.M. (1984). Cloning of an *X. laevis* gene expressed during embryogenesis coding for a peptide region homologous to *Drosophila* homoeotic genes. *Cell*, **37**, 409–14.

Carroll, S.B. & Scott, M.P. (1985). Localisation of the *fushi tarazu* protein during *Drosophila* embryogenesis. *Cell*, **43**, 47–57.

Carroll, S.B. & Scott, M.P. (1986). Zygotically active genes that affect the spatial expression fo the *fushi tarazu* segmentation gene during early *Drosophila* embryogenesis. *Cell*, **45**, 113–26.

Carroll, S.B., Laymon, R.A., McCutcheon, M.A., Riley, P.D. & Scott, M.P. (1986a). The localisation and regulation of *Antennapedia* protein expression in *Drosophila* embryos. *Cell*, **47**, 113–22.

Carroll, S.B., Winslow, G.M., Schupbach, T. & Scott, M.P. (1986b). Maternal control of *Drosophila* segmentation gene expression. *Nature (Lond.)* **323**, 278–80.

Carroll, S.B., DiNardo, S., O'Farrell, P.H., White, R.A.H. & Scott, M.P. (1988a). Temporal and spatial relationships between segmentation and homeotic gene expression in *Drosophila* embryos: distributions of the *fushi tarazu, engrailed, Sex combs reduced, Antennapedia,* and *Ultrabithorax* proteins. *Genes & Dev.* **2**, 350–60.

Carroll, S.B., Laughon, A. & Thalley, B.S. (1988b). Expression, function and regulation of the *hairy* segmentation protein in the *Drosophila* embryo. *Genes & Dev.* **2**, 883–90.

Casanova, J. & White, R.A.H. (1987). *Trans*-regulatory functions in the *Abdominal-B* gene of the bithorax complex. *Development*, **101**, 117–22.

Casanova, J., Sanchez-Herrero, E. & Morata, G. (1985a).*Contrabithorax* and the control of spatial expression of the bithorax complex genes of *Drosophila*. *J. Embryol. Exp. Morphol.* **90**, 179–96.

Casanova, J., Sanchez-Herrero, E. & Morata, G. (1985b). Prothoracic transformation and functional structure of the *Ultrabithorax* gene of *Drosophila*. *Cell*, **42**, 663–9.

Casanova, J., Sanchez-Herrero, E., Busturia, A. & Morata, G. (1987). Double and triple mutant combinations of the bithorax complex of *Drosophila*. *EMBO J.* **6**, 3130–9.

Casanova J., Sanchez-Herrero, E. & Morata, G. (1988). Developmental analysis of a hybrid gene composed of parts of the *Ubx* and *abd-A* genes of *Drosophila*. *EMBO J.* **7**, 1097–105.

Castrillo, J.-L., Bodner, M. & Karin, M. (1989). Purification of growth hormone-specific transcription factor GHF-1 containing homeobox. *Science*, **243**, 814–7.

Catalano, G., Eilbeck, C., Monroy, A. & Parisi, E. (1979). A model for early segregation of territories in the ascidean egg. In *Cell Lineage, Stem Cells and Cell Determination*, ed. N. le Douarin, INSERM Symp. **10**, 15–28. North Holland, Amsterdam.

Cather, J.N., Verdonk, N.H. & Dohmen, M.R. (1976). Role of the vegetal body in the regulation of development in *Bithynia tentaculata* (Prosobranchia, Gastropoda). *Amer. Zool.* **16**, 455–68.

Catterall, J.F., Stein, J.P., Lai, E.C., Woo, S.L.S., Dugaiczyk, A., Mace, M.L., Means, A.R. & O'Malley, B.W. (1979). The chick ovomucoid gene contains at least six intervening sequences. *Nature (Lond.)* **278**, 323–7.

Caudy, M., Grell, E.H., Dambly-Chaudière, C., Ghysen, A., Jan, L.Y. & Jan, Y.N. (1988). The maternal sex determination gene *daughterless* has zygotic activity necessary for the formation of peripheral neurons in *Drosophila*. *Genes & Dev.* **2**, 843–52.

Cech, T.R. (1983). RNA splicing: three themes with variations. *Cell*, **34**, 713–16.

Cedar, H. (1988). DNA methylation and gene activity. *Cell*, **53**, 3–4.

Celniker, S.E. & Lewis E.B. (1987). *Transabdominal*, a dominant mutant of the bithorax complex, produces a sexually dimorphic segemental transformation in *Drosophila*. *Genes & Dev.* **1**, 111–23.

Chadwick, R. & McGinnis, W. (1987). Temporal and spatial distribution of transcripts from the *Deformed* gene of *Drosophila*. *EMBO J.* **6**, 779–89.

Chalfie, M. & Au, M. (1989). Genetic control of differentiation of the *Caenorhabditis elegans* touch receptor neurons. *Science*, **243**, 1027–33.

Chalfie, M. & Sulston, J. (1981). Developmental genetics of the mechanosensory neurons of *Caenorhabditis elegans*. *Devl. Biol.* **82**, 358–70.

Chalfie, M.& White, J. (1988). The nervous system. In Wood, W.B. (ed.) *The nematode* Caenorhabditis elegans, ed. W.B. Wood, CSH Monograph, **17**, 337–91.

Chalfie, M., Horvitz, H.R. & Sulston, J.E. (1981). Mutations that lead to reiterations in the cell lineages of *C. elegans*. *Cell*, **24**, 59–69.

Chalfie, M., Sulston, J.E., White, J.G., Southgate, E., Thomson, J.M. & Brenner, S. (1985). The neural circuit for touch sensitivity in *Caenorhabditis elegans*. *J. Neurosci.* **5**, 956–64.

Chapman, B.S. & Tobin, A.J. (1979). Distribution of developmentally-regulated haemoglobins in embryonic erythroid populations. *Devl. Biol.* **69**, 375–87.

Charnay, P., Treisman, R., Mellon, P., Chao, M., Axel, R. & Maniatis, T. (1984). Differences in human α- and β-globin gene expression in mouse erythroleukaemia cells; the role of intragenic sequences. *Cell*, **38**, 251–63.

Chasan, R. & Anderson, K.V. (1989). The role of *easter*, an apparent serine protease, in organising the dorsal–ventral pattern of the *Drosophila* embryo. *Cell*,

56, 391–400.

Chavrier, P., Zerial, M., Lemaire, P., Almendral, J., Bravo, R. & Charnay, P. (1988). A gene encoding a protein with zinc fingers is activated during G_0/G_1 transition in cultured cells. *EMBO J.* **7**, 29–35.

Cho, C.W.Y., Goetz, J., Wright, C.V.E., Fritz, A., Hardwicke, J. & de Robertis, E.M. (1988). Differential utilisation of the same reading frame in a *Xenopus* homeobox gene encodes two related proteins sharing the same DNA-binding activity. *EMBO J.* **7**, 2139–49.

Chodosh, L.A., Baldwin, A.S., Carthew, R.W. & Sharp, P.A. (1988). Human CCAAT-binding proteins have heterologous subunits. *Cell*, **53**, 11–24.

Choi, O.-R. & Engel, J.D. (1986). A 3' enhancer is required for temporal and tissue-specific transcriptional activation of the chicken adult β globin gene. *Nature (Lond.)* **323** 731–4.

Choi, O.-R.B. & Engel, J.D. (1988). Developmental regulation of β-globin gene switching. *Cell*, **55**, 17–26.

Chowdhury, K., Deutsch, A. & Gruss, P. (1987). A multigene family encoding several 'finger' structures is present and differentially active in mammalian genomes. *Cell*, **48**, 771–8.

Chowdhury, K., Dressler, G., Breier, G., Deutsch, U. & Gruss, P. (1988). The primary structure of the murine multifinger gene *mKr2* and its specific expression in developing and adult neurons. *EMBO J.* **7**, 1345–53.

Ciejek, E.M., Tsai, M.-J. & O'Malley, B.W. (1983). Actively-transcribed genes are associated with the nuclear matrix. *Nature (Lond.)* **306**, 607–9.

Ciliberto, G., Raugei, G., Costanzo, F., Dente, L. & Cortese, R. (1983). Common and interchangeable elements in the promoters of genes transcribed by RNA polymerase III. *Cell*, **32**, 725–33.

Clark, W.C., Doctor, J., Fristrom, J.W. & Hodgetts, R.B. (1986). Differential responses of the dopa decarboxylase gene to 20-OH-ecdysone in *Drosophila melanogaster*. *Devl. Biol.* **114**, 141–50.

Clement, A.C. (1962). Development of *Ilyanassa* following removal of the D macromere at successive cleavage stages. *J. exp. Zool.* **149**, 193–215.

Clement, A.C. (1976). Cell determination and organogenesis in molluscan development: a reappraisal based on deletion experiments in *Ilyanassa*. *Amer. Zool.* **16**, 447–53.

Cline, T.W. (1976). A sex-specific, temperature-sensitive maternal effect of the *daughterless* mutation of *Drosophila melanogaster*. *Genetics*, **84**, 723–42.

Cline, T.W. (1978). Two closely-linked mutations in *Drosophila melanogaster* that are lethal to opposite sexes and interact with *daughterless*. *Genetics*, **90**, 683–98.

Cline, T.W. (1980). Maternal and zygotic sex-specific gene interactions in *Drosophila melanogaster*. *Genetics*, **96**, 903–26.

Cline, T.W. (1984). Autoregulatory functioning of a *Drosophila* gene product that establishes and maintains the sexually determined state. *Genetics*, **107**, 231–77.

Cline, T.W. (1986). A female-specific lethal lesion in an X-linked positive regulator of the *Drosophila* sex determination gene, *Sex-lethal*. *Genetics*, **113**, 641–63.

Cline, T.W. (1989). The affairs of *daughterless* and the promiscuity of developmental regulators. *Cell*, **59**, 231–4.

Cochet, M., Gannon, F., Hen, R. Maroteaux, L., Perrin, F. & Chambon, P. (1979). Organisation and sequence studies of the 17 piece chicken conalbumin gene. *Nature (Lond.)* **282**, 5657–74.

Cohen, S., Bronner, G., Kuttner, F., Jurgens, G. & Jackle, H. (1989). *Distal-less* encodes a homeodomain protein required for limb development in *Drosophila*. *Nature (Lond.)* **338**, 432–4.

Cohen, S.M. & Jurgens. G. (1989). Proximal–distal pattern formation in *Drosophila*: cell autonomous requirement for *Distal-less* gene activity in limb development. *EMBO J.* **8**, 2045–55.

Colberg-Poley, A.M., Voss, S.D., Chowdhury, K. & Gruss, P. (1985a). Structural analysis of murine genes containing homeobox sequences and their expression in embryonal carcinoma cells. *Nature (Lond.)* **314**, 713–6.

Colberg-Poley, A.M., Voss, S.D., Chowdhury K., Stewart, C.L., Wagner, E.F. & Gruss, P. (1985b). Clustered homeoboxes are differentially expressed during murine development. *Cell*, **43**, 39–45.

Coleman, K.G., Poole, S.J., Weir, M.P., Seller, W.C. & Kornberg, T. (1987). The *invected* gene of *Drosophila*: sequence analysis and expression studies reveal a close kinship to the *engrailed* gene. *Genes & Dev.* **1**, 19–28.

Colin, S.M. & Hille, M.B. (1986). Injected mRNA does not increase protein synthesis in unfertilised, fertilised or ammonia-activated sea urchin eggs. *Devl. Biol.* **115**, 184–93.

Collier, J.R. (1983). Protein synthesis during *Ilyanassa* organogenesis. *Devl. Biol.* **100**, 256–9.

Collins, F.S., Stoeckert, C.J., Sergeant, G.R., Forget, B.G. & Weissman, S.M. (1984). $^G\gamma\beta^+$ hereditary persistence of foetal haemoglobin: cosmid cloning and identification of a specific mutation 5' to the $^G\gamma$ gene. *Proc. Natl. Acad. Sci. (USA)* **81**, 4894–8.

Condie, B.G. & Harland, R.M. (1987). Posterior expression of a homeobox gene in early *Xenopus* embryos. *Development*, **101**, 93–105.

Conklin, E.G., (1905). The organisation and cell lineage of the ascidean egg. *J. Acad. Natl. Sci. Philadelphia*, **13**, 5–118.

Cooke, J. (1989). Mesoderm-inducing factors and Spemann's organiser phenomenon in amphibian development. *Development*, **107**, 229–41.

Cooke, J., Smith, J.C., Smith, & Yaqoob, M. (1987). The organisation of mesodermal pattern in *Xenopus laevis*: experiments using a *Xenopus* mesoderm-inducing factor. *Development*, **101**, 893–908.

Cory, S., Jackson, J. & Adam, J.M. (1980). Deletions in the constant region locus can account for switches in immunoglobulin heavy chain expression. *Nature (Lond.)* **284**, 450–3.

Costa, M., Weir, M., Coulson, A., Sulston, J. & Kenyon, C. (1988). Posterior pattern formation in *C. elegans* involves position-specific expression of a gene containing a homeobox. *Cell*, **55**, 747–56.

Costa, R.H., Lai, E. & Darnell, J.E. (1986). Transcriptional control of the mouse prealbumin (transthyretin) gene: both promoter sequences and a distinct enhancer are cell specific. *Mol. & Cell. Biol.* **6**, 4697–708.

Costa, R.H., Grayson, D.R., Xanthopoulos, K.G. & Darnell, J.E. (1988). A liver-specific DNA-binding protein recognises multiple regulatory regions of transthyretin, α_1-antitrypsin, albumin and simian virus 40 genes. *Proc. Natl.*

Acad. Sci. (USA) **85**, 3840–4.

Costantini, F.D., Scheller, R.H., Britten, R.J. & Davidson, E.H. (1978). Repetitive sequence transcripts in the mature sea urchin oocyte. *Cell*, **15**, 173–87.

Côté, S., Preiss, A., Haller, J., Schuh, R., Kienlin, A., Seifert, E. & Jackle, H. (1987). The *gooseberry-zipper* region of *Drosophila*: five genes encode different spatially regulated transcripts in the embryo. *EMBO J.* **6**, 2793–801.

Cowan, A.E. & McIntosh. (1985). Mapping the distribution of differentiation potential for intestine, muscle and hypodermis during early development in *Caenorhabditis elegans*. *Cell*, **41**, 923–32.

Cowie, A. & Myers, R.M. (1988). DNA sequences involved in transcriptional regulation of the mouse. β-globin promoter in murine erythroleukaemia cells. *Mol. & Cell. Biol.* **8**, 3122–8.

Cox, G.N. & Hirsh, D. (1985). Stage-specific patterns of collagen gene expression during development of *Caenorhabditis elegans*. *Mol. & Cell. Biol.* **5**, 363–72.

Cox, K.H., Angerer, L.M., Lee, J.J., Davidson, E.H. & Angerer, R.C. (1986). Cell lineage-specific programs of expression of multiple action genes during sea urchin embryogenesis. *J. Mol. Biol.* **188**, 159–72.

Crenshaw, E. B. III, Kalla, K., Simmons, D.M., Swanson, L.W. & Rosenfeld, M.G. (1989). Cell-specific expression of the prolactin gene in transgenic mice is controlled by synergistic interactions between promoter and enhancer elements. *Genes & Dev.* **3**, 959–72.

Crews, S.T., Thomas, J.B. & Goodman, C.S. (1988). The *Drosophila single-minded* gene encodes a nuclear protein with sequence homology to the *per* gene product. *Cell*, **52**, 143–51.

Crick, F.H.C. & Lawrence, P.A. (1975). Compartments and polyclones in insect development. *Science*, **189**, 340–7.

Cronmiller, C. & Cline, T.W. (1987). The *Drosophila* sex determination gene *daughterless* has different functions in the germ line versus the soma. *Cell*, **48**, 479–87.

Crowley, T.E., Bond, M.W., & Meyerowitz, E.M. (1983). The structural genes for three *Drosophila* glue proteins reside at a single polytene chromosome puff locus. *Mol. & Cell. Biol.* **3**, 623–34.

Crowther, R.J. & Whittaker, J.R. (1986). Differentation without cleavage: multiple cytospecific ultrastructural expressions in individual one-celled ascidean embryos. *Devl. Biol.* **117**, 114–26.

Cudennec, C.A., Thiery, J.-P. & le Douarin, N.M. (1981). *In vitro* induction of adult erythropoiesis in early mouse yolk sac. *Proc. Natl. Acad. Sci. (USA)* **78**, 241–6.

Curran, T. & Franza, B.R., Jr (1988). Fos and Jun: the AP-1 connection. *Cell*, **55**, 395–7.

Curtis, P.J., Mantei, N., Van den Berg, J. & Weissman, C. (1977). Presence of a putative 15S precursor to β-globin mRNA but not to α-globin mRNA in Friend cells. *Proc. Natl. Acad. Sci. (USA)* **74**, 3184–8.

Cyert, M.S. & Kirschner, M.W. (1988). Regulation of MPF activity *in vitro*. *Cell*, **53**, 185–9.

Dale, L. & Slack, J.M.W. (1987). Regional specification within the mesoderm of early embryos of *Xenopus laevis*. *Development*, **100**, 279–95.

Dale, L., Smith, J.C. & Slack, J.M.W. (1985). Mesoderm induction in *Xenopus laevis*; a quantitative study using a cell-lineage label and tissue-specific antibodies. *J. Embryol. Exp. Morphol.* **89**, 289–312.

Daneholt, B., Andersson, K. & Fagerlind, M. (1977). Large sized polysomes in *Chironomus tentans* salivary glands and their relation to Balbiani ring 75S RNA. *J.Cell Biol.* **73** , 149–60.

Daneholt, B., Case, S.T., Derksen, J., Lamb, M.M., Nelson, L.G. & Wieslander, L. (1979). The transcription unit in Balbiani ring 2 and its relation to the chromomeric subdivision of the polytene chromosome. In *Specific Eukaryotic Genes*, Alfred Bentzon Symp. **13**, 39–51. Munksgaard, Copenhagen.

Danilchik, M.V. & Hille, M.B. (1981). Sea urchin egg and embryo ribosomes: differences in translational activity in a cell-free system. *Devl. Biol.* **84**, 291–8.

Danner, D. & Leder, P. (1985). Role of an RNA cleavage/poly(A) addition site in the production of membrane-bound and secreted IgM mRNA. *Proc. Natl. Acad. Sci. (USA)* **82**, 8658–62.

Dan-Sohkawa, M. & Satoh, N. (1978). Studies on dwarf larvae from isolated blastomeres of the starfish *Asterina pectenifera*. *J. Embryol. Exp. Morphol.* **46**, 171–85.

Darby, M.K., Andrews, M.T. & Brown, D.D. (1988). Transcription complexes that program *Xenopus* 5S RNA genes are stable *in vitro*. *Proc. Natl. Acad. Sci. (USA)* **85**, 5516–20.

Darnell, J., Lodish, H. & Baltimore, D. (1986). *Molecular Cell Biology*. Scientific American Books, W.H. Freeman & Co., New York.

Davidson, D., Graham, E., Sime, C. & Hill, R. (1988). A gene with sequence similarity to *Drosophila engrailed* is expressed during development of the neural tube and vertebrae in the mouse. *Development*, **104**, 305–16.

Davidson, E.H. (1986). *Gene Activity in Early Development*, 3rd ed. Academic Press, New York.

Davidson, E.H. (1989). Lineage-specific gene expression and the regulative capacities of the sea urchin embryo: a proposed model. *Development*, **105**, 421–5.

Davidson, E.H., Hough-Evans, B.R. & Britten, R.J. (1982). Molecular biology of the sea urchin embryo. *Science*, **217**, 17–26.

Davidson, E.H., Flytzanis, C.N., Lee, J.J., Robinson, J.J., Rose, S.J. III, & Sucov, H.M. (1985). Lineage-specific gene expression in the sea urchin embryo. *Cold Spring Harbor Symp. Quant. Biol.* **50**, 321–30.

Davis, C.A., Noble-Topham, S.E., Rossant, J. & Joyner, A.L. (1988). Expression of the homeobox-containing gene *En-2* delineates a specific region of the developing mouse brain. *Genes & Dev.* **2**, 361–71.

Davis, R.L., Weintraub, H. & Lassar, A.B. (1987). Expression of a single transfected cDNA converts fibroblasts to myoblasts. *Cell*, **51**, 987–1000.

Dean, D.C., Knoll, B.J., Riser, M.E. & O'Malley, B.W. (1983). A 5′ flanking sequence essential for progesterone regulation of an ovalbumin fusion gene. *Nature (Lond.)* **305**, 551–3.

Dean, D.C., Gope, R., Knoll, B.J., Riser, M.E. & O'Malley, B.W. (1984). A similar 5′-flanking region is required for oestrogen and progesterone induction of ovalbumin gene expression. *J. Biol. Chem.* **259**, 9967–70.

Dearolf, C.R., Topol, J. & Parker, C.S. (1989). The *caudal* gene is a direct activator of *fushi tarazu* transcription during *Drosophila* embryogenesis. *Nature (Lond.)* **341**, 340–2.

de Boer, E., Antoniou, M., Mignotte, V., Wall, L. & Grosveld, F. (1988). The human β-globin promoter: nuclear protein factors and erythroid specific induction of transcription. *EMBO J.* **7**, 4203–12.

de Cicco, D.V. & Spradling, A.C. (1984). Localisation of a *cis*-acting element responsible for the developmentally regulated amplification of *Drosophila* chorion genes. *Cell*, **28**, 45–54.

De Leon, D.V., Cox, K.H., Angerer, L.M. & Angerer, R.C. (1983). Most early-variant histone mRNA is contained in the pronucleus of sea urchin eggs. *Devl. Biol.* **100**, 197–206.

DeLorenzi, M., Ali, N., Saari, G., Henry, C., Wilcox, M. & Bienz, M. (1988). Evidence that the *Abdominal-B* r element function is conferred by a *trans*-regulatory homeoprotein. *EMBO J.* **7**, 3223–31.

deLotto, R. & Spierer, P. (1986). A gene required for the specification of dorsal-ventral pattern in *Drosophila* appears to encode a serine protease. *Nature (Lond.)* **323**, 688–92.

del Pino, E.M. & Elinson, RE.P. (1983). A novel development pattern for frogs: gastrulation produces an embryonic disc. *Nature (Lond.)* **306**, 589–92.

del Pino, E.M., Steinbeiser, H., Hofman, A., Dreyer, C., Campos, M. & Trendelburg, M.F. (1986). Oogenesis in the egg-brooding frog *Gastrotheca riobambae* produces large oocytes with fewer nucleoli and low RNA content in comparison to *Xenopus laevis*. *Differentiation*, **32**, 24–33.

De Ponti-Zilli, L., Seiler-Tuyns, A. & Paterson, B.M. (1988). A 40-base-pair sequence in the 3′ end of the β-actin gene regulates β-actin mRNA transcription during myogenesis. *Proc. Natl. Acad. Sci. (USA)* **85**, 1389–93.

Derksen, J., Wieslander, L., van der Ploeg, M. & Danholt, B. (1980). Identification of the Balbiani ring 2 chromomere and determination of the content and compaction of its DNA. *Chromosoma*, **81**, 65–84.

Desai, C., Garriga, G., McIntire, S.L. & Horvitz, H.R. (1988). A genetic pathway for the development of the *Caenorhabditis elegans* HSN motor neurons. *Nature (Lond.)* **336**, 639–46.

de Robertis, E.M., Oliver, G. & Wright, C.V.E. (1989). Determination of axial polarity in the vertebrate embryo: homeodomain proteins and homeo-genetic induction. *Cell*, **57**, 189–91.

Deschamps, J., de Laaf, R., Joosen, L., Meijlink, F. & Destrée O. (1987). Abundant expression of homeobox genes in mouse embryonal carcinoma cells correlates with chemically induced differentiation. *Proc. Natl. Acad. Sci. (USA)* **84**, 1304–8.

Desplan, C., Theis, J. & O'Farrell, P.H. (1985). The *Drosophila* developmental gene, *engrailed*, encodes a sequence-specific DNA binding activity. *Nature (Lond.)* **318**, 630–4.

Desplan, C., Theis, J. & O'Farrell, P.H. (1988). The sequence specificity of homeodomain–DNA interaction. *Cell*, **54**, 1081–90.

Deutsch, U., Dressler, G. & Gruss, P. (1988). *Pax-1*, a member of a paired box homologous murine gene family, is expressed in segmented structures during development. *Cell*, **53**, 617–25.

Diaz, M.O., Barasacchi-Pilone, G., Mahon, K.A. & Gall, J.G. (1981). Transcripts from both strands of a satellite DNA occur on lampbrush chromosome loops of the newt *Notophthalmus*. *Cell*, 24, 649–59.

Diaz, M.O. & Gall, J.G. (1985). Giant readthrough transcription units at the histone loci on lampbrush chromosomes of the newt *Notophthalmus*. *Chromosoma*, 92, 243–53.

Di Berardino, M.A. & Hoffner, N.J. (1983). Gene reactivation in erythrocytes: nuclear transplantation in oocytes and eggs of *Rana*. *Science*, 219, 862–4.

Di Berardino, M.A., Hoffner-Orr, N. & McKinnell, R.G. (1986). Feeding tadpoles cloned from *Rana* erythrocyte nuclei. *Proc. Natl. Acad. Sci. (USA)* 83, 8231–4.

Dierich, A., Gaub, M.-P., Le Pennec, J.-P., Astinotti, D. & Chambon, P. (1987). Cell-specificity of the chicken ovalbumin and conalbumin promoters. *EMBO J.* 6, 2305–12.

Dierks, P., van Ooyen, A., Cochran, M.D., Dobkin, C., Reiser, J. & Weissman, C. (1983). Three regions upstream from the cap site are required for efficient and accurate transcription of the rabbit β globin gene in mouse 3T6 cells. *Cell*, 32, 695–706.

Diesseroth, A., Nienhuis, A., Turner, P., Velez, R., Anderson, W.F., Ruddle, F., Lawrence, J., Creagan, R. & Kucherlapati, R. (1977). Localisation of the human α globin structural gene to chromosome 16 in somatic cell hybrids by molecular hybridisation assay. *Cell*, 12, 205–18.

Diesseroth, A., Nienhuis, A., Lawrence, J., Giles, R., Turner, P. & Ruddle, F.H. (1978). Chromosomal localisation of human β globin gene on human chromosome 11 in somatic cell hybrids. *Proc. Natl. Acad. Sci. (USA)* 75, 1456–60.

DiLiberto, M., Lai, Z.-C., Fei, H. & Childs, G. (1989). Developmental control of promoter-specific factors responsible for the embryonic activation and inactivation of the sea urchin early histone H3 gene. *Genes & Dev.* 3, 973–85.

DiNardo, S. & O'Farrell, P.H. (1987). Establishment and refinement of segmental pattern in the *Drosophila* embryo: spatial control of *engrailed* expression by pair-rule genes. *Genes & Dev.* 1, 1212–25.

DiNardo, S., Kuner, J.M., Theis, J. & O'Farrell, P.H. (1985). Development of embryonic pattern in *D. melanogaster* as revealed by accumulation of the nuclear *engrailed* protein. *Cell*, 43, 59–69.

DiNardo, S., Sher, E., Heemskerk-Jongens, J., Kassis, J.A. & O'Farrell, P.H. (1988). Two-tiered regulation of spatially patterned *engrailed* gene expression during *Drosophila* embryogenesis. *Nature (Lond.)* 332, 604–9.

Dobbeling, U., Rob, K., Klein-Hitpaβ, L. Morley, C., Wagner, V. & Ryfell, G.V. (1988). A cell-specific activator in the *Xenopus* A2 vitellogenin gene: promoter elements functioning with rat liver nuclear extracts. *EMBO J.* 7, 2495–501.

Doe, C.Q. & Goodman, C.S. (1985a). Early events in insect neurogenesis: I Development and segmental differences in the pattern of neuronal precursor cells. *Devl. Biol.* 111, 193–205.

Doe, C.Q. & Goodman, C.S. (1985b). Early events in the insect neurogenesis: II The role of cell interactions and cell lineage in the determination of neuronal precursor cells. *Devl. Biol.* 111, 206–19.

Doe, C.Q. & Scott, M.P. (1988). Segmentation and homoeotic gene function in the developing nervous system of *Drosophila*. *Trends Neurosci.* 11, 101–6.

Doe, C.Q., Kuwada, J.Y. & Goodman, C.S. (1985). From epithelium to neuroblasts to neurons: the role of cell interaction and cell lineage during insect neurogenesis. *Phil. Trans. Roy. Soc. Lond. B*, **312**, 67–81.

Doe, C.Q., Hiromi, Y., Gehring, W.J. & Goodman, C.S. (1988a). Expression and function of the segmentation gene *fushi tarazu* during *Drosophila* neurogenesis. *Science*, **239**, 170–4.

Doe, C.Q., Smouse, D. & Goodman, C.S. (1988b). Control of neuronal fate by the *Drosophila* segmentation gene *even-skipped*. *Nature (Lond.)* **333**, 376–9.

Dolan, M., Sugarman, B.M., Dodgson, J.B. & Engel, J.D. (1981). Chromosomal arrangement of the chicken β type globin genes. *Cell*, **24**, 669–77.

Dolecki, G.J., Wannakrairoj, S., Lum, R., Wang, G., Riley, H.D., Carlos, R., Wang, A. & Humphreys, H. (1986). Stage-specific expression of a homeobox-containing gene in the non-segmented sea-urchin embryo. *EMBO J.* **5**, 925–30.

Dony, C & Gruss, P. (1987). Specific expression of the *Hox 1.3* homeobox gene in murine embryonic structures originating from or induced by the mesoderm. *EMBO J.* **6**, 2965–75.

Dony, C. & Gruss, P. (1988). Expression of a murine homeobox gene precedes the induction of *c-fos* during mesodermal differentiation of P19 teratocarcinoma cells. *Differentiation*, **37**, 115–22.

Dorn, A., Bollekens, J., Staub, A., Benoist, C. & Mathis, D. (1987). A multiplicity of CCAAT box-binding proteins. *Cell*, **50**, 863–72.

Doyle, H.J., Harding, K., Hoey, T. & Levine, M. (1986). Transcripts encoded by a homeobox gene are restricted to dorsal tissues of *Drosophila* embryos. *Nature (Lond.)* **323**, 76–9.

Draetta, G., Luca, F., Westendorf, J., Brizuela, L., Ruderman, J. & Beach, D. (1989). *cdc-2* protein kinase is complexed with both cyclin A and B: evidence for proteolytic inactivation of MPF. *Cell*, **56**, 829–38.

Drager, B.J., Harkey, M.A., Iwata, M. & Whitely, A.H. (1989). The expression of primary mesenchyme genes of the sea urchin, *Strongylocentrotus purpuratus*, in the adult skeletogenic tissues of this and other species of echinoderms. *Devl. Biol.* **133**, 14–23.

Drees, B., Ali, Z., Soeller, W.C., Coleman, K.G., Poole, S.J. & Kornberg, T. (1987). The transcription unit of the *Drosophila engrailed* locus: an unusually small portion of a 70,000 bp gene. *EMBO J.* **6**, 2803–9.

Driesch, H. (1898). Uber rein-mütterliche charaktere und bastardlarven von Echiniden. *Arch. Entwickelungsmech. Org.* **7**, 5–102.

Driever, W. & Nusslein-Volhard, C. (1988a). A gradient of *bicoid* protein in *Drosophila* embryos. *Cell*, **54**, 83–93.

Driever, W. & Nusslien-Volhard, C. (1988b). The bicoid protein determines position in the *Drosophila* embryo in a concentration-dependent manner. *Cell*, **54**, 95–104.

Driever, W. & Nusslein-Volhard, C. (1989). The bicoid protein is a positive regulator of *hunchback* transcription in the early *Drosophila* embryo. *Nature (Lond.)* **337**, 138–43.

Driever, W., Ma, J., Nusslein-Volhard, C. & Ptashne, M. (1989a). Rescue of *bicoid* mutant embryos by Bicoid fusion proteins containing heterologous activating sequences. *Nature (Lond.)* **342**, 149–53.

Driever, W., Thoma, G. & Nusslein-Volhard, C. (1989b). Determination of spatial domains of zygotic gene expression in the *Drosophila* embryo by the affinity of binding sites for the bicoid morphogen. *Nature (Lond.)* **340**, 363–6.

Duboule, D., Baron, A., Mahl, P. & Galliot, B. (1986). A new homeobox is present in overlapping cosmid clones which define the mouse Hox-1 locus. *EMBO J.* **5**, 1973–80.

Dubrovsky, E.B. & Zhimulev, I.F. (1988). *Trans* regulation of ecdysterone-induced protein synthesis in *Drosophila melanogaster* salivary glands. *Devl. Biol.* **127**, 33–44.

Dugaiczyk, A., Woo, S.L.C., Colbert, D.A., Lai, E.C., Mace Jr., M.L. & O'Malley, B.W. (1979). The ovalbumin gene: cloning and molecular organisation of the entire natural gene. *Proc. Natl. Acad. Sci. (USA)* **76**, 2253–7.

Dunaway, M. & Droge, P. (1989). Transactivation of the *Xenopus* rRNA gene promoter by its enhancer. *Nature (Lond.)* **341**, 657–9.

Duncan, I. (1982). *Polycomblike*: a gene that appears to be required for the normal expression of the bithorax and Antennapedia gene complexes of *Drosophila melanogaster. Genetics*, **102**, 49–70.

Duncan, I. (1986). Control of bithorax complex functions by the segmentation gene *fushi tarazu* of *D. melanogaster. Cell*, **47**, 297–309.

Duncan, I. (1987). The bithorax complex. *Ann. Rev. Genet.* **21**, 285–319.

Dunphy, W.G. & Newport, J.W. (1988). Unravelling of mitotic control mechanisms. *Cell*, **55**, 925–8.

Dunphy, W.G., Brizuela, L., Beach, D. & Newport, J. (1988). The *Xenopus cdc2* protein is a component of MPF, a cytoplasmic regulator of mitosis. *Cell*, **54**, 423–31.

Duprey, P., Chowdhury, K., Dressler, G.R., Balling, R., Simon, D., Guenet, J.-L. & Gruss, P. (1988). A mouse gene homologous to the *Drosophila* gene *caudal* is expressed in epithelial cells from the embryonic intestine. *Genes & Dev.* **2**, 1647–54.

Dura, J.-M. & Ingham, P. (1988). Tissue- and stage-specific control of homeotic and segmentation gene expression in *Drosophila* embryos by the *polyhomeotic* gene. *Development*, **103**, 733–41.

Dura, J.-M., Randsholt, N.B., Deatrick, J., Erk, I., Santamaria, P., Freeman, S.J., Weddell, D. & Brock, H.W. (1987). A complex genetic locus, *polyhomeotic*, is required for segmental specification and epidermal development in *D. melanogaster. Cell*, **51**, 829–39.

Dworkin, M.B., Shrutkowski, A. & Dworkin-Rastl, E. (1985). Mobilisation of specific maternal RNA species into polysomes after fertilisation in *Xenopus laevis. Proc. Natl. Acad. Sci. (USA)* **82**, 7636–40.

Dworniczak, B., Seidel, R. & Pongs, O. (1983). Puffing activities and binding of ecdysteriod to polytene chromosomes of *Drosophila melanogaster. EMBO J.* **2**, 1323–30.

Dynan, W.S. (1989). Modularity in promoters and enhancers. *Cell*, **58**, 1–4.

Dynan, W.S., Saffer, J.D., Lee, W.S. & Tijan, R. (1985). Transcription factor Sp1 recognises promoter sequences from the monkey genome that are similar to the simian virus 40 promoter. *Proc. Natl. Acad. Sci. (USA)* **82**, 4915–9.

Dynan, W.S., Sazer, S., Tijan, R. & Schimke, R.T. (1986). Transcription factor Sp1 recognises a DNA sequence in the mouse dihydrofolate reductase promoter. *Nature (Lond.)* **319**, 246–9.

Dzierzak, E.A., Papayannopoulou, T. & Mulligan, R.C. (1988). Lineage-specific expression of a human β globin gene in murine bonemarrow recipients reconstituted with retrovirus-transduced stem cells. *Nature (Lond.)* **331**, 35-8.

Early, P., Huang, H., Davis, M., Calame, K. & Hood, L. (1980a). An immunoglobulin heavy chain variable region is generated from three segments of DNA; V_H, D and J_H. *Cell*, **19**, 981-92.

Early, P., Rogers, J., Davis, M., Calame, K., Bond, M., Wall, R & Hood, L. (1980b). Two mRNAs can be produced from a single immunoglobulin μ gene by alternative RNA processing pathways. *Cell*, **20**, 313-9.

Ede, D.A. (1978). *An Introduction to Developmental Biology.* Blackie, London & Glasgow.

Edgar, B.A. & O'Farrell, P. (1989). Genetic control fo cell division patterns in the *Drosophila* embryo. *Cell*, **57**, 177-87.

Edgar, B.A. & Schubiger, G. (1986). Parameters controlling transcriptional activation during early *Drosophila* development. *Cell*, **44**, 871-7.

Edgar B.A., Weir, M.P., Schubiger, G. & Kornberg, T. (1986). Repression and turnover pattern of *fushi tarazu* RNA in the early *Drosophila* embryo. *Cell*, **47**, 747-54.

Edgar, B.A., Odell, G.M., & Schubiger, G. (1987). Cytoarchitecture and the patterning of *fushi tarazu* expression in the *Drosophila* blastoderm. *Genes & Dev.* **1**, 1226-37.

Edgar, L.G. & McGhee, J.D. (1986). Embryonic expression of a gut-specific esterase in *Caenorhabditis elegans*. *Devl. Biol.* **114**, 109-18.

Edgar, L.G. & McGhee, J.D. (1988). DNA synthesis and the control of embryonic gene expression in *C. elegans*. *Cell*, **53**, 589-99.

Edwards, J.S., Milner, M.J. & Chen, S.W. (1978). Integument and sensory nerve differentiation of *Drosophila* leg and wing imaginal discs *in vitro*. *Roux' Arch. Devl. Biol.* **185**, 59-77.

Efstratiadis, A., Posakony, J. W., Maniatis, T., Lawn, R.M., O'Connell, C., Spritz, R.A., de Riel, J.K., Forget, B.G., Weissmann, S.M., Slightom, J.L., Blechl, A.E., Smithies, O., Baralle, F.E., Shoulders, C.C. & Proudfoot, N.J. (1980). The structure and evolution of the human β globin gene cluster. *Cell*, **21**, 653-68.

Eide, D. & Anderson, P. (1985). Transposition of Tc1 in the nematode *Caenorhabditis elegans*. *Proc. Natl. Acad. Sci. (USA)* **82**, 1756-60.

Eide, E. & Anderson, P. (1988). Insertion and excision of *Caenorhabditis elegans* transposable element Tc1. *Mol. Cell. Biol.* **8**, 737-46.

Elgin, S.C.R. (1984). Anatomy of hypersensitive sites. *Nature (Lond.)* **309**, 213-4.

Ellis, H.M. & Horvitz, H.R. (1986). Genetic control of programmed cell death in the nematode *C. elegans*. *Cell*, **44**, 817-29.

Engelke, D.R., Ng, S.-Y., Shastry, B.S. & Roeder, R.G. (1980). Specific interation of a purified transcription factor with an internal control region of 5S RNA genes. *Cell*, **19**, 717-28.

Engels, W.R. (1988). P elements in *Drosophila*. In *Mobile DNA*, ed. D. Berg & M. Howe, pp. 437-484. Amer. Soc. Microbiol. publications, Washington.

Enver, T. (1985). Gene transcription: a pulling out of fingers. *Nature (Lond.)* **317**, 385-6.

Enver, T., Zhang, J.-V., Papayannopoulou, T. & Stamatoyannopoulos, G. (1988). DNA methylation: a secondary event in globin gene switching? *Genes & Dev.* **2**, 698-706.

Epper, F. & Nothiger, R. (1982). Genetic and developmental evidence for a repressed genital primordium in *Drosophila melanogaster*. *Devl. Biol.* **94**, 163–75.

Ernst, S.G., Britten, R.J. & Davidson, E.H. (1979). Distinct single copy sequence sets in sea urchin nuclear RNAs. *Proc. Natl. Acad. Sci. (USA)* **76**, 2209–12.

Ernst, S.G., Hough-Evans, B.R., Britten, R.J. & Davidson, E.H. (1980). Limited complexity of the RNA in micromeres of sixteen cell sea urchin embryos. *Devl. Biol.* **79**, 119–27.

Ettensohn, C.A. (1985). Gastrulation in the sea urchin embryo is accompanied by the rearrangement of invaginating epithelial cells. *Devl. Biol.* **112**, 383–90.

Ettensohn, C.A. & McClay, D.R. (1988). Cell lineage conversion in the sea urchin embryo. *Devl. Biol.* **125**, 396–409.

Evans, R.M., & Hollenberg, S.M. (1988). Zinc fingers: gilt by association. *Cell*, **52**, 1–3.

Evans, T. & Felsenfeld, G. (1989). The erythroid-specific transcription factor Eryf1: a new finger protein. *Cell*, **58**, 877–85.

Evans, T., Rosenthal, E.T., Youngblom, J., Distel, D. & Hunt, T. (1983). Cyclin: a protein specified by maternal mRNA in sea urchin eggs that is destroyed at each cleavage division. *Cell*, **33**, 389–96.

Evans, T., Reitman, M. & Felsenfeld, G. (1988). An erythrocyte-specific DNA-binding factor recognises a regulatory sequence common to all chicken globin genes. *Proc. Natl. Acad. Sci. (USA)* **85**, 5976–80.

Fainsod, A., Bogorad, L.A., Ruusala, T., Lubin, M., Crothers, D.M. & Ruddle, F.H. (1986). The homeodomain of a murine protein binds 5′ to its own homeobox. *Proc. Natl. Acad. Sci. (USA)* **83**, 9532–6.

Fainsod, A., Awgulewitsch, A. & Ruddle, F.H. (1987). Expression of the murine homeobox gene *Hox 1.5* during embryogenesis. *Devl. Biol.* **124**, 125–33.

Falkner, F.G. & Zachau, H.G. (1984). Correct transcription of an immuno-globin κ gene requires an upstream fragment containing conserved sequence elements. *Nature (Lond.)* **310**, 71–3.

Feavers, I.M., Jiricny, J., Montcharmant, B., Saluz, H.P. & Jost, J.P. (1987). Interaction of two non histone proteins with the oestradiol response element of the avian vitellogenin gene modulates the binding of oestradiol–receptor complex. *Proc. Natl. Acad. Sci. (USA)* **84**, 7453–7.

Feiler, R., Harris, W.A., Kirschfeld, K., Wehrhahn, C. & Zuker, C.S. (1988). Targeted misexpression of a *Drosophila* opsin gene leads to altered visual function. *Nature (Lond.)* **333**, 731–41.

Ferguson, E.L., Sternberg, P.W. & Horvitz, H.R. (1987). A genetic pathway for the specification of the vulval cell lineages of *Caenorhabditis elegans*. *Nature (Lond.)* **326**, 259–68.

Fienberg, A.A., Utset, M.F., Bogorad, L.D., Hart, C.P., Awgulewitsch, A., Ferguson-Smith, A., Fainsod, A., Rabin, M. & Ruddle, F.H. (1987). Homeobox genes in murine development. *Curr. Topics Devl. Biol.* **23**, 233–56.

Files, J.G., Carr, S. & Hirsh, D. (1983). Actin gene family of *Caenorhabditis elegans*. *J. Mol. Biol.* **164**, 355–75.

Fink, R.D. & McClay, D.R. (1985). Three cell recognition changes accompany the ingression of sea urchin primary mesenchyme cells. *Devl. Biol.* **107**, 66–74.

Finney, M., Ruvkun, G. & Horvitz, H.R. (1988). The *C.elegans* cell lineage and differentiation gene *unc-86* encodes a protein with a homeodomain and extended similarity to transcription factors. *Cell*, **55**, 757–69.

Fire, A. (1986). Integrative transformation of *Caenorhabditis elegans*. *EMBO J.* **5**, 267–80.

Fire, A. & Waterston, R.H. (1989). Proper expression of myosin genes in transgenic nematodes. *EMBO J.* **8**, 3419–28.

Fjose, A., McGinnis, W.J. & Gehring, W.J. (1985). Isolation of a homeobox-containing gene from the *engrailed* region of *Drosophila* and the spatial distribution of its transcripts. *Nature (Lond.)* **313**, 284–8.

Flavell, R.A. (1982). The mystery of the mouse α globin pseudogene. *Nature (Lond.)* **295**, 370.

Flenniken, A.M. & Newrock, K.M. (1987). H1 subtypes and subtype synthesis switches of normal and delobed embryos of *Ilyanassa obsoleta*. *Devl. Biol.* **124**, 457–68.

Fletcher, C., Heintz, N. & Roeder, R.G. (1987). Purification and characterisation of OTF-1, a transcription factor regulating cell-cycle-specific expression of a human histone H2b gene. *Cell*, **51**, 773–81.

Floyd, E.E., Gong, S., Brandhorst, B.P. & Klein, W.H. (1986). Calmodulin gene expression during sea urchin development: persistence of a prevalent maternal protein. *Devl. Biol.* **113**, 501–11.

Flytzanis, C.N., Britten, R.J. & Davidson, E.H. (1987). Ontogenic activation of a fusion gene introduced into sea urchin eggs. *Proc. Natl. Acad. Sci. (USA)* **84**, 151–5.

Foe, V.E. (1989). Mitotic domains reveal early commitment of cells in *Drosophila* embryos. *Development*, **107**, 1–22.

Forrester, W.C., Novak, U., Gelinas, R. & Groudine, M. (1989). Molecular analysis of the β-globin locus activation region. *Proc. Natl. Acad. Sci. (USA)* **86**, 5439–43.

Frain, M., Swart, G., Monaci, P., Nicosia, A., Stampfli, S., Frank, R. & Cortese, R. (1989). The liver-specific transcription factor LF-B1 contains a highly diverged homeobox DNA-binding domain. *Cell*, **59**, 145–57.

Francke, C., Edstrom, J.-E., McDowall, A.W. & Miller, O.L. Jr (1982). Electron microscopic visualisation of a discrete class of giant translation units in salivary gland cells of *Chironomus tentans*. *EMBO J.* **1**, 59–62.

Franks, R.R., Hough-Evans, B.R., Britten, R.J. & Davidson, E.H. (1988). Spatially deranged though temporally correct expression of a *Strongylocentrotus purpuratus* actin gene fusion in transgenic embryos of a different sea urchin family. *Genes & Dev.* **2**, 1–12.

Frasch, M. & Levine, M. (1987). Complementary patterns of *even-skipped* and *fushi tarazu* expression involve their differential regulation by a common set of segmentation genes in *Drosophila*. *Genes & Dev.* **1**, 981–95.

Frasch, M., Hoey, T., Rushlow, C., Doyle, H. & Levine, M. (1987). Characterisation and localisation of the *even-skipped* protein of *Drosophila*. *EMBO J.* **6**, 749–59.

Freeman, G. (1976). The role of cleavage in the localisation of developmental potential in the ctenophore *Mnemiopsis leidyi*. *Devl. Biol.* **49**, 143–77.

Freeman, G. & Reynolds, G.T. (1973). The development of bioluminescence in the ctenophore *Mnemiopsis leidyi*. *Devl. Biol.* **31**, 61–100.

Freeman, G. & Lundelius, J.W. (1982). The developmental genetics of dextrality and sinistrality in the gastropod *Lymnaea peregra. Roux' Arch. Devl. Biol.* **191**, 69–83.

Frey, A., Sander, K. & Gutzeit, H. (1984). The spatial arrangement of germ line cells in ovarian follicles of the mutant *dicephalic* in *Drosophila melanogaster. Roux' Arch. Devl. Biol.* **193**, 388–93.

Frigiero, G., Burri, M., Bopp, D., Baumgartner, S. & Noll, M. (1986). Structure of the segmentation gene *paired* and the *Drosophila* PRD gene set as part of a gene network. *Cell*, **47**, 735–46.

Fritsch, E.F., Lawn, R.M. & Maniatis, T. (1980). Molecular cloning and characterisation of the human β-like globin gene cluster. *Cell*, **19**, 959–72.

Frischer, L.E., Hagen, F.S. & Garber, R.L. (1986). An inversion that disrupts the *Antennapedia* gene causes abnormal structure and localisation of RNAs. *Cell*, **47**, 1017–23.

Fritton, H.P., Igo-Kemenes, T., Nowock, J., Strech-Jurk, U., Thiesen, M. & Sippel, A.E. (1984). Alternative sets of DNase I-hypersensitive sites characterise the various functional states of the chicken lysozyme gene. *Nature (Lond.)* **311**, 163–6.

Frohnhofer, H.G. & Nusslein-Volhard, C. (1986). Organisation of anterior pattern in the *Drosophila* embryo by the maternal gene *bicoid. Nature (Lond.)* **324**, 120–6.

Frohnhofer, H.G. & Nusslein-Volhard, C. (1987). Maternal genes required for the anterior localisation of *bicoid* activity in the embryo of *Drosophila. Genes & Dev.* **1**, 880–90.

Galau, G.A., Klein, W.H., Davis, M.M., Wold, B.J., Britten, R.J. & Davidson, E.H. (1976). Structural gene sets active in embryos and adult tissues of the sea urchin. *Cell*, **7**, 487–505.

Galau, G.A., Klein, W.H., Britten, R.J. & Davidson, R.J. (1977). Significance of rare mRNA sequences in liver. *Arch. Biochem. Biophys.* **179**, 584–99.

Galli, G., Hofstetter, H. & Birnstiel, M.L. (1981). Two conserved sequence blocks within eucaryotic tRNA genes are major promoter elements. *Nature (Lond.)* **294**, 626–9.

Galli, G., Hofstetter, H., Stunnenberg, H.G. & Birnstiel, M.L. (1983). Biochemical complementation with RNA in the *Xenopus* oocyte: a small RNA is required for the generation of 3' histone mRNA termini. *Cell*, **34**, 823–32.

Garber, R.L., Kuroiwa, A. & Gehring, W.J. (1983). Genomic and cDNA clones of the homeotic locus *Antennapaedia* in *Drosophila. EMBO J.* **2**, 2027–36.

Garcia, A.D., O'Connell, A.M.& Sharp, S.J. (1987). Formation of an active transcription complex in the *Drosophila melanogaster* 5S RNA gene is dependent on an upstream region. *Mol. & Cell. Biol.* **7**, 2046–51.

Garcia-Bellido, A. & Merriam, J.R. (1969). Cell lineage of the imaginal discs in *Drosophila* gynandromorphs. *J. Exp. Zool.* **170**, 61–75.

Garcia-Bellido, A., Ripoll, P. & Morata, G. (1973). Developmental compartments of the wing disk of *Drosophila. Nature New Biol.* **245**, 251–3.

Garcia-Bellido, A., Lawrence, P.A., & Morata, G. (1979). Compartments in animal development. *Scient. Amer.*, **241**, 102–10.

Garcia-Blanco, M.A., Clerc, R.G. & Sharp, P.A. (1989). The DNA-binding homeodomain of the Oct-2 protein. *Genes & Dev.* **3**, 739–45.

Garel, A. & Axel, R. (1976). Selective digestion of transcriptionally-active ovalbumin genes from oviduct nuclei. *Proc. Natl. Acad. Sci. (USA)* **73**, 70.

Gaub, M.-P., Dierich, A., Astinotti, D., Touitou, I. & Chambon, P. (1987). The chicken ovalbumin promoter is under negative control which is relieved by steroid hormones. *EMBO J.* **6**, 2313–20.

Gaul, U. & Jackle, H. (1987). Pole region-dependent repression of the *Drosophila* gap gene *Kruppel* by maternal gene products. *Cell*, **51**, 549–55.

Gaul, U., Seifert, E., Schuh, R. & Jackle, H. (1987). Analysis of *Kruppel* protein distribution during early *Drosophila* development reveals post-transcriptional regulation. *Cell*, **50**, 639–47.

Gaunt, S.J. (1987). Homeobox gene *Hox 1.5* expression in mouse embryos: earliest detection by *in situ* hybridisation is during gastrulation. *Development*, **101**, 51–60.

Gaunt, S.J., Krumlauf, R. & Duboule, D. (1989). Mouse homeo-genes within a subfamily, *Hox-1.4*, *-2.6* and *-5.1*, display similar anteroposterior domains of expression in the embryo, but show stage- and tissue-dependent differences in their regulation. *Development*, **107**, 131–41.

Gautier, J., Norbury, C., Lohka, M., Nurse, P. & Maller, J. (1988). Purified maturation-promoting factor contains the product of a *Xenopus* homologue of the fission yeast cell cycle control gene *cdc 2+*. *Cell*, **54**, 433–9.

Gay, N.J., Poole, S. & Kornberg, T. (1988). Association of the *Drosophila melanogaster engrailed* protein with specific soluble nuclear protein complexes. *EMBO J.* **7**, 4291–7.

Gehring, W.J. & Hiromi, Y. (1986). Homoeotic genes and the homeobox. *Ann. Rev. Genet.* **20**, 147–73.

Gelinas, R., Endlich, B., Pfeiffer, C., Yagi, M. & Stamatoyannopoulos, G. (1985). A substitution in the distal CCAT box of the $^A\gamma$-globin gene in Greek hereditary persistence of foetal haemoglobin. *Nature (Lond.)* **313**, 323–6.

Gerasimova, T.I. & Smirnova, S.G. (1979). Maternal effect for genes encoding 6-phosphogluconate dehydrogenase and glucose 6-phosphate dehydrogenase in *Drosophila melanogaster*. *Dev. Genet.* **1**, 97–107.

Gergen, J.P. & Butler, B.A. (1988). Isolation of the *Drosophila* segmentation gene *runt* and analysis of its expression during embryogenesis. *Genes & Dev.* **2**, 1179–93.

Gergen, J.P. & Wieschaus, E. (1986). Dosage requirements for *runt* in the segmentation of *Drosophila* embryos. *Cell*, **45**, 289–99.

Gerhart, J., Wu, M. & Kirschner, M. (1984). Cell cycle dynamics of an M-phase-specific cytoplasmic factor in *Xenopus laevis* oocytes and eggs. *J. Cell Biol.* **98**, 1247–55.

Geyer-Duszynska, I. (1966). Genetic factors in oogenesis and spermatogenesis in *Cecidomyiidae*. *Chromosomes Today*, **1**, 174–8.

Ghysen, A. & Dambly-Chaudière, C. (1988). From DNA to form: the *achaete-scute* complex. *Genes & Dev.* **2**, 495–501.

Ghysen, A., Dambly-Chaudière, C., Jan, L.Y. & Jan, Y.N. (1982). Segmental differences in the protein content of imaginal discs. *EMBO J.* **1**, 1373–9.

Ghysen, A., Jan, L.Y. & Jan, Y.N. (1985). Segmental determination in *Drosophila* central nervous system. *Cell*, **40**, 943–8.

Gibson, A.W. & Burke, R.D. (1985). The origin of pigment cells in embryos of the sea urchin *Strongylocentrotus purpuratus*. *Devl. Biol.* **107**, 414–22.

Giebelhaus, D.H, Heikkala, J.J. & Schultz, G.A. (1983). Changes in the quantity of histone and actin messenger RNA during the development of preimplantation mouse embryos. *Devl. Biol.* **98**, 148–57.

Giglioni, G., Gianni, A.M., Comi, P., Ottolenghi, S. & Runnger, D. (1973). Translational control of globin syntheses by haemin in *Xenopus* oocytes. *Nature New Biol.* **246**, 99–102.

Giguère, V., Ong, E.S., Segui, P. & Evans, R.M. (1987). Identification of a receptor for the morphogen retinoic acid. *Nature (Lond.)* **330**, 624–8.

Gillies, S.D., Morrison, S.L., Oi, V.T. & Tonegawa, S. (1983). A tissue-specific transcription enhancer element is located in the major intron of a rearranged immunoglobulin heavy-chain gene. *Cell*, **33**, 717–28.

Gilmartin, G.M., McDevitt, M.A. & Nevins, J.R. (1988). Multiple factors are required for specific RNA cleavage at a poly(A) addition site. *Genes & Dev.* **2**, 278–87.

Gilmour, D.S., Pflugfelder, G., Wang, J.C. & Lis, J.T. (1986). Topoisomerase I interacts with transcribed regions in *Drosophila* cells. *Cell*, **44**, 401–7.

Ginsberg, A.M., King, B.O. & Roeder, R.G. (1984). *Xenopus* 5S gene transcription factor, TFIIIA: characterisation of a cDNA clone and measurement of RNA levels throughout development. *Cell*, **39**, 479–89.

Glass, C.K., Holloway, J.M., Devary, O.V. & Rosenfeld, M.G. (1988). The thyroid hormone receptor binds with opposite transcriptional effects to a common sequence motif in thyroid hormone and oestrogen response elements. *Cell*, **55**, 313–23.

Glicksman, M.A. & Brower, D.L. (1988). Expression of the *Sex combs reduced* protein in *Drosophila* larvae. *Dev. Biol.* **127**, 113–8.

Glover, D.M., Zaha, A., Stocker, A.J., Santelli, R.V., Pueyo, M.T., de Toledo, S.M. & Lara, F.J.S. (1982). Gene amplification in *Rhynochosciara* salivary gland chromosomes. *Proc. Natl. Acad. Sci. (USA)* **79**, 2947–51.

Goldberg, D., Posakony, J. & Maniatis, T. (1983). Correct developmental expression of a cloned alcohol dehydrogenase gene transduced into the *Drosophila* germ line. *Cell*, **34**, 59–73.

Golden, J.W. & Riddle, D.L. (1985). A gene affecting production of the *Caenorhabditis elegans* dauer-inducing pheromone. *Mol. & Gen. Genet.* **198**, 534–6.

Golden, L., Schafer, U. & Rosbash, M. (1980). Accumulation of individual poly A$^+$ RNAs during oogenesis of *Xenopus laevis*. *Cell*, **22**, 835–44.

Golub, E.S. (1987). Somatic mutation: diversity and regulation of the immune response. *Cell*, **48**, 723–4.

Goodbourn, S.E.Y., Higgs, D.R., Clegg, J.B. & Weatherall, D.J. (1983). Molecular basis of length polymorphism in the human ζ_2 globin gene complex. *Proc. Natl. Acad. Sci. (USA)* **80**, 5022–6.

Goralski, T.J., Edstrom, J.-E. & Baker, B.S. (1989). The sex determination locus *transformer-2* of *Drosophila* encodes a polypeptide with similarity to RNA-binding proteins. *Cell*, **56**, 101–18.

Goto, T., Macdonald, P. & Maniatis, T. (1989). Early and late patterns of *even-skipped* expression are controlled by distinct regulatory elements that respond to different spatial cues. *Cell*, **57**, 413–22.

Gossett, L.A. & Hecht, R.M. (1982). Muscle differentiation in normal and cleavage-arrested mutant embryos of *Caenorhabditis elegans*. *Cell*, **30**, 193–204.

Gough, N.M., Kemp, D.J., Tyler, B.M., Adam, J.M. & Cory, S. (1980). Intervening sequences divide the gene for the constant region of mouse immunoglobulin μ chains into segments, each encoding a domain. *Proc. Natl. Acad. Sci. (USA)* **77**, 554–8.

Grabowski, P.J., Seiler, S.R. & Sharp, P.A. (1985). A multicomponent complex is involved in the splicing of messenger RNA precursors. *Cell*, **42**, 345–53.

Graham, A., Papalopulu, N., Lorimer, J., McVey, J.H., Tuddenham, E.G.D. & Krumlauf, R. (1988). Characterisation of a murine homeobox gene, *Hox-2.6*, related to the *Drosophila Deformed* gene. *Genes & Dev.* **2**, 1424–38.

Graham, A., Papalopulu, N. & Krumlauf, R. (1989). The murine and *Drosophila* homeobox gene clusters have common features of organisation and expression. *Cell*, **57**, 367–78.

Grainger, R.M., Henry, J.J. & Henderson, R.A. (1988). Reinvestigation of the role of the optic vesicle in embryonic lens induction. *Development*, **102**, 517–26.

Green, S., Kumar, V., Theulaz, I., Wahli, W. & Chambon, P. (1988). The N-terminal DNA-binding 'zinc-finger' of the oestrogen and glucocorticoid receptors determines target gene specificity. *EMBO J.* **7**, 3037–44.

Greenberg, R.M. & Adler, P.N. (1982). Protein synthesis and accumulation in *Drosophila melanogaster* imaginal discs: identification of a protein with a nonrandom spatial distribution. *Devl. Biol.* **89**, 273–86.

Greenwald, I. (1985). *lin-12*, a nematode homoeotic gene, is homologous to a set of mammalian proteins that includes epidermal growth factor. *Cell*, **43**, 583–90.

Greenwald, I.S., Sternberg, P.W. & Horvitz, H.R. (1983). The *lin-12* locus specifies cell fates in *Caenorhabditis elegans*. *Cell*, **34**, 435–44.

Griffin-Shea, R., Thireos, G. & Kafatos, F.C. (1982). Organisation of a cluster of four chorion genes in *Drosophila* and its relationship to developmental expression and amplification. *Devl. Biol.* **91**, 325–36.

Gronemeyer, H. & Pongs, O. (1980). Localisation of ecdysterone on polytene chromosomes of *Drosophila melanogaster*. *Proc. Natl. Acad. Sci. (USA)* **77**, 2108–12.

Gronemeyer, H., Turcotte, B., Quirin-Stricker, C., Bocquel, M.T., Meyer, M.E., Krozowski, Z., Lerouge, T., Garnier, J.M. & Chambon, P. (1987). The chicken progesterone receptor: sequence, expression and functional analysis. *EMBO J.* **6**, 3985–94.

Grossbach, U. (1973). Chromosome puffs and gene expression in polytene cells. *Cold Spring Harbor Symp. Quant. Biol.* **38**, 619–27.

Grosschedl, R. & Baltimore, D. (1985). Cell-type specificity of immunoglobulin gene expression is regulated by at least three DNA sequence elements. *Cell*, **41**, 885–97.

Grosschedl, R. & Birnstiel, M.L. (1980a). Identification of regulatory sequences in the prelude sequences of an H2A histone gene by the study of specific deletion mutants *in vivo*. *Proc. Natl. Acad. Sci. (USA)* **77**, 1432–6.

Grosschedl, R. & Birnstiel, M.L. (1980b). Spacer DNA sequences upstream of the TATAAATA sequence are essential for promotion of H2A histone gene transcription *in vivo. Proc. Natl. Acad. Sci. (USA)* **77**, 7102–6.

Grosschedl, R., Machler, M., Rohrer, U. & Birnstiel, M.L. (1983). A functional component of the sea urchin H2A gene modulator contains an extended sequence homology to a viral enhancer. *Nucleic Acids Res.* **11**, 8123–9.

Grosveld, G.C. Shewmaker, C.M., Jat, P. & Flavell, R.A. (1981). Localisation of DNA sequences necessary for transcription of the rabbit β globin gene *in vitro. Cell*, **25**, 215–26.

Groudine, M. & Weintraub, H. (1981). Activation of globin genes during chicken development. *Cell*, **24**, 393–401.

Groudine, M. & Weintraub, H. (1982). Propagation of globin DNaseI-hypersensitive sites in the absence of factors required for induction: a possible mechanism for determination. *Cell*, **30**, 131–9.

Groudine, M., Peretz, M. & Weintraub, H. (1981). Transcriptional regulation of haemoglobin switching in chickens. *Mol. Cell. Biol.* **1**, 281–8.

Groudine, M., Kohwi-Shigematsu, T., Gelinas, R., Stamatoyannopoulos, G. & Papayannopoulou, T. (1983). Human foetal to adult haemoglobin switching: changes in chromatin structure of the β globin gene locus. *Proc. Natl. Acad. Sci. (USA)* **80**, 7551–5.

Gruenbaum, Y., Cedar, H, & Razin, A. (1982). Substrate and sequence specificity of a eucaryotic DNA methylase. *Nature (Lond.)* **295**, 620–2.

Grummt, I., Rosenbauer, H., Niedermeyer, I., Maier, U. & Ohrlein, A. (1986). A repeated 18bp sequence in the mouse rDNA spacer mediates binding of a nuclear factor and transcriptional termination. *Cell*, **45**, 837–46.

Gruskin, K.D., Smith, T.F. & Goodman, M. (1987). Possible origin of a calmodulin gene that lacks intervening sequences. *Proc. Natl. Acad. Sci. (USA)* **84**, 1605–8.

Guerrier, P., van der Biggelaar, J.A.M., van Dongen, C.A.M. & Verdonk, N.H. (1978). Significance of the polar lobe for the determination of dorso-ventral polarity in *Dentalium vulgare* (da Costa). *Devl. Biol.* **63**, 233–42.

Gurdon, J.B. (1962). Adult frogs derived from the nuclei of single somatic cells. *Devl. Biol.* **4**, 256–73.

Gurdon, J.B. (1987). Embryonic induction – molecular prospects. *Development*, **99**, 285–306.

Gurdon, J.B. & Laskey, R.A. (1970). The transplantation of nuclei from single cultured cells into enucleated frogs' eggs. *J. Embryol. Exp. Morphol.* **24**, 227–48.

Gurdon, J.B. & Uehlinger, V. (1966). 'Fertile' intestinal nuclei. *Nature (Lond.)* **210**, 1240–1.

Gurdon, J.B., Fairman, S., Mohun, T.J. & Brennan, S. (1985a). Activation of muscle-specific actin genes in *Xenopus* development by an induction between animal and vegetal cells of a blastula. *Cell*, **41**, 913–22.

Gurdon, J.B., Mohun, T.J., Fairman, S. & Brennan, S. (1985b). All components required for the eventual activation of muscle-specific actin genes are localised in the subequatorial region of an uncleaved amphibian egg. *Proc. Natl. Acad. Sci. (USA)* **82**, 139–43.

Hadorn, E. (1978). Transdetermination, In *The Genetics and Biology of Drosophila*, Vol. 2C, ed. M. Ashburner & T.R.F. Wright, pp 555–617. Academic Press, New York & London.

Hafen, E., Kuroiwa, A & Gehring, W.J. (1984). Spatial distribution of transcripts from the segmentation gene *fushi tarazu* during *Drosophila* embryonic development. *Cell*, 37, 833–41.

Hafen, E., Basler, K. Edstroem, J.E. & Rubin, G.M. (1987). *sevenless*, a cell-specific homoeotic gene of *Drosophila*, encodes a putative trans-membrane receptor with a tyrosine kinase domain. *Science*, 236, 55–62.

Hai, T., Horikoshi, M., Roeder, R.G. & Green, M.R. (1988a). Analysis of the role of the transcription factor ATF in the assembly of a functional preinitiation complex. *Cell*, 54, 1043–51.

Hai, T., Liu, F., Allegretto, E.A., Karin, M. & Green, M.R. (1988b). A family of immunologically related transcription factors that includes multiple forms of ATF and AP-1. *Genes & Dev.* 2,.1216–26.

Hall, J.C., Gelbart, W.M. & Kankel, D.R. (1976). Mosaic systems. In *The Genetics and Biology of* Drosophila, Vol. 1a, ed. M. Ashburner & E. Novitski, pp. 265–314. Academic Press, New York.

Hall, P.A. & Watt, F.M. (1989). Stem cells: the generation and maintenance of cellular diversity. *Development*, 106, 619–33.

Hannah-Alava, A. & Stern, C. (1957). The sex-combs in males and intersexes of *Drosophila melanogaster*. *J. Exp. Zool.* 134, 533–56.

Hanneman, E., Trevarrow, B., Metcalfe, W.K., Kimmel, C.B. & Westerfield, M. (1988). Segmental pattern of development of the hindbrain and spinal cord of the zebrafish embryo. *Development*, 103, 49–58.

Harbecke, R. & Janning, W. (1989). The segmentation gene *Kruppel* of *Drosophila melanogaster* has homoeotic properties. *Genes & Dev.* 3, 114–22.

Harding, K. & Levine, M. (1988). Gap genes define the limits of Antennapedia and bithorax gene expression during early development in *Drosophila*. *EMBO J.* 7, 205–14.

Harding, K., Hoey, T., Warrior, R. & Levine, M. (1989). Autoregulatory and gap gene response elements of the *even-skipped* promoter of *Drosophila*. *EMBO J.* 8, 1205–12.

Harkey, M.A. & Whiteley, A.H. (1983). The program of protein synthesis during the development of the micromere–primary mesenchyme cell line in the sea urchin embryo. *Devl. Biol.* 100, 12–28.

Harkey, M.A., Whiteley, H.R. & Whiteley, A.R. (1988). Coordinate accumulation of five transcripts in the primary mesenchyme during skeletogenesis in the sea urchin embryo. *Devl. Biol.* 125, 381–95.

Harland, R. & Misher, L. (1988). Stability of RNA in developing *Xenopus* oocytes and identification of a destabilising sequence in TFIIIA messenger RNA. *Development*, 102, 837–52.

Harlow, P. & Nemer, M. (1987). Developmental and tissue-specific regulation of β-tubulin gene expression in the embryo of the sea urchin *Strongylocentrotus purpuratus*. *Genes & Dev.* 1, 147–60.

Harper, M.L. & Monk, M. (1983). Evidence for translation of HPRT enzyme on maternal mRNA in early mouse embryos. *J. Embryol. Exp. Morphol.* 74, 15–28.

Hart, C.P., Awgulewitsch, A., Fainsod, A., McGinnis, W. & Ruddle, F.H. (1985). Homeobox gene complex on mouse chromosome 11: molecular cloning, expression in embryogenesis, and homology to a human homeobox locus. *Cell*, 43, 9–18.

Hart, C.P., Fainsod, A. & Ruddle, F.H. (1987). Sequence analysis of the murine *Hox 2.2, 2.3* and *2.4* homeoboxes: evolutionary and structural comparisons. *Genomics*, **1**, 182–95.

Hartley, D.A., Xu, T. & Artavanis-Tsakonas, S. (1987). The embryonic expression of the *Notch* locus of *Drosophila melanogaster* and the implications of point mutations in the extracellular EGF-like domain of the predicted protein. *EMBO J.* **6**, 3407–17.

Hartley, D.A., Preiss, A. & Artavanis-Tsakonas, S. (1988). A deduced gene product from the *Drosophila* neurogenic locus, *Enhancer of split*, shows homology to mammalian G-protein β subunit. *Cell*, **55**, 785–95.

Harvey, E.B. (1936). Parthenogeneic merogony or cleavage without nuclei in *Arbacia punctulata. Biol. Bull.* **71**, 101–21.

Harvey, R.P. & Melton, D.A. (1988). Microinjection of synthetic *Xhox-1A* homeobox mRNA disrupts somite formation in developing *Xenopus* embryos. *Cell*, **53**, 687–97.

Harvey, R.P., Tabin, C.J. & Melton, D.A. (1986). Embryonic expression and nuclear localisation of *Xenopus* homeobox (*Xhox*) gene products. *EMBO J.* **5**, 1237–44.

Hashimoto, C., Hudson, K.L. & Anderson, K.V. (1988). The *Toll* gene of *Drosophila*, required for dorsal-ventral embryonic polarity, appears to encode a transmembrane protein. *Cell*, **52**, 269–79.

Hauser, C.A,. Joyner, A.L., Klein, R.D., Learned, T.K., Martin, G.R. & Tijan, R. (1985). Expression of homologous homeobox-containing genes in differentiated human teratocarcinoma cells and mouse embryos. *Cell*, **43**, 19–28.

Hawley, R.J. & Waring, G.L. (1988). Cloning and analysis of the *dec-1* female sterile locus, a gene required for proper assembly of the *Drosophila* eggshell. *Genes & Dev.* **2**, 341–9.

Hay, B., Jan, L.Y. & Jan, Y.N. (1988). A protein component of *Drosophila* polar granules is encoded by *vasa* and has extensive sequence similarity to ATP-dependent helicases. *Cell*, **55**, 577–87.

Hayes, P.H., Sato, T. & Dennell, R.E. (1984). Homoeosis in *Drosophila:* the *Ultrabithorax* larval syndrome. *Proc. Natl. Acad. Sci. (USA)* **81**, 545–9.

Haynie, J.L. (1983). The maternal and zygotic roles of the gene *Polycomb* in embryonic determination in *Drosophila melanogaster. Devl. Biol.* **100**, 399–411.

He, X., Treacy, M.N., Simmons, D.M., Ingraham, H.A., Swanson, L.W. & Rosenfeld, M.G. (1989). Expression of a large family of POU-domain regulatory genes in mammalian brain development. *Nature (Lond.)* **340**, 35–9.

Hecht, R.M., Gossett, L.A. & Jeffery, W.R. (1981). Ontogeny of maternal and newly-transcribed mRNA analysed by *in situ* hybridisation during development of *Caenorhabditis elegans. Devl. Biol.* **83**, 374–9.

Hedgecock, E.M. (1985). Cell lineage mutants in the nematode *Caenorhabditis elegans. Trends Neurosci.* **8**, 288–93.

Hedgecock, E.M., Sulston, J.E. & Thomson, J.N. (1983). Mutations affecting programmed cell deaths in the nematode *Caenorhabditis elegans. Science*, **220**, 1277–9.

Heine, U. & Blumenthal, T. (1986). Characterisation of regions of the *Caenorhabditis elegans* X chromosome containing vitellogenin genes. *J. Mol. Biol.* **188**, 301–12.

Herman, R.K. (1987). Mosaic analysis of two genes that affect nervous system structure in *Caenorhabditis elegans*. *Genetics*, **116**, 377–88.

Herr, W., Sturm, R.A., Clerc, R.G., Corcoran, L.M., Baltimore, D., Sharp, P.A., Ingraham, H.A., Rosenfeld, M.G., Finney, M., Ruvkun, G. & Horvitz, H.R., (1988). The POU domain: a large conserved region in the mammalian *pit-1*, *oct-1* and *oct-2* and *Caenorhabditis elegans unc-86* gene products. *Genes & Dev.* **2**, 1513–16.

Hicks, J.B., (1987). Mechanisms of differentiation. *(Nature (Lond.)* **326**, 444–5.

Hill, D.P., Shakes, D.C., Ward, S. & Strome, S. (1989). A sperm-supplied product essential for initiation of normal embryogenesis in *Caenorhabditis elegans* is encoded by the paternal-effect embryonic-lethal gene, *spe-11*. *Devl. Biol.* **136**, 154–66.

Hill, R.E., Hall, A.E., Sime, C.M. & Hastie, N.D. (1987). A mouse homeobox-containing gene maps near a developmental mutation. *Cytogenet. Cell Genet.* **44**, 171–4.

Hill, R.E., Jones, P.F., Rees, A.R., Sime, C.M., Justice, M.J., Copeland, N.G., Jenkins, N.A., Graham, E., & Davidson, D. R. (1989). A new family of mouse homeobox-containing genes: molecular structure, chromosomal location and developmental expression of *Hox 7.1*. *Genes & Dev.* **3**, 26–37.

Hill, R.S. & MacGregor, H.C. (1980). The development of lampbrush chromosome-type transcription in the early diplotene oocytes of *Xenopus laevis*: an electron-microscope analysis. *J. Cell Sci.* **44**, 87–101.

Hiromi, Y. & Gehring, W.J. (1987). Regulation and function of the *Drosophila* segmentation gene *fushi tarazu*. *Cell*, **50**, 963–74.

Hiromi, Y., Kuroiwa, A. & Gehring, W.J. (1985). Control elements of the *Drosophila* segmentation gene *fushi tarazu*. *Cell*, **43**, 603–13.

Hirsh, D. & Vanderslice, R. (1976). Temperature-sensitive developmental mutants of *Caenorhabditis elegans*. *Devl. Biol.* **49**, 220–35.

Hochman, B. (1973). Analysis of a whole chromosome in *Drosophila*. *Cold Spring Harbor Symp. Quant. Biol.* **38**, 581–9.

Hodgkin, J. (1985). Novel nematode amber suppressors. *Genetics*, **111**, 287–310.

Hodgkin, J. (1987a). Sex determination and dosage compensation in *Caenorhabditis elegans*. *Ann. Rev. Genet.* **21**, 133–54.

Hodgkin, J. (1987b). A genetic analysis of the sex-determining gene *tra-1*, in the nematode *Caenorhabditis elegans*. *Genes & Dev.* **1**, 731–45.

Hodgkin, J. (1988a). Sexual dimorphism and sex determination. In *The Nematode* Caenorhabditis elegans, ed. W.B. Wood. CSH Monograph **17**, 243–79.

Hodgkin, J. (1988b). Everything you always wanted to know about sex . . . *Nature (Lond.)* **331**, 300–1.

Hodgkin, J. (1989). *Drosophila* sex determination: a cascade of regulated splicing. *Cell*, **56**, 905–6.

Hoey, T. & Levine, M. (1988). Divergent homeobox proteins recognise similar DNA sequences in *Drosophila*. *Nature (Lond.)* **332**, 858–61.

Hoey, T., & Doyle, H.J., Harding, K., Wedeen, C. & Levine, M. (1986). Homeobox gene expression in anterior and posterior regions of the *Drosophila* embryo. *Proc. Natl. Acad. Sci. (USA)* **83**, 4809–13.

Hofstetter, H., Kressmann, A. & Birnstiel, M.L. (1981). A split promoter for a eucaryotic tRNA gene. *Cell*, **24**, 573–85.

Hogan, B., Holland, P. & Schofield, P. (1985). How is the mouse segmented? *Trends Genet.* **1**, 67–74.

Holland, P.W.H. & Hogan, B.L.M. (1986). Phylogenetic distribution of *Antennapedia*-like homeoboxes. *Nature (Lond.)* **321**, 251–3.

Holland, P.W.H. & Hogan, B.L.M. (1988a). Expression of homeobox genes during mouse development: a review. *Genes & Dev.* **2**, 773–82.

Holland, P.W.H. & Hogan, B.L.M. (1988b). Spatially restricted patterns of expression of the homeobox-containing gene *Hox 2.1* during mouse embryogenesis. *Development*, **102**, 159–74.

Hollis, G.F., Hieter, P.A., McBride, O.W., Swan, D. & Leder, P. (1982). Processed genes: a dispersed human immunoglobulin gene bearing evidence of RNA-type processing. *Nature (Lond.)* **296**, 321–6.

Honda, B.M. & Roeder, R.G. (1980). Association of a 5S gene transcription factor with 5S RNA and altered levels of the factor during cell differentiation. *Cell*, **22**, 119–26.

Hooper, J.E. (1986). Homoeotic gene function in the muscles of *Drosophila* larvae. *EMBO J.* **5**, 2321–9.

Hoppe, P.E. & Greenspan, R.J. (1986). Local function of the *Notch* gene for embryonic ectodermal pathway choice in *Drosophila*. *Cell*, **46**, 773–83.

Hopwood, N.D., Pluck, A. & Gurdon, J.B. (1989a). Myo D expression in forming somites is an early response to mesoderm induction in *Xenopus* embryos. *EMBO J.* **8**, 3409–17.

Horikoshi, M., Hai, T., Lin, Y.-S., Green, M.R. & Roeder, R.G. (1988). Transcription factor ATF interacts with the TATA factor to facilitate establishment of a preinitiation complex. *Cell*, **54**, 1033–42.

Hörstadius, S. (1928). Über die determination des keimes der Echinodermen. *Acta. Zool. Stockholm* **9**, 1–192.

Hörstadius, S, (1973). *Experimental Embryology of Echinoderms*. Oxford University Press, London and New York.

Horvitz, H.R. (1988). Genetics of cell lineage. In *The Nematode* Caenorhabditis elegans, ed. W.B. Wood. CSH Monographs **17**, 157–90.

Horvitz, H.R., Sternberg, P.W., Greenwald, I.S., Fixen, W. & Ellis, H.M. (1983). Mutations that affect neural cell lineage and cell fates during the development of the nematode *Caenorhabditis elegans*. *Cold Spring Harbor Symp. Quant. Biol.* **48**, 453–63.

Hough-Evans, B.R., Wold, B.J., Ernst, S.G., Britten, R.J. & Davidson, E.H. (1977). Appearance and persistence of maternal RNA sequences in sea urchin development. *Devl. Biol.* **60**, 258–77.

Hough-Evans, B.R., Ernst, S.G., Britten, R.J., & Davidson, E.H. (1979). RNA complexity in developing sea urchin oocytes. *Devl. Biol.* **69**, 258–69.

Hourcade, D., Dressler, D. & Wolfson, J. (1973). The nucleolus and the rolling circle. *Cold Spring Harbor Symp. Quant. Biol.* **38**, 537–50.

Howard, K.R. & Ingham, P.W. (1986). Regulatory interactions between the segmentation genes *fushi tarazu*, *hairy* and *engrailed* in the *Drosophila* blastoderm. *Cell*, **44**, 949–57.

Howard, K., Ingham, P. & Rushlow, C. (1988). Region-specific alleles of the *Drosophila* segmentation gene *hairy*. *Genes & Dev.* **2**, 1037–46.

Hsu, S.-L., Marks, J., Shaw, J.-P., Tam, M., Higgs, D.R., Shen, C.C. & Shen C.-

K.J. (1988). Structure and expression of the human θ_1 globin gene. *Nature (Lond.)* **331**, 94–6.

Huber, P.W. & Wool, I.G. (1986). Identification of the binding site on 5S rRNA for the transcription factor IIIA: proposed structure of a common binding site on 5S rRNA and on the gene. *Proc. Natl. Acad. Sci. (USA)* **83**, 1593–7.

Huckaby, C.S., Conneely, O.M., Beattie, W.G., Dobson, A.D.W., Tasai, M.-J. & O'Malley, B.W. (1987). Structure of the chromosomal chicken progesterone receptor gene. *Proc. Natl. Acad. Sci. (USA)* **84**, 8380–4.

Hülskamp, M., Schroder, C., Pfeifle, C., Jackle, H. & Tautz, D. (1989). Posterior segmentation of the *Drosophila* embryo in the absence of a maternal posterior organiser gene. *Nature (Lond.)* **338**, 629–33.

Humphrey, T.H. & Proudfoot, N.J. (1988). A beginning to the biochemistry of polyadenylation. *Trends Genet.* **4**, 243–5.

Hutchinson, N. & Weintraub, H. (1985). Localisation of DNAaseI-sensitive sequences to specific regions in interphase nuclei. *Cell*, **43**, 471–82.

Hyer, B.J. & Chan, L.-N.L. (1978). Initial synthesis of globin peptide chains in differentiating embryonic red blood cells. *Devl. Biol.* **66**, 279–84.

Iguchi-Ariga, S.M.M. & Schaffner, W. (1989). CpG methylation of the cAMP-responsive enhancer/promoter sequence TGACGTCA abolishes specific factor binding as well as transcriptional activation. *Genes & Dev.* **3**, 612–9.

Illmensee, K. (1972). Developmental potencies of nuclei from cleavage, preblastoderm and syncitial blastoderm transplanted into unfertilised eggs of *Drosophila melanogaster*. *Roux' Arch. Devl. Biol.* **170**, 267–298.

Illmensee, K. & Hoppe, P.C. (1981). Nuclear transplantation in *Mus musculus*: developmental potencies of nuclei from preimplantation embryos. *Cell*, **25**, 9–18.

Illmensee, K. & Mahowald, A.P. (1974). Transplantation of posterior pole plasm in *Drosophila*: induction of germ cells at the anterior pole of the egg. *Proc. Natl. Acad. Sci. (USA)* **71**, 1016–20.

Imaizumi-Scherrer, M.T., Maundrell, K., Civelli, O. & Scherrer, K. (1982). Transcriptional and posttranscriptional regulation in duck erythroblasts. *Devl. Biol.* **93**, 126–38.

Ingham, P.W. (1983). Differential expression of bithorax complex genes in the absence of the *extra sex combs* and *trithorax* genes. *Nature (Lond.)* **306**, 591–3.

Ingham, P.W. (1984). A gene that regulates the bithorax complex differentially in larval and adult cells of *Drosophila*. *Cell*, **37**, 815–23.

Ingham, P.W. (1988). The molecular genetics of embryonic pattern in *Drosophila*. *Nature (Lond.)* **335**, 25–34.

Ingham, P.W. & Martinez-Arias, A. (1986). The correct activation of Antennapedia and bithorax complex genes requires the *fushi tarazu* gene. *Nature (Lond.)* **324**, 592–7.

Ingham, P.W., Howard, K.R., & Ish-Horowicz, D. (1985a). Transcription pattern of the *Drosophila* segmentation gene *hairy*. *Nature (Lond.)* **318**, 437–40.

Ingham, P., Martinez-Arias, A., Lawrence, P.A., & Howard, K.R. (1985b). Expression of *engrailed* in the parasegment of *Drosophila*. *Nature (Lond.)* **317**, 634–6.

Ingham, P.W., Ish-Horowicz, D. & Howard, K.R. (1986). Correlative changes in homoeotic and segmentation gene expression in *Kruppel* mutant embryos of *Drosophila. EMBO J.* **5**, 1659–65.

Ingham, P.W., Baker, N.E. & Martinez-Arias, A. (1988). Regulation of segment polarity genes in the *Drosophila* blastoderm by *fushi tarazu* and *even-skipped*. *Nature (Lond.)* **3331**, 73–5.

Ingraham, H.A., Chen, R., Mangalam, H.J., Elsholotz, H.P., Flynn, S.E., Lin, C.R., Simmons, D.M., Swanson, L. & Rosenfeld, M.G. (1988). A tissue-specific transcription factor containing a homeodomain specifies a pituitary phenotype. *Cell*, **55**, 519–29.

Irish, V., Lehmann, R. & Akam, M. (1989a). The *Drosophila* posterior-group gene *nanos* functions by repressing *hunchback* activity. *Nature (Lond.)* **338**, 646–8.

Irish, V.F., Martinez-Arias, A. & Akam, M. (1989b). Spatial regulation of the *Antennapedia* and *Ultrabithorax* homeotic genes during *Drosophila* early development. *EMBO J.* **8**, 1539–48.

Ish-Horowicz, D. & Pinchin, S.M. (1987). Pattern abnormalities induced by ectopic expression of the *Drosophila* gene *hairy* are associated with the repression of *ftz* transcription. *Cell*, **51**, 405–15.

Jack, T., Regulski, M. & McGinnis, W. (1988). Pair-rule segmentation genes regulate the expression of the homoeotic selector gene *Deformed*. *Genes & Dev.* **2**, 645–51.

Jackle, H., Tautz, D., Schuh, R., Seifert, E. & Lehmann, R. (1986). Cross-regulatory interactions among the gap genes of *Drosophila. Nature (Lond.)* **324**, 668–70.

Jackson, D.A. & Cook, P.R. (1985). Transcription occurs at a nucleoskeleton. *EMBO J.* **4**, 919–25.

Jackson, D.A., McReady, S.J. & Cook, P.R. (1981). RNA is synthesised at the nuclear cage. *Nature (Lond.)* **292**, 552–4.

Jacobson, A.G. & Sater, A.K. (1988). Features of embryonic induction. *Development*, **104**, 341–59.

Jamrich, M., Greenleaf, A.L. & Bautz, E.K.F. (1977a). Localisation of RNA polymerase in polytene chromosomes of *Drosophila. Proc. Natl. Acad. Sci. (USA)* **74**, 2079–83.

Jamrich, M., Greenleaf, A.L., Bautz, F.A. & Bautz, E.K.F. (1977b). Functional organisation of polytene chromosomes. *Cold Spring Harbor Symp Quant. Biol.* **42**, 389–96.

Jamrich, M., Sargent, T.D. & David, I.B. (1987). Cell-type-specific expression of epidermal cytokeratin genes during gastrulation of *Xenopus laevis. Genes & Dev.* **1**, 124–32.

Jaynes, J.B., Chamberlin, J.S., Buskin, J.N., Johnson, J.E. & Hauschka, S.D. (1986). Transcriptional regulation of the muscle creatine kinase gene and regulated expression in transfected muscle myoblasts. *Mol. & Cell. Biol.* **6**, 2855–64.

Jeffery, W.R. (1985a). Identification of proteins and mRNAs in isolated yellow crescents of ascidean eggs. *J. Embryol. Exp. Morophol.* **89**, 275–87.

Jeffery, W.R. (1985b). Specification of cell fate by cytoplasmic determinants in ascidean embryos. *Cell*, **41**, 11–12.

Jeffery, W.R., Tomlinson, C.R. & Brodeur, R.D. (1983). Localisation of actin messenger RNA during early ascidean development. *Devl. Biol.* **99**, 408–17.

Jeffreys, A.J. (1982). Evolution of globin genes. In *Genome Evolution*, ed. G.A. Dover & R.B. Flavell, pp. 157–76. Academic Press, New York & London.

Jimenez, F. & Campos-Ortega, J.A. (1982). Maternal effects of zygotic mutants affecting early neurogenesis in *Drosophila*. *Roux' Arch. Devl. Biol.* **191**, 191–201.

John, H.A., Patrinou-Georgopoulos, M. & Jones, K.W. (1977). Detection of myosin heavy chain mRNA during myogenesis in tissue culture by *in vitro* and *in situ* hybridisation. *Cell*, **12**, 501–8.

Johnson, M.H. & Pratt, H.P.M. (1983). Cytoplasmic localisations and cell interactions in the formation of the mouse blastocyst. In *Time Space and Pattern in Embryonic Development*, ed. W.R. Jeffery & R.A. Raff, pp. 287–312. A.R. Liss Inc., New York.

Johnson, M.H. & Ziomek, C.A. (1981a). The foundation of two distinct cell lineages within the mouse morula. *Cell*, **24**, 71–80.

Johnson, M.H. & Ziomek, C.A. (1981b). Induction of polarity in mouse 8-cell blastomeres: specificity, geometry and stability. *J. Cell Biol.* **91**, 303–12.

Johnson, M.H. & Ziomek, C.A. (1983). Cell interactions influence the fate of mouse blastomeres undergoing the transition from the 16- to 32-cell stage. *Devl. Biol.* **95**, 211–8.

Johnson, W.A., McCormick, C.A., Bray, S.J. & Hirsh, J. (1989). A neuron-specific enhancer of the *Drosophila* dopa decarboxylase gene. *Genes & Dev.* **3**, 676–86.

Johnston, R.N., Beverley, S.M. & Schimke, R.T. (1983). Rapid spontaneous dihydrofolate gene amplification shown by fluorescence-activated cell sorting. *Proc. Natl. Acad. Sci. (USA)* **80**, 3711–5.

Jones, K.A., Kadonaga, J.T., Rosenfeld, P.J., Kelly, T.J. & Tijan, R. (1987). A cellular DNA-binding protein that activates eucaryotic transcription and DNA replication. *Cell*, **48**, 79–89.

Jongens, T.A., Fowler, T., Shermoen, A.W. & Beckendorf, S.K. (1988). Functional redundancy in the tissue-specific enhancer of the *Drosophila Sgs-4* gene. *EMBO J.* **7**, 2559–67.

Jorgensen, E.M. & Garber, R.L. (1987). Function and misfunction of the two promoters of the *Drosophila Antennapedia* gene. *Genes & Dev.* **1**, 544–54.

Jost, J.-P., Moncharmant, B., Jiricny, J., Saluz, H. & Hertner, T. (1986), *In vitro* secondary activation (memory effect) of avian vitellogenin II gene in isolated liver nuclei. *Proc. Natl. Acad. Sci. (USA)* **83**, 43–7.

Joyner, A.L. & Martin, G.R. (1987). *En-1* and *En-2*, two mouse genes with sequence homology to the *Drosophila engrailed* gene: expression during embryogenesis. *Genes & Dev.* **1**, 29–38.

Joyner, A.L. Kornberg, T., Coleman, K.G., Cox, D.R. & Martin, G.R. (1985). Expression during embryogenesis of a mouse gene with sequence homology to the *Drosophila engrailed* gene. *Cell*, **43**, 29–37.

Judd, B.H. & Young, M.W. (1973). An examination of the one cistron: one chromomere concept. *Cold Spring Harbor Symp. Quant. Biol.* **38**, 573–9.

Jung, A., Sippel, A.E., Grez, M. & Schütz, G. (1980). Exons encode functional and structural units of chicken lysozyme. *Proc. Natl. Acad. Sci. (USA)* **77**, 5759–63.

Jurgens, G. (1985). A group of genes controlling the spatial expression of the bithorax complex in *Drosophila*. *Nature (Lond.)* **316**, 153–5.

Jurgens, G. (1988). Head and tail development of the *Drosophila* embryo involves *spalt*, a novel homoeotic gene. *EMBO J.* 7, 189–96.

Karch, F., Wieffenbach, W., Peifer, M., Bender, W., Duncan, I., Celniker, S., Crosby, M. & Lewis, E.B. (1985). The abdominal region of the bithorax complex. *Cell*, 43, 81–96.

Karin, M. (1989). Complexities of gene regulation by cAMP. *Trends Genet.* 5, 65–7.

Karlik, C.C. & Fyrberg, E.A. (1985). An insertion within a variably spliced *Drosophila* tropomyosin gene blocks accumulation of only one encoded isoform. *Cell*, 41, 57–66.

Karn, J., Brenner, S. & Barnett, L. (1983). Protein structural domains in the *Caenorhabditis elegans unc-54* myosin heavy chain gene are not separated by introns. *Proc. Natl. Acad. Sci. (USA)* 80, 4253–7.

Karr, T.L. & Kornberg, T.B. (1989). *fushi tarazu* protein expression in the cellular blastoderm of *Drosophila* detected using a novel imaging technique. *Development*, 105, 95–103.

Karr, T.L., Ali, Z., Drees, B. & Kornberg, T. (1985). The *engrailed* locus of *D. melanogaster* provides an essential zygotic function in precellular embryos. *Cell*, 43, 591–601.

Kato, H., Nagamine, M., Kominami, R. & Muramatsu, M. (1986). Formation of the transcription initiation complex on mammalian rDNA. *Mol. & Cell. Biol.* 6, 3418–27.

Katula, K.S., Hough-Evans, B.R., Britten, R.J. & Davidson, E.H. (1987). Ontogenic expression of a CyI actin fusion gene injected into sea urchin eggs. *Development*, 101, 437–47.

Kaufman, R.J., Brown, P.C. & Schimke, R.T. (1979). Amplified dihydrofolate reductase genes in unstably methotrexate resistant cells are associated with double minute chromosomes. *Proc. Natl. Acad. Sci. (USA)* 76, 5669–73.

Kaufman, S.J. & Foster, R.F. (1988). Replicating myoblasts express a muscle-specific phenotype. *Proc. Natl. Acad. Sci. (USA)* 85, 9606–10.

Kaye, J.S., Pratt-Kaye, S., Bellard, M., Dretzen, G., Bellard, F. & Chambon, P. (1986). Steroid hormone dependence of four DNase I-hypersensitive regions located within the 7000 bp 5′ -flanking segment of the ovalbumin gene. *EMBO J.* 5, 277–85.

Kedes, L. & Maxson, R. (1981). Histone gene organisation: paradigm lost. *Nature (Lond.)* 294, 11–12.

Keller, W. (1984). The RNA lariat: a new ring to the splicing of mRNA precursors. *Cell*, 39, 423–5.

Kemphues, K.J., Priess, J.R., Morton, D.G. & Cheng, N. (1988). Identification of genes required for cytoplasmic localisation in early *C. elegans* embryos. *Cell*, 52, 311–20.

Keynes, R.J. & Stern, C.D. (1984). Segmentation in the vertebrate nervous system. *Nature (Lond.)* 310, 786–8.

Kidder, G.M. & McLachlin, J.R. (1985). Timing of transcription and protein synthesis underlying morphogenesis in preimplantation mouse embryos. *Devl. Biol.* 112, 265–75.

Kilcherr, F., Baumgartner, S., Bopp, D., Frei, E. & Noll, M. (1986). Isolation of the *paired* gene of *Drosophila* and its spatial expression during early embryogenesis. *Nature (Lond.)* **321**, 493–7.

Killian, C.E. & Wilt, F.H. (1989). The accumulation and translation of a spicule matrix protein mRNA during sea urchin embryo development. *Devl. Biol.* **133**, 148–56.

Kimble, J. (1981a). Alterations in cell lineage following laser ablation of cells in the somatic gonad of *Caenorhabditis elegans*, *Devl. Biol.* **87**, 286–300.

Kimble, J.E. (1981b). Strategies for control of pattern formation in *Caenorhabditis elegans*. *Phil. Trans. Roy. Soc. Lond. B*, **295**, 539–51.

Kimble, J. & Hirsh, D. (1979). The postembryonic cell lineages of the hermaphrodite and male gonads of *Caenorhabditis elegans*. *Devl. Biol.* **70**, 396–417.

Kimble, J.E. & White, J.G. (1981). On the control of germ cell development in *Caenorhabditis elegans*. *Devl. Biol.* **81**, 208–19.

Kimble, J., Hodgkin, J., Smith, T. & Smith, J. (1982). Suppression of an amber mutation by microinjection of suppressor tRNA in *Caenorhabditis elegans*. *Nature (Lond.)* **299**, 456–8.

Kimelman, D. & Kirschner, M. (1987). Synergistic induction of mesoderm by FGF and TGF-β and the identification of an mRNA coding for FGF in the early *Xenopus* embryo. *Cell*, **51**, 869–77.

Kimelman, D., Kirschner, M. & Scherson, T. (1987). The events of the midblastula transition in *Xenopus* are regulated by changes in the cell cycle. *Cell*, **48**, 399–407.

Kimelman, D., Abraham, J.A., Haaparanta, T., Palisi, T.M. & Kirschner, M.W. (1988). The presence of fibroblast growth factor in the frog egg: its role as a natural mesoderm inducer. *Science*, **242**, 1053–6.

King, C.R. & Piatigorsky, J. (1983). Alternative RNA splicing of the murine αA crystallin gene: protein coding information within an intron. *Cell*, **32**, 707–12.

King, M.L. & Barkliss, E. (1985). Regional distribution of maternal messenger RNA in the amphibian oocyte. *Devl. Biol.* **112**, 203–12.

King, R.C. & Aggarwal, S.K. (1965). Oogenesis in *Hyalophora cecropia*. *Growth*, **29**, 17–83.

King, W.J. & Green, G.L. (1984). Monoclonal antibodies localise oestrogen receptor in the nuclei of target cells. *Nature (Lond.)* **307**, 745–7.

Kioussis, D., Wilson, F., Khazaie, K. & Grosveld, F. (1985). Differential expression of human globin genes introduced in K562 cells. *EMBO J.* **4**, 927–31.

Klass, M., Ammons, D. & Ward, S. (1988). Conservation of the 5′ flanking sequences of transcribed members of the *Caenorhabditis elegans* major sperm protein gene family. *J. Mol. Biol.* **199**, 14–22.

Kleene, K.C. & Humphries, T. (1977). Similarity of hnRNA sequences in blastula and pluteus stage sea urchin embryos. *Cell*, **12**, 143–55.

Klein-Hitpaβ, L., Schorpp, M., Wagner, U. & Ryffel, G.U. (1986). An estrogen-responsive element derived from the 5′ flanking region of the *Xenopus* vitellogenin A2 gene functions in transfected human cells. *Cell*, **46**, 1053–61.

Klingler, M., Erdelyi, M., Szabad, J. & Nusslein-Volhard, C. (1988). Function of *torso* in determining the terminal anlagen of the *Drosophila* embryo. *Nature (Lond.)* **335**, 275–8.

Klock, G., Strahle, U. & Schutz, G. (1987). Oestrogen and glucocorticoid responsive elements are closely related but distinct. *Nature (Lond.)* **329**, 734–6.

Knipple, D.C., Seifert, E., Rosenberg, U.B., Preiss, A. & Jackle, H. (1985). Spatial and temporal patterns of *Kruppel* gene expression in early *Drosophila* embryos. *Nature (Lond.)* 317, 40–4.

Knust, E., Dietrich, U., Tepass, U., Bremer, K.A., Weigel, D., Vassin, H. & Campos-Ortega, J.A. (1987a). EGF-homologous sequences encoded in the genome of *Drosophila melanogaster*, and their relation to neurogenic genes. *EMBO J.* 6, 761–6.

Knust, E., Tietze, K. & Campus-Ortega, J.A. (1987b). Molecular analysis of the neurogenic locus *Enhancer of Split* of *Drosophila melanogaster*. *EMBO J.* 6, 4113–23.

Ko, H.-S., Fast, P., McBride, W. & Staudt, L.M. (1988). A human protein specific for the immunoglobin octamer DNA motif contains a functional homeobox doman. *Cell*, 55, 135–44.

Kominami, T. (1984). Allocation of mesendodermal cells during early embryogenesis in the starfish, *Asterina pectenifera*. *J. Embryol. Exp. Morphol.* 84, 177–90.

Konarska, M.M. & Sharp, P.A. (1987). Interactions between small nuclear ribonucleoprotein particles in formation of spliceosomes. *Cell*, 49, 763–74.

Konarska, M.M., Grabowski, P.J., Padgett, R.A. & Sharp, P.A. (1985). Characterisation of the branch site in lariat RNAs produced by splicing of mRNA precursors. *Nature (Lond.)* 313, 552–6.

Kongsuwan, K., Webb, E., Houiaux, P. & Adams, J.M. (1988). Expression of multiple homeobox genes within diverse mammalian haematopoietic lineages. *EMBO J.* 7, 2131–8.

Konrad, K.D. & Marsh, J.L. (1987). Developmental expression and spatial distribution of dopa decarboxylase in *Drosophila*. *Dev. Biol.* 122, 172–85.

Kornberg, T. (1981). *engrailed*, a gene controlling compartment and segment formation in *Drosophila*. *Proc. Natl. Acad. Sci. USA* 78, 1095–9.

Kornberg, T., Siden, I., O'Farrell, P.H. & Simon, M. (1985). The *engrailed* locus of *Drosophila*: *in situ* localisation of transcripts reveals compartment-specific expression. *Cell*, 40, 45–53.

Kornfeld, K., Saint, R.B., Beachy, P.A., Harte, P.J., Peattie, D.A. & Hogness, D.A. (1989). Structure and expression of a family of *Ultrabithorax* mRNAs generated by alternative splicing and polyadenylation in *Drosophila*. *Genes & Dev.* 3, 243–58.

Kramer, J.M., Cox, G.N. & Hirsh, D. (1985). Expression of the *Caenorhabditis elegans* collagen genes *col-1* and *col-2* is developmentally regulated. *J. Biol. Chem.* 260, 1945–51.

Kramer, J.M., Johnson, J.J., Edgar, R.S., Basch, C. & Roberts, C. (1988). The *sqt-1* gene of *C. elegans* encodes a collagen critical for organismal morphogenesis. *Cell*, 55, 555–65.

Kraminsky, G.P., Clark, W.C., Estelle, M.A., Gietz, R.D., Sage, B.D., O'Connor, J.D. & Hodgetts, R.B. (1980). Induction of translatable mRNA for dopa decarboxylase in *Drosophila*: an early response to ecdysone. *Proc. Natl. Acad. Sci. (USA)* 77, 4175–9.

Krasnow, M.A., Saffman, E.E., Kornfeld, K. & Hogness, D.S. (1989). Transcriptional activation and repression by *Ultrabithorax* proteins in cultured *Drosophila* cells. *Cell*, 7, 1031–43.

Krause, M. & Hirsh, D. (1987). A *trans*-spliced leader sequence on actin mRNA in *C. elegans*. *Cell*, 49, 753–61.

Krause, H.M., Klemenz, R. & Gehring, W.J. (1988). Expression, modification and localisation of the *fushi tarazu* protein in *Drosophila* embryos. *Genes & Dev.* 2, 1021–36.

Krieg, P.A. & Melton, D.A. (1987). An enhancer responsible for activating transcription at the midblastula transition in *Xenopus* development. *Proc. Natl. Acad. Sci. (USA)*, 84, 2331–5.

Krumlauf, R., Holland, P.W.H., McVey, J.H. & Hogan, B.L.M. (1987). Developmental and spatial patterns of expression of the mouse homeobox gene, *Hox 2.1*. *Development*, 99, 603–17.

Krumm, A., Roth, G.E. & Korge, G. (1985). Transformation of salivary gland secretion protein gene *sgs-4* in *Drosophila*: stage- and tissue-specific regulation, dosage compensation and position effect. *Proc. Natl. Acad. Sci. (USA)* 82, 5055–9.

Kuhn, A. & Grummt, I. (1987). A novel promoter in the mouse rDNA spacer is active *in vivo* and *in vitro*. *EMBO J.* 6, 3487–92.

Kuhn, D.T. & Packert, G. (1988). Tumorous-head type mutants of the distal bithorax complex cause dominant gain and recessive loss of function in *Drosophila melanogaster*. *Devl. Biol.* 125, 8–18.

Kuner, J.M., Nakanishi, M., Ali, Z., Drees, B., Gustavson, E., Theis, J., Kauvar, L., Kornberg, T. & O'Farrell, P.H. (1985). Molecular cloning of *engrailed*: a gene involved in the development of pattern in *Drosophila melanogaster*. *Cell*, 42, 309–16.

Kunz, W., Trepte, H.-H. & Bier, K. (1970). On the function of the germ line chromosomes in the oogenesis of *Wachtliella persicariae* (Cecidomyiidae). *Chromosoma*, 30, 180–91.

Kuroiwa, A., Hafen, E. & Gehring, W.J. (1984). Cloning and transcriptional analysis of the segmentation gene *fushi tarazu* of *Drosophila*. *Cell*, 37, 825–31.

Kuroiwa, A., Kloter, U., Baumgartner, P. & Gehring, W.J. (1985). Cloning of the homoeotic *Sex combs reduced* gene in *Drosophila* and *in situ* localisation of its transcripts. *EMBO J.* 4, 3757–64.

Kuziora, M.A. & McGinnis, W. (1988a). Different transcripts of the *Drosophila Abd-B* gene correlate with distinct genetic sub-functions. *EMBO J.* 7, 3233–44.

Kuziora, M.A. & McGinnis, W. (1988b). Auto-regulation of a *Drosophila* homeotic selector gene. *Cell*, 55, 477–85.

Kuziora, M.A. & McGinnis, W. (1989). A homeodomain substitution changes the regulatory specificity of the *Deformed* protein in *Drosophila* embryos. *Cell*, 59, 563–71.

Labhart, P. & Reeder, R.H. (1989). High initiation rates at the ribosmal gene promoter do not depend upon spacer transcription. *Proc. Natl. Acad. Sci. (USA)* 86, 3155–8.

LaBonne, S.G. & Mahowald, A. P. (1985). Partial rescue of embryos from two maternal-effect neurogenic mutants by transplantation of wild-type ooplasm. *Devl. Biol.* 110, 264–74.

Lacy, E. & Maniatis, T. (1980). The nucleotide sequence of a rabbit β globin pseudogene. *Cell*, 21, 545–53.

Laimins, L., Holmgren-Konig, M. & Khoury, G. (1986). Transcriptional 'silencer' element in rat repetitive sequences associated with the rat insulin 1 gene locus. *Proc. Natl. Acad. Sci. (USA)* **83**, 3151–5.

Lamb, M.M. & Daneholt, B. (1979). Characterisation of active transcription units in Balbiani rings of *Chironomus tentans*. *Cell*, **17**, 835–48.

Lambert, B. (1972). Repeated sequences in a Balbiani ring. *J. Mol. Biol.* **72**, 65–75.

Landschulz, W.H., Johnson, P.F. & McKnight, S.L. (1988). The leucine zipper: a hypothetical structure common to a new class of DNA binding proteins. *Science*, **240**, 1759–64.

Lane, C.D., Gurdon, J.B. & Woodland, H.R. (1974). Control of translation of globin mRNA in embryonic cells. *Nature (Lond.)* **251**, 436–7.

Laski, F.A., Rio, D.C. & Rubin, G.M. (1986). Tissue specificity of *Drosophila* P element transposition is regulated at the level of mRNA splicing. *Cell*, **44**, 7–19.

Laski, F.A. & Rubin, G.M. (1989). Analysis of *cis*-acting requirements for germ-line-specific splicing of the P-element ORF2-ORF3 intron. *Genes & Dev.* **3**, 720–8.

Lasko, P.F. & Ashburner, M. (1988). The product of the *Drosophila* gene *vasa* is very similar to eucaryotic initiation factor-4A. *Nature (Lond.)* **335**, 611–6.

Lassar, A.B., Martin, P.L. & Roeder, R.G. (1983). Transcription of class III genes: formation of preinitiation complexes. *Science*, **222**, 740–3.

Lassar, A.B., Paterson, B.M. & Weintraub, H. (1986). Transfection of a DNA locus that mediates the conversion of 10T1/2 fibroblasts to myoblasts. *Cell*, **47**, 649–56.

Lassar, A.B., Buskin, J.N., Lockshon, D., Davis, R.L., Apone, S., Hauschka, S.D. & Weintraub, H. (1989). Myo D is a sequence-specific DNA binding protein requiring a region of *myc* homology to bind to the muscle creatine kinase enhancer. *Cell*, **58**, 823–31.

Lauer, J., Shen, C.-K.J. & Maniatis, T. (1980). The chromosomal arrangement of human β-like globin genes; sequence homology and β globin gene deletions. *Cell*, **20**, 119–30.

Laufer, J.S., Bazzicalupo, P. & Wood, W.B. (1980). Segregation of developmental potential in early embryos of *Caenorhabditis elegans*. *Cell*, **19**, 569–77.

Laughon, A. & Scott, M.P. (1984). Sequence of a *Drosophila* segmentation gene: protein structure homology with DNA-binding proteins. *Nature (Lond.)* **310**, 25–8.

Laughon, A., Boulet, A.M., Bermingham, J.R., Laymon, R.A. & Scott, M.P. (1986). Structure of transcripts from the homoeotic *Antennapedia* gene of *Drosophila melanogaster*; two promoters control the major protein-coding region. *Mol. Cell. Biol.* **6**, 4676–89.

Lawn, R.M., Efstratiadis, A., O'Connell, C. & Maniatis, T. (1980). The nucleotide sequence of the human β globin gene. *Cell*, **21**, 647–51.

Lawrence, J.B., Taneja, K. & Singer, R. (1989). Temporal resolution and sequential expression of muscle-specific genes revealed by *in situ* hybridisation. *Devl. Biol.* **133**, 235–46.

Lawrence, P.A. (1981). A general cell marker for clonal analysis of *Drosophila* development. *J. Embryol. Exp. Morphol.* **64**, 321–32.

Lawrence, P.A. (1982). Cell lineage of the thoracic muscles of *Drosophila*. *Cell*, **129**, 493–503.

Lawrence, P.A. (1983). A new homoeotic gene. *Nature (Lond.)* **306**, 643.

Lawrence, P.A. (1985). Notes on the genetics of pattern formation in the internal organs of *Drosophila*. *Trends Neurosci.* **8**, 267–9.

Lawrence, P.A. (1988). Background to *bicoid*. *Cell*, **54**, 1–2.

Lawrence, P.A. & Brower, D.L. (1982). Myoblasts from *Drosophila* wing disks can contribute to developing muscles throughout the fly. *Nature (Lond.)* **295**, 55–7.

Lawrence, P.A. & Johnston, P. (1984a). On the role of the *engrailed*+ gene in the internal organs of *Drosophila*. *EMBO* J. **3**, 2889–44.

Lawrence, P.A. & Johnston, P. (1984b). The genetic specification of pattern in a *Drosophila* muscle. *Cell*, **36**, 775–82.

Lawrence, P. & Johnston, P. (1986). The muscle pattern of a segment of *Drosophila* may be determined by neurons and not by contributing myoblasts. *Cell*, **45**, 505–13.

Lawrence, P.A. & Johnston, P. (1989). Pattern formation in the *Drosophila* embryo: allocation of cells to parasegments by *even skipped* and *fushi tarazu*. *Development* **105**, 761–7.

Lawrence, P.A. & Morata, G. (1976). Compartments in the wing of *Drosophila*: a study of the *engrailed* gene. *Devl. Biol.* **50**, 321–37.

Lawrence, P.A. & Morata, G. (1983). The elements of the bithorax complex *Cell*, **35**, 595–601.

Lawrence, P.A. & Struhl, G. (1982). Further studies of the *engrailed* phenotype in *Drosophila EMBO J.* **1**, 827–33.

Lawrence, P.A., Johnston, P. & Struhl, G. (1983). Different requirements for homoeotic genes in the soma and germ line of *Drosophila*. *Cell*, **35**, 27–34.

Lawrence, P.A., Johnston, P., Macdonald, P. & Struhl, G. (1987). Borders of parasegments in *Drosophila* embryos are delimited by the *fushi tarazu* and *even skipped* genes. *Nature (Lond.)* **328**, 440–2.

Learned, R.M., Learned, T.K., Haltiner, M.M. & Tijan, R.T. (1986). Human rRNA transcription is modulated by the coordinate binding of two factors to an upstream control element. *Cell*, **45**, 847–57.

LeBowitz, J.H., Kobayashi, T., Staudt, L., Baltimore, D. & Sharp, P. (1988). Octamer-binding proteins from B or Hela cells stimulate transcription of the immunoglobulin heavy-chain promoter *in vitro*. *Genes & Dev.* **2**, 1227–37.

Leder, A., Swan, D., Ruddle, F., D'Eustachio, P. & Leder, P. (1981). Dispersion of α-like globin genes of the mouse to three different chromosomes. *Nature (Lond.)* **293**, 196–200.

Lehmann, R. & Nusslein-Volhard, C. (1986). Abdominal segmentation, pole cell formation and embryonic polarity require the localised activity of *oskar*, a maternal gene in *Drosophila*. *Cell*, **47**, 141–52.

Lehmann, R. & Nusslein-Volhard, C. (1987a). Involvement of the *pumilio* gene in the transport of an abdominal signal in the *Drosophila* embryo. *Nature (Lond.)* **329**, 167–9.

Lehmann, R. & Nusslein-Volhard, C. (1987b). *hunchback*, a gene required for segmentation of anterior and posterior region of the *Drosophila* embryo. *Devl. Biol.* **119**, 402–17.

Lehmann, R., Dietrich, U., Jimenez, F. & Campos-Ortega, J.A. (1981). Mutations of early neurogenesis in *Drosophila*. *Roux' Arch. Devl. Biol.* **190**, 226–9.

Lehmann, R., Jimenez, F., Dietrich, U. & Campos-Ortega, J.A. (1983). On the phenotype and development of mutants of early neurogenesis in *Drosophila melanogaster. Roux' Arch. Devl. Biol.* **192**, 62–74.

LeMeur, M., Glanville, N., Mandel, J.L., Gerlinger, P., Palmiter, R. & Chambon, P. (1981). The ovalbumin gene family: hormonal control of X and Y gene transcription and mRNA accumulation. *Cell,* **23**, 561–71.

Lemischka, I.R., Raulet, D.H. & Mulligan, R.C. (1986). Developmental potential and dynamic behaviour of haematopoietic stem cells. *Cell,* **45**, 917–27.

Le Mouellic, H., Condamine, H. & Brulet, P. (1988). Pattern of transcription of the homeobox gene *Hox 3.1* in the mouse embryo. *Genes & Dev.* **2**, 125–35.

Lenardo, M.J. & Baltimore, D. (1989). NF-κB: a pleiotropic mediator of inducible and tissue-specific gene control. *Cell,* **58**, 227–9.

Lendahl, U. & Wieslander, L. (1987). Balbiani ring (BR) genes exhibit different patterns of expression during development. *Devl. Biol.* **121**, 130–8.

Leung, S.-O., Proudfoot, N.J. & Whitelaw, E. (1987). The gene for θ globin is transcribed in human foetal erythroid tissues. *(Nature Lond.)* **329**, 551–3.

Levine, M. (1988). Molecular analysis of dorsal–ventral polarity in *Drosophila. Cell,* **52**, 785–6.

Levine, M., Hafen, E., Garber, R.L. & Gehring, W.J. (1983). Spatial distribution of *Antennapaedia* transcripts during *Drosophila* development. *EMBO J.* **2**, 2037–46.

Levine, M. & Hoey, T. (1988). Homeobox proteins as sequence-specific transcription factors. *Cell,* **55**, 537–40.

Lewin, B. (1987). *Genes,* 3rd ed. Wiley, New York.

Lewis, E.B. (1978). A gene complex controlling segmentation in *Drosophila. Nature (Lond.)* **276**, 565–70.

Lewis, J. (1989). Genes and segmentation. *Nature (Lond.)* **341**, 382–3.

Lieber, M.R., Hesse, J.E., Lewis, S., Bosman, G.C., Rosenberg, N., Mizuuchi, K., Bosman, M.J. & Gellert, M. (1988). The defect in murine severe combined immune deficiency: joining of signal sequences but not coding segments in V(D)J recombination. *Cell,* **55**, 7–16.

Lifton, R.P. & Kedes, L.H. (1976). Size and sequence homology of masked maternal and embryonic histone messenger RNAs. *Devl. Biol.* **48**, 47–55.

Lifton, R.P., Goldberg, M.L., Karp, R.W. & Hogness, D.S. (1977). The organisation of the histone genes in *D. melanogaster:* functional and evolutionary implications. *Cold Spring Harbor Symp. Quant. Biol.* **42**, 1047–51.

Lillie, J.W. & Green, M.R. (1989). Activator's target in sight. *Nature (Lond.)* **341**, 279–80.

Lin, F.-K., Suggs, S., Lin, G.-H., Browne, J.K., Smalling, R., Egrie, J.C., Chen, K.K., Fox, G.M., Martin, F., Stabinsky, Z., Badrawi, S.M., Lai, P.-H., & Goldwasser, E. (1985). Cloning and expression of the human erythropoietin gene. *Proc. Natl. Acad. Sci. (USA)* **82**, 7580–4.

Lindenmaier, E., Nguyen-Huu, M.C., Lurz, R., Stratmann, M., Blin, N., Wurtz, T., Hauser, H.J., Sippel, A.E. & Schutz, G. (1979). Arrangement of coding and intervening sequences in the chicken lysozyme gene. *Proc. Natl. Acad. Sci. (USA)* **76**, 6196–200.

Lipshitz, H.D., Peattie, D.A. & Hogness, D. (1987). Novel transcripts from the *Ultrabithorax* domain of the bithorax complex. *Genes & Dev.* **1**, 307–22.

Little, P.F.R. (1982). Globin pseudogenes. *Cell,* **28**, 683–4.

Liu, J.-K., Bergman, Y. & Zaret, K.S. (1988). The mouse albumin promoter and a distal upstream site are simultanteously DNase I hypersensitive in liver chromatin and bind similar liver-abundant factors *in vitro. Genes & Dev.* **2**, 528–41.

Livingston, B.T. & Wilt, F.H. (1989). Lithium evokes expression of vegetal-specific molecules in the animal blastomeres of sea urchin embryos. *Proc. Natl. Acad. Sci. (USA)* **86**, 3669–73.

Lonai, P., Arman, E., Czosnek, H., Ruddle, F.H. & Blatt, C. (1987). New murine homeoboxes: structure, chromosomal assignment and differential expression in adult erythropoiesis. *DNA,* **6**, 409–18.

London, C., Akers, R. & Phillips, C. (1988). Expression of Epi 1, an epidermis-specific marker in *Xenopus laevis* embryos, is specified prior to gastrulation. *Devl. Biol.* **129**, 380–9.

McCarrey, J.R. & Thomas, K. (1987). Human testis-specific PGK gene lacks introns and possesses characteristics of a processed gene. *Nature (Lond.)* **326**, 501–4.

McConkey, G.A. & Bogenhagen, D.F. (1988). TFIIIA binds with equal affinity to somatic and major oocyte 5S RNA genes. *Genes & Dev.* **2**, 205–14.

McCormick, A., Wu, D., Castrillo, J.-L., Dana, S., Strobl, J., Thompson, E.B. & Karin, M. (1988). Extinction of growth hormone expression in somatic cell hybrids involves repression of the specific *trans*-activator GHF-1. *Cell,* **55**, 379–89.

McCoubrey, W.K., Nordstrom, K.D. & Meneely, P.M. (1988). Microinjected DNA from the X chromosome affects sex determination in *Caenorhabditis elegans. Science,* **242**, 1146–50.

McDevitt, M.A., Gilmartin, G.M. Reeves, W.H. & Nevins, J.R. (1988). Multiple factors are required for poly(A) addition to a mRNA 3' end. *Genes & Dev.* **2**, 588–97.

Macdonald, P.M. & Struhl, G. (1986). A molecular gradient in early *Drosophila* embryos and its role in specifying the body pattern. *Nature (Lond.)* **324**, 537–45.

Macdonald, P.M. & Struhl, G. (1988). *Cis*-acting sequences responsible for anterior localisation of *bicoid* mRNA in *Drosophila* embryos. *Nature (Lond.)* **336**, 595–7.

Macdonald, P.M., Ingham, P. & Struhl, G. (1986). Isolation, structure and expression of *even skipped:* a second pair-rule gene of *Drosophila* containing a homeo box. *Cell,* **47**, 721–34.

McGinnis, W., Levine, M.S., Hafen, E., Kuroiwa, A. & Gehring, W.J. (1984a). A conserved DNA sequence in homeotic genes of the *Drosophila* Antennapedia and bithorax complexes. *Nature (Lond.)* **308**, 428–33.

McGinnis, W., Garber, R.L., Wirtz, J., Kuroiwa, A. & Gehring, W.J. (1984b). homologous protein-coding sequence in *Drosophila* homeotic genes and its conservation in other metazoans. *Cell,* **37**, 403–8.

McKeown, M., Belote, J.M. & Baker, B.S. (1987). A molecular analysis of *trans-former*, a gene in *Drosophila melanogaster* that controls female sexual differentiation. *Cell,* **48**, 489–99.

McKeown, M., Belote, J.M. & Boggs, R.T. (1988). Ectopic expression of the female *transformer* gene product leads to female differentiation of chromosomally male *Drosophila*. *Cell*, **53**, 887–95.

McKnight, G.S. (1978). The induction of ovalbumin and conalbumin mRNA by oestrogen and progesterone in chick oviduct explant cultures. *Cell*, **14**, 403–13.

Macleod, A.R., Waterson, R.H., Fishpool, R.M. & Brenner, S. (1977). Identification of the structural gene for a myosin heavy-chain in *Caenorhabditis elegans*. *J. Mol. Biol.* **114**, 133–40.

McMahon, A.P. & Moon, R.T. (1989). Ectopic expression of the proto-oncogene *int-1* in *Xenopus* embryos leads to duplication of the embryonic axis. *Cell*, **58**, 1075–84.

McReynolds, L., O'Malley, B.W., Nisbet, A.D., Fothergill, J.E., Givol, D., Fields, S., Robertson, M. & Brownlee, G.G. (1978). Sequence of chicken ovalbumin mRNA. *Nature (Lond.)* **273**, 723–8.

McStay, B. & Reeder, R.H. (1986). A termination site for *Xenopus* RNA polymerase I also acts as an element of an adjacent promoter. *Cell*, **47**, 913–20.

Magram, J., Chada, K. & Costantini, F. (1985). Developmental regulation of a cloned adult β globin gene in transgenic mice. *Nature (Lond.)* **315**, 338–40.

Magrassi, L. & Lawrence, P.A. (1988). The pattern of cell death in *fushi tarazu*, a segmentation gene of *Drosophila*. *Development*, **104**, 447–51.

Mahaffey, J.W. & Kaufman, T.C. (1987). Distribution of the *Sex combs reduced* gene products in *Drosophila melanogaster*. *Genetics* **117**, 51–60.

Mahaffey, J.W., Diederich, R.J. & Kaufman, T.C. (1989). Novel patterns of homeotic protein accumulation in the head of the *Drosophila* embryo. *Development*, **105**, 167–74.

Mahoney, P.A. & Lengyel, J.A. (1987). The zygotic segmentation mutant *tailless* alters the blastoderm fate map of the *Drosophila* embryo. *Devl. Biol.* **122**, 464–70.

Maine, E. & Kimble, J. (1989). Identification of genes that interact with *glp-1*, a gene required for inductive cell interactions in *Caenorhabditis elegans*. *Development* **105**, 133–43.

Malynn, B.A., Blackwell, T.K., Fulop, G.M., Rathbun, G.A., Furley, A.J.W., Ferrier, P., Heinke, L.B., Phillips, R.A., Yancopoulos, G.D. & Alt, F.W. (1988). The *scid* defect affects the final step of the immunoglobulin VDJ recombinase mechanism. *Cell*, **54**, 453–60.

Maniatis, T. & Reed, R. (1987). The role of small nuclear ribonucleoprotein particles in pre-mRNA splicing. *Nature (Lond.)* **325**, 673–8.

Mangalam, H.J., Albert, V.R., Ingraham, H.A., Kapiloff, M., Wilson, L., Nelson, C., Elsholtz, H. & Rosenfeld, M.G. (1989). A pituitary POU-domain protein, Pit-1, activates both growth hormone and prolactin promoters transcriptionally. *Genes & Dev.* **3**, 946–58.

Marcu, K.B. & Cooper, M.D. (1982). New views of the immunoglobulin heavy chain switch. *Nature (Lond.)* **298**, 327–8.

Martindale, M.Q., Doe, C.Q. & Morrill, J.B. (1985). The role of animal–vegetal interaction with respect to the determination of dorsoventral polarity in the equal-cleaving spiralian, *Lymnaea palustris*. *Roux' Arch. Devl. Biol.* **194**, 281–95.

Martinez, E., Givel, F. & Wahli, W. (1987). The oestrogen responsive element as an inducible enhancer: DNA sequence requirements and conversion to a glucocorticoid response element. *EMBO J.* **6**, 3719–27.

Martinez-Arias, A. (1986). The *Antennapedia* gene is required and expressed in parasegments 4 and 5 of the *Drosophila* embryo. *EMBO J.* **5**, 135–41.

Martinez-Arias, A. & Lawrence, P.A. (1985). Parasegements and compartments in the *Drosophila* embryo *Nature (Lond.)* **313**, 639–42.

Martinez-Arias, A. & White, R.A.H. (1988). *Ultrabithorax* and *engrailed* expression in *Drosophila* embryos mutant for segmentation genes of the pair-rule class. *Development*, **102**, 325–8.

Martinez-Arias, A., Ingham, P.W., Scott, M.P. & Akam, M.E. (1987). The spatial and temporal deployment of *Dfd* and *Scr* transcripts throughout development of *Drosophila*. *Development*, **100**, 673–83.

Martinez-Arias, A., Baker, N.E. & Ingham, P.W. (1988). Role of segment polarity genes in the definition and maintenance of cell states in the *Drosophila* embryo. *Development*, **103**, 157–70.

Maruyama, Y.-K., Nakaseko, Y. & Yagi, S. (1985). Localisation of cytoplasmic determinants responsible for primary mesenchyme formation and gastrulation in the unfertilised egg of the sea urchin *Hemicentrotus pulcherrimus*. *J. Exp. Zool.* **236**, 155–63.

Maser, R.L. & Calvet, J.P. (1989). U3 small nuclear RNA can be psoralen-crosslinked *in vivo* to the 5' external transcribed spacer of pre-ribosmal RNA. *Proc. Natl. Acad. Sci. (USA)* **86**, 6523–7.

Mather, E.L., Nelson, K.J., Haimovitch, J. & Perry, R.P. (1984). Mode of regulation of immunoglobulin μ- and δ-chain expression varies during B lymphocyte maturation. *Cell*, **36**, 329–38.

Mauron, A., Levy, S., Childs, G. & Kedes, L. (1981). Monocistronic transcription is the physiological mechanism of sea urchin embryonic histone gene expression. *Mol. & Cell. Biol.* **1**, 661–8.

Mavilio, F., Giampaolo, A., Carè, A., Migliaccio, G., Calandrini, M., Russo, G., Pagliardi, G.L., Mastroberardino, G., Marinucci, M. & Peschle, C. (1983). Molecular mechanisms of human haemoglobin switching; selective undermethylation and expression of globin genes in embryonic, fetal and adult erythroblasts. *Proc. Natl. Acad. Sci. (USA)* **80**, 6907–11.

Mavilio, F., Simeone, A., Giampaolo, A., Faiella, A., Zappavigna, V., Acampora, D., Poiana, G., Russo, G., Peschle, C. & Boncinelli, E. (1986). Differential and stage-related expression in embryonic tissues of a new human homoeobox gene. *Nature (Lond.)* **324**, 664–7.

Mavilio, F., Simeone, A., Boncinelli, E. & Andrews, P.W. (1988). Activation of four homeobox gene clusters in human embryonal carcinoma cells induced to differentiate by retinoic acid. *Differentiation*, **37**, 73–9.

Maxson, R.E. & Egrie, J.C. (1980). Expression of maternal and paternal histone genes during early cleavage stages of the echinoderm hybrid *Strongylocentrotus purpuratus* × *Lytechninus pictus*. *Devl. Biol.* **74**, 335–43.

Maxson, R., Mohun, T., Gormezano, G., Childs, G. & Kedes, L. (1983). Distinct organisation and patterns of expression of early and late histone gene sets in the sea urchin. *Nature (Lond.)* **301**, 120–4.

Mayer, U. & Nusslein-Volhard, C. (1988). A group of genes required for pattern formation in the ventral ectoderm of the *Drosophila* embryo. *Genes & Dev.* **2**, 1496–1511.

Meedel, T.H. (1983). Myosin expression in the developing ascidean embryo. *J. Exp. Zool,* **227**, 203–11.

Meedel, T.H. & Whittaker, J.R. (1983). Development of translationally active mRNA for larval muscle acetylcholinesterase during ascidean embryogenesis. *Proc. Natl. Acad. Sci. (USA)* **80**, 4761–5.

Meedel, T.H. & Whittaker, J.R. (1984). Lineage segregation and developmental autonomy in expression of functional acetylcholinesterase mRNA in the ascidean embryo. *Devl. Biol.* **105**, 479–87.

Meehan, R.R., Lewis, J.D., McKay, S., Kleiner, E.L. & Bird, A.P. (1989). Identification of a mammalian protein that binds specifically to DNA containing methylated CpGs. *Cell,* **58**, 499–507.

Meinhart, H. (1986). Hierarchical inductions of cell states: a model for segmentation in *Drosophila. J. Cell Sci. Suppl.* **4**, 357–81.

Melton, D.A. (1987). Translocation of a localised maternal RNA to the vegetal pole of *Xenopus* oocytes. *Nature (Lond.)* **328**, 80–2.

Melton, D.W., McEwan, C., McKie, A.B. & Reid, A.M. (1986). Expression of the mouse HPRT gene: deletional analysis of the promoter of an X-chromosome-linked housekeeping gene. *Cell,* **44**, 319–28.

Merriam, R.W. & Clark, T.G. (1978). Actin in *Xenopus* oocytes: II intracellular distribution and polymerisability. *J. Cell Biol.* **77**, 439–46.

Merriam, R.W. & Sauterer, R.A. (1983). Localisation of a pigment-containing structure near the surface of *Xenopus* eggs which contracts in response to calcium. *J. Embryol. Exp. Morphol.* **76**, 51–65.

Merrill, V.K.L., Turner, F.R. & Kaufman, T.C. (1987). A genetic and developmental analysis of mutations in the *Deformed* locus in *Drosophila melanogaster. Devl. Biol.* **122**, 379–95.

Metcalf, D. (1985). The granulocyte-macrophage colony stimulating factors. *Cell,* **43**, 5–6.

Metcalf, D. (1989). The molecular control of cell division, differentiation commitment and maturation in haematopoietic cells. *Nature (Lond.)* **339**, 27–30.

Meyer, B.J. (1988). Primary events in *C. elegans* sex determination and dosage compensation. *Trends Genet.* **4**, 337–42.

Meyer, M., Gronemeyer, H., Turcotte, B. Bocquel, M.-T., Tasset, D. & Chambon, P. (1989). Steroid hormones compete for factors that mediate their enhancer function. *Cell,* **57**, 433–42.

Meyerowitz, E.M. & Hogness, D.S. (1982). Molecular organisation of a *Drosophila* puff site that responds to ecdysone. *Cell,* **28**, 165–76.

Mihara, H. & Kaiser, E.T. (1988). A chemically synthesised *Antennapedia* homeodomain binds to a specific DNA sequence. *Science,* **242**, 925–8.

Milbrandt, J.D., Heintz, N.H., White, W.C., Rothman, S.M. & Hamlin, J.I. (1981). Methotrexate-resistant Chinese hamster ovary cells have amplified a 135 kilobase pair region that includes the dihydrofolate reductase gene. *Proc. Natl. Acad. Sci. (USA).* **78**, 6043–7.

Miller, D.M., Stockdale, F.E. & Karn, J. (1986). Immunological identifcation of the genes encoding the four myosin heavy chain isoforms in *Caenorhabditis elegans. Proc. Natl. Acad. Sci. (USA)* **83**, 2305–9.

Miller, L.M., Plenefisch, J.D., Casson, L.P. & Meyer, B.J. (1988). *xol-1*: a gene that controls the male modes of both sex determination and X chromosome dosage compensation in *C. elegans*. *Cell*, 55, 167–83.

Miller, O.L. & Bakken, A.H. (1972). Morphological studies of transcription. *Acta Endocrinol. Suppl.* 168, 155–73.

Mita-Miyazawa, I., Nishikata, T. & Satoh, N. (1987). Cell- and tissue-specific monoclonal antibodies in eggs and embryos of the ascidean *Halocynthia roretzi*. *Development*, 99, 155–62.

Mitchell, P.J., Carothers, A.M., Han, J.H., Harding, J.D., Kas, E., Venolia, L. & Chasin, L.A. (1986). Multiple transcription start sites, DNase I-hypersensitive sites, and an opposite-strand exon in the 5′ region of the CHO *dhfr* gene. *Mol. & Cell. Biol.* 6, 425–40.

Miwa, J., Schierenberg, E., Miwa, S. & von Ehrenstein, G. (1980). Genetics and mode of expression of temperature-sensitive mutations arresting embryonic development in *Caenorhabditis elegans*. *Devl. Biol.* 76, 160–74.

Mizushima-Sugano, J. & Roeder, R.G. (1986). Cell-type-specific transcription of an immunoglobulin κ light chain gene *in vitro*. *Proc. Natl. Acad. Sci. (USA)* 83, 8511–5.

Mlodzik, M. & Gehring, W.J. (1987). Expression of the *caudal* gene in the germ line of *Drosophila*: formation of an RNA and protein gradient during early embryogenesis. *Cell*, 48, 465–78.

Mlodzik, M., Fjose, A & Gehring, W.J. (1985). Isolation of *caudal*, a *Drosophila* homeobox-containing gene with maternal expression, whose transcripts form a concentration gradient at the pre-blastoderm stage. *EMBO J.* 4, 2961–9.

Mlodzik, M., De Montrion, C.M., Hiromi, Y., Krause, H.M. & Gehring, W.J. (1987). The influence on the blastoderm fate map of maternal-effect genes that affect the antero-posterior pattern in *Drosophila*. *Genes & Dev.* 1, 603–14.

Mlodzik, M., Fjose, A. & Gehring, W.J. (1988). Molecular structure and spatial expression of a homeobox gene from the *labial* region of the Antennapedia-complex, *EMBO J.* 7, 2569–78.

Mock, B.A., D'Hoostelaere, L.A., Matthai, R. & Huppi, K. (1987). A mouse homeobox gene, *Hox-1.5* and the morphological locus, *Hd*, map to within 1 cM. *Genetics*, 116, 607–12.

Moerman, D.G., Benian, G.M., Barstead, R.J., Schriefer, L.A. & Waterston, R.H. (1988). Identification and intracellular localisation of the *unc-22* gene product of *Caenorhabditis elegans*. *Genes & Dev.* 2, 93–105.

Mohler, J. & Wieschaus, E. (1985). *Bicaudal* mutations of *Drosophila melanogaster*: alteration of blastoderm cell fate. *Cold Spring Harbor Symp. Quant. Biol.* 50, 105–11.

Mohler, J., Eldon, E.D. & Pirrotta, V. (1989). A novel spatial transcription pattern associated with the segmentation gene, *giant*, of *Drosophila*. *EMBO J.* 8, 1539–48.

Mohun, T.J. & Garrett, N. (1987). An amphibian cytoskeletal-type actin gene is expressed exclusively in muscle tissue. *Development*, 101, 393–402.

Mohun, T.J., Brennan, S., Dathan, N., Fairman, S. & Gurdon, J.B. (1984). Cell type-specific activation of actin genes in the early amphibian embryo. *Nature (Lond.)* 311, 716–9.

Mohun, T., Maxson, R., Gormezano, G. & Kedes, L. (1985). Differential regulation of individual late histone genes during development of the sea urchin (*Strongylocentrotus purpuratus*). *Devl. Biol.* 108, 491–9.

Molgaard, H.V. (1980). Assembly of immunoglobulin heavy chain genes.

Nature (Lond.) **286**, 657–9.

Monk, M. (1988). Genomic imprinting. *Genes & Dev.* **2**, 921–5.

Monk, M. & Harper, M. (1978). X chromosome activity in preimplantation embryos from XX and XO mothers. *J. Embryol. Exp. Morphol.* **46**, 53–64.

Montminy, M.R. & Bilezkjian, L.M. (1987). Binding of a nuclear protein to the cyclic-AMP response element of the somatostatin gene. *Nature (Lond.)* **328**, 175–8.

Moon, R.T., Danilchik, M.V. & Hille, M.B. (1982). An assessment of the masked messenger hypothesis: sea urchin egg messenger ribonucleoprotein complexes are efficient templates for *in vitro* protein synthesis. *Devl. Biol.* **93**, 389–402.

Morata, G. & Kerridge, S. (1981). Sequential functions of the bithorax complex of *Drosophila*. *Nature (Lond.)* **290**, 778–81.

Morata, G. & Ripoll, P. (1975). *Minutes*: mutants of *Drosophila* autonomously affecting cell division rate. *Devl. Biol.* **42**, 211–21.

Morata, G., Sanchez-Herrero, E. & Casanova, J. (1986). The bithorax complex of *Drosophila*: an overview. *Cell Differ.* **18**, 67–78.

Morgan, T.H. (1934). *Embryology and Genetics*. Columbia University Press, New York.

Moritz, K.B. & Roth, G.E. (1976). Complexity of germline and somatic DNA in *Ascaris. Nature (Lond.)* **259**, 55–7.

Moses, K., Ellis, M.C. & Rubin, G.M. (1989). The *glass* gene encodes a zinc-finger protein required by *Drosophila* photoreceptor cells. *Nature (Lond.)* **340**, 521–4.

Mount, S. & Steitz, J. (1983). Lessons from mutant globins. *Nature (Lond.)* **303**, 380–1.

Mous, J., Stunnenberg, H., Georgiev, O. & Birnstiel, M.L. (1985). Stimulation of sea urchin H2b histone gene transcription by a chromatin-associated protein fraction depends on gene sequences downstream of the transcription start site. *Mol. & Cell. Biol.* **5**, 2764–83.

Muller, M., Affolter, M., Leupin, W., Otting, G., Wuthrich, K. & Gehring, W.J. (1988a). Isolation and sequence-specific binding of the *Antennapedia* homeodomain. *EMBO J.* **7**, 4299–304.

Muller, M.M., Carrasco, A.E. & de Robertis, E.M. (1984). A homeobox-containing gene expressed during oogenesis in *Xenopus. Cell*, **39**, 157–62.

Muller, M.M., Ruppert, S., Schaffner, W. & Matthias, P. (1988b). A cloned octamer transcription factor stimulates transcription from lymphoid-specific promoters in non-B cells. *Nature (Lond.)* **336**, 544–8.

Muller-Holtkamp, D., Knipple, D.C., Seifert, E. & Jackle, H. (1985). An early role of maternal mRNA in establishing the dorsoventral pattern in *pelle* mutant *Drosophila* embryos. *Devl. Biol.* **110**, 238–46.

Muller-Storm, H.-P., Sogo, J.M. & Schaffner, W. (1989). An enhancer stimulates transcription in *trans* when attached via a protein bridge. *Cell*, **58**, 767–77.

Munke, M., Cox, D.R., Jackson, I.J., Hogan, B.L.M. & Francke, U. (1986). The murine Hox-2 cluster of homeobox-containing genes maps distal on chromosome 11 near the *tail-short* (*Ts*) locus. *Cytogenet. Cell Genet.* **42**, 236–40.

Murphy, P., Davidson, D.R. & Hill, R.E. (1989). Segment-specific expression of a homeobox-containing gene in the mouse hindbrain. *Nature (Lond.)* 341, 156–8.

Murphy, S.P., Garbera, J., Odenwald, W.F., Lazzarini, R.A. & Linney, E. (1988). Differential expression of the homeobox gene *Hox-1.3* in F9 embryonal carcinoma cells. *Proc. Natl. Acad. Sci. (USA)* 85, 5587–91.

Murray, A.W. (1987). Cyclins in meiosis and mitosis. *Nature (Lond.)* 326, 542–3.

Murray, A.W. (1988). A mitotic inducer matures. *Nature (Lond.)* 335, 207–8.

Murray, A.W. (1989). The cell cycle as a *cdc-2* cycle. *Nature (Lond.)* 342, 14–15.

Murray, A.W. & Kirschner, M.W. (1989). Cyclin synthesis drives the early embryonic cell cycle. *Nature (Lond.)* 339, 275–8.

Murre, C., Schonleber McCaw, P. & Baltimore, D. (1989). A new DNA-binding and dimerisation motif in immunoglobulin enhancer-binding, *daughterless*, *MyoD* and *myc* proteins. *Cell*, 56, 777–83.

Nagoshi, R.N., McKeown, M., Burtis, K.C., Belote, J.M. & Baker, B.S. (1988). The control of alternative splicing at genes regulating sexual differentiation in *Drosophila melanogaster*. *Cell.* 53, 229–36.

Nakano, Y., Guerrero, I., Hidalgo, A., Taylor, A., Whittle, J.R.S. & Ingham, P.W. (1989). A protein with several possible membrane-spanning domains encoded by the *Drosophila* segment polarity gene *patched*. *Nature (Lond.)* 341, 508–12.

Nandi, A.K., Roginski, R.S., Gregg, R.G., Smithies, O. & Skoultchi, A.I. (1988). Regulated expression of genes inserted at the human chromosomal β globin locus by homologous recombination. *Proc. Natl. Acad. Sci. (USA)* 85, 3845–9.

Natzle, J.E., Fristrom, D.K. & Fristrom, J.W. (1988). Genes expressed during imaginal disc morphogenesis: IMP-E, a gene associated with epithelial cell rearrangement. *Devl. Biol.* 129, 428–38.

Nauber, U., Pankratz, M.J., Kienlin, A., Seifert, E., Klemm, U. & Jackle, H. (1988). Abdominal segmentation of the *Drosophila* embryo requires a hormone receptor-like protein encoded by the gap gene *knirps*. *Nature (Lond.)* 336, 489–92.

Nelson, C., Albert, V.R., Elsholtz, H.P., Liu, L.I.-W. & Rosenfeld, M.R. (1988). Activation of cell-specific expression of rat growth hormone and prolactin genes by a common transcription factor. *Science.* 239, 1400–3.

Nemer, M. (1986). An altered series of ectodermal gene expression accompanying the reversible suspension of differentiation in the zinc-animalised sea urchin embryo. *Devl. Biol.* 114, 214–24.

Newport. J. & Kirschner, M. (1982a). A major developmental transition in early *Xenopus* embryos: I Characterisation and timing of cellular changes at the midblastula stage. *Cell*, 30, 675–86.

Newport. J. & Kirschner, M. (1982b). A major developmental transition in early *Xenopus* embryos: II Control of the onset of transcription. *Cell*, 30, 687–96.

Newport, J. & Kirschner, M.W. (1984). Regulation of the cell cycle during early *Xenopus* development. *Cell*, 37, 731–42.

Nienhuis, A.W. & Stamatoyannopoulos, G. (1978). Haemoglobin switching. *Cell*, 15, 307–15.

Nishida, H. (1987). Cell lineage analysis in ascidean embryos by intracellular injection of a tracer enzyme: III Up to the tissue restricted stage. *Devl. Biol.* 121, 526–41.

Nishida, H. & Satoh, N. (1989). Determination and regulation in the pigment cell lineage of the ascidean embryo. *Devl. Biol.* **132**, 355–67.

Nishikata, T., Mita-Miyazawa, I., Deno, T., Takamura, K. & Satoh, N. (1987a). Expression of epidermis-specific antigens during embryogenesis of the ascidean *Halocynthia roretzi*. *Devl. Biol.* **121**, 408–16.

Nishikata, T., Mita-Muyazawa, I., Deno, T. & Satoh, N. (1987b). Muscle cell differentiation in ascidean embryos analysed with a tissue-specific mono-colonal antibody. *Development*, **99**, 163–71.

Nishikata, T., Mita-Miyazawa, I., Deno, T. & Satoh, N. (1987c). Monoclonal antibodies against components of the myoplasm of eggs of the ascidean *Ciona intestinalis* partially block the development of muscle-specific acetylcholinesterase. *Development*, **100**, 577–86.

Nishioka, Y., Leder, A. & Leder, P. (1980). Unusual α-globin like gene that has cleanly lost both globin intervening sequences. *Proc. Natl. Acad. Sci. (USA)* **77**, 2806–9.

Nocente-McGrath, C., Brenner, C.A. & Ernst, S.G. (1989). Endo-16, a lineage-specific protein of the sea urchin embryo, is first expressed just prior to gastrulation. *Devl. Biol.* **136**, 264–72.

Nomura, S., Yamagoe, S., Kamiya, T. & Oishi, M. (1986). An intracellular factor that induces erythroid differentiation in mouse erythroleukaemia (Friend) cells. *Cell*, **44**, 663–9.

Nordstrom, J.L., Roop, D.R., Tsai, M.J. & O'Malley, B.W. (1979). Identification of potential ovomucoid mRNA precursors in chick oviduct nuclei. *Nature (Lond.)* **278**, 328–33.

North, G. (1988). Pattern formation: in my beginning is my end. *Nature (Lond.)* **332**, 785–6.

Nothiger, R., Dubendorfer, A. & Epper, F. (1977). Gynandromorphs reveal two separate primordia for male and femal genitalia in *Drosophila melanogaster*. *Roux' Arch. Devl. Biol.* **181**, 367–73.

Nothiger, R. & Steinmann-Zwicky, M. (1985). A single principle for sex determination in insects. *Cold Spring Harbor Symp. Quant. Biol.*, **50**, 615–20.

Nudel, U., Greenberg, D., Ordahl, C.P., Saxel, O., Neuman, S. & Yaffe, D. (1985). Developmentally regulated expression of a chicken muscle-specific gene in stably transfected rat myogenic cells. *Proc. Natl. Acad. Sci. (USA)* **82**, 3106–9.

Nunberg, J.H., Kaufman, R.J., Schimke, R.T., Urlaub, G. & Chasin, L.A. (1978). Amplified dihydrofolate reductase genes are localised to a homogeneously staining region of a single chromosome in a methotrexate-resistant Chinese hamster ovary cell line. *Proc. Natl. Acad Sci. (USA)* **65**, 5553–6.

Nusslein-Volhard, C. (1979). Maternal effect mutations that alter the spatial coordinates of the embryo of *Drosophila melanogaster*. In *Determinants of Spatial Organisation*, ed. S. Subtelny & I.R. Konigsberg, Symp. Soc. Devl. Biol. 37, 185–211. Academic Press, New York.

Nusslein-Volhard, C. & Wieschaus, E. (1980). Mutations affecting segment number and polarity in *Drosophila*. *Nature (Lond.)* **287**, 795–9.

Nusslien-Volhard, C., Frohnhofer, H.G. & Lehmann, R. (1987). Determination of antero-posterior polarity in *Drosophila*. *Science*, **238**, 1675–81.

O'Brochta, D.A. & Bryant, P.J. (1985). A zone of non-proliferating cells at a lineage restriction boundary in *Drosophila. Nature (Lond.)* **313**, 138–40.

O'Connor, M.B., Binari, R., Perkins, L.A. & Bender, W. (1988). Alternative RNA products from the *Ultrabithorax* domain of the bithorax complex. *EMBO J.* **7**, 435–45.

Odenwald, W.F., Garbern, J., Arnheiter, H., Tournier-Lasserve, E. & Lazzarini, R.A. (1989). The *Hox-1.3* homeobox protein is a sequence-specific DNA-binding phosphoprotein. *Genes & Dev.* **3**, 156–72.

O'Hare, K. & Rubin, G.M. (1983). Structure of P transposable elements and their sites of insertion and excision in the *Drosophila melanogaster* genome. *Cell*, **34**, 25–35.

Okada, M. & Togashi, S. (1985). Isolation of a factor inducing pole-cell formation from *Drosophila* embryos. *Int. J. Invert. Reprod. & Dev.* **8**, 207–17.

Okazaki, K. (1975). Spicule formation by isolated micromeres of the sea urchin embryo. *Amer. Zool.* **15**, 567–581.

Oliver, G., Wright, C.V.E., Hardwicke, J. & De Robertis, E.M. (1988a). Differential antero-posterior expression of two proteins encoded by a homeobox gene in *Xenopus* and mouse embryos. *EMBO J.* **7**, 3199–209.

Oliver, G., Wright, C.V.E., Hardwicke, J. & De Robertis, E.M. (1988b). A gradient of homeodomain protein in developing forelimbs of *Xenopus* and mouse embryos. *Cell*, **55**, 1017–24.

Oliver, G., Sidell, N., Fiske, W., Heinzmann, C., Mohandas, T., Sparkes, R.S. & de Robertis, E.M. (1989). Complementary homeoprotein gradients in developing limb buds. *Genes & Dev.* **3**, 641–50.

Orkin, S.H., Kazazian, H.H. Jr., Antonarakis, S.E., Ostrer, H., Goff, S.C. & Sexton, J.P. (1982). Abnormal RNA processing due to the exon mutation of β^E-globin gene. *Nature (Lond.)* **300**, 768–70.

Ornitz, D.M., Palmiter, R.D., Hammer, R.E., Brinster, R.L., Swift, G.H. & Mac-Donald, R.J. (1985). Specific expression of an elastase–human growth hormone fusion gene in pancreatic acinar cells of transgenic mice. *Nature (Lond.)* **313**, 600–3.

Oro, A.E., Ong, E.S., Margolis, J.S., Posakony, J.W., McKeown, M. & Evans, R.M. (1988). The *Drosophila* gene *knirps-related* is a member of the steroid-receptor gene superfamily. *Nature (Lond.)* **336**, 493–6.

Orr, W., Komitopoulou, K. & Kafatos, F.C. (1984). Mutants suppressing in *trans* chorion gene amplification in *Drosophila. Proc. Natl. Acad. Sci. (USA)* **8**, 3773–7.

Osheim, Y.N. & Miller, O.L. Jr. (1983). Novel amplification and transcriptional activity of chorion genes in *Drosophila melanogaster* follicle cells. *Cell*, **33**, 543–53.

Osterbur, D.L., Fristrom, D.K., Natzle, J.E., Tojo, S.J. & Fristrom, J.W. (1988). Genes expressed during imaginal disc morphogenesis: IMP-L2, a gene expressed during imaginal disc and imaginal histoblast morphogenesis. *Devl. Biol.* **129**, 439–48.

Padgett, R.W., St. Johnston, R.D. & Gelbart, W.M. (1987). A transcript from a *Drosophila* pattern gene predicts a protein homologous to the transforming growth factor β family. *Nature (Lond.)* **325**, 81–4.

Palmiter, R.D. (1975). Quantitation of the parameters that determine the rate of ovalbumin synthesis. *Cell*, **4**, 189–97.

Palmiter, R.D. & Brinster, R.L. (1985). Transgenic mice. *Cell*, **41**, 343–5.

Palmiter, R.D. & Carey, N.H. (1974). Rapid inactivation of ovalbumin messenger ribonucleic acid after acute withdrawal of oestrogen. *Proc. Natl. Acad. Sci. (USA)* **71**, 2357–61.

Palmiter, R.D., Mulvihill, E.R., McKnight, G.S. & Senear, A.W. (1977). Regulation of gene expression in the chick oviduct by steroid hormones. *Cold Spring Harbor Symp. Quant. Biol.* **42**, 639–47.

Palmiter, R.D., Behringer, R.R., Quaife, C.J., Maxwell, F., Maxwell, I.H. & Brinster, R.H. (1987). Cell lineage ablation in transgenic mice by cell-specific expression of a toxin gene. *Cell*, **50**, 435–43.

Pancratz, M.J., Hoch, M., Seifert, E. & Jackle, H. (1989). *Kruppel* requirement for *knirps* enhancement reflects overlapping gap gene activities in the *Drosophila* embryo. *Nature (Lond.)* **341**, 337–40.

Pankow, W., Leszzi, M. & Holdregger-Mähling, I. (1976). Correlated changes of Balbiana ring expression and secretory protein synthesis in larval salivary glands of *Chironomus tentans*. *Chromosoma*, **58**, 137–53.

Papayannopoulou, T., Bryce, M. & Stamatoyannopoulos, G. (1986). Analysis of human haemoglobin switching in MEL × human foetal erythroblast cell hybrids. *Cell*, **46**, 469–76.

Parks, S. & Spradling, A.C. (1987). Spatially regulated expression of chorion genes during *Drosophila* oogenesis. *Genes & Dev.* **1**, 497–509.

Parks, S., Wakimoto, B. & Spradling, A.C. (1986). Replication and expression of an X-linked cluster of *Drosophila* chorion genes. *Devl. Biol.* **117**, 294–305.

Pastorcic, M., Wang, H., Elbrecht, A., Tsai, S.Y., Tsai, M.-J. & O'Malley, B.W. (1986). Control of transcription initiation *in vitro* requires binding of a transcription factor to the distal promoter of the ovalbumin gene. *Mol. & Cell. Biol.* **6**, 2784–91.

Patel, N.H., Martin-Blanco, E., Coleman, K.G., Poole, S.J., Ellis, M.C., Kornberg, T.B. & Goodman, C.S. (1989). Expression of *engrailed* proteins in arthropods, annelids and chordates. *Cell*, **58**, 955–68.

Peifer, M. & Bender, W. (1986). The *anterobithorax* and *bithorax* mutations of the bithorax complex. *EMBO J.* **5**, 2293–303.

Peifer, M., Karch, F. & Bender, W. (1987). The bithorax complex: control of segmental identity. *Genes & Dev.* **1**, 891–8.

Pelham, H.R.B. & Brown, D.D. (1980). A specific transcription factor that can bind either the 5S RNA gene or 5S RNA. *Proc. Natl. Acad. Sci. (USA)* **77**, 4170–4.

Perkowska, E., MacGregor, H.C. & Birnstiel, M.L. (1968). Gene amplification in the oocyte nucleus of mutant and wild-type *Xenopus laevis*. *Nature (Lond.)* **217**, 649–50.

Perrimon, N. & Mahowald, A.P. (1987). Multiple functions of segment polarity genes in *Drosophila*. *Devl. Biol.* **119**, 587–600.

Peschle, C., Mavilio, F., Carè, A., Migliaccio, G., Migliaccio, A.R., Salvo, G., Samoggia, P., Petti, S., Guerriero, R., Marinucci, M., Lazzaro, D., Russo, G. & Mastroberardino, G. (1985). Haemoglobin synthesis in human embryos: asynchrony of ζ→α and ε→γ globin switches in primitive and definitive erythropoietic lineage. *Nature (Lond.)* **313**, 235–7.

Pestell, R.Q.W. (1975). Microtubule protein synthesis during oogenesis and early embryogenesis in *Xenopus laevis. Biochem. J.* **145**, 527–34.

Peterson, C.L., Orth, K. & Calame, K. (1986). Binding *in vitro* of multiple cellular proteins to immunoglobulin heavy chain enhancer DNA. *Mol. & Cell. Biol.* **6**, 4168–78.

Petkovich, M., Brand, N.J., Krust, A. & Chambon, P. (1987). A human retinoic acid receptor which belongs to the family of nuclear receptors. *Nature (Lond.)* **330**, 444–7.

Picard, D. & Schaffner, W. (1984). A lymphocyte-specific enhancer in the mouse immunoglobulin κ gene. *Nature (Lond.)* **307**, 80–3.

Picard, D. & Yamamoto, K.R. (1987). Two signals mediate hormone-dependent nuclear localisation of the glucocortoid receptor. *EMBO J.* **6**, 3333–40.

Pierandrei-Amaldi, P., Campioni, N., Beccari, E., Bozzoni, I. & Amaldi, F. (1982). Expression of ribosomal-protein genes in *Xenopus laevis* development. *Cell*, **30**, 163–71.

Pierandrei-Amaldi, P., Beccari, E., Bozzoni, I. & Amaldi, F. (1985). Ribosomal protein production in normal and anucleate *Xenopus* embryos: regulation at the post-transcriptional and translational levels. *Cell*, **42**, 317–233.

Pinkert, C.A., Ornitz, D.M., Brinster, R.L. & Palmiter, R.D. (1987). An albumin enhancer located 10kb upstream functions along with its promoter to direct efficient liver-specific expression in transgenic mice. *Genes & Dev.* **1**, 268–76.

Pinney, D.F., Pearson-White, S.H., Konieczny, S.F., Latham, K.E. & Emerson, C.P. (1988). Myogenic lineage determination and differentiation: evidence for a regulatory gene pathway. *Cell*, **53**, 781–91.

Poccia, D., Salik, J. & Krystal, G. (1981). Transitions in histone variants of the male pronucleus following fertilisation and evidence for a matenal store of cleavage-stage histones in the sea urchin egg. *Devl. Biol.* **82**, 287–97.

Poellinger, L., Yoza, B.K. & Roeder, R.G. (1989). Functional cooperativity between protein molecules bound at two distinct sequence elements of the immunoglobulin heavy-chain promoter. *Nature (Lond.)* **337**, 573–6.

Politz, J.C. & Edgar, R.S. (1984). Overlapping stage-specific sets of numerous small collagenous polypeptides are translated *in vitro* from *Caenorhabditis elegans* RNA. *Cell*, **37**, 853–60.

Pollock, J.A. & Benzer, S. (1988). Transcript localisation of four opsin genes in the three visual organs of *Drosophila*; RH2 is ocellus specific. *Nature (Lond.)* **333**, 779–81.

Ponglikitmongkol, M., Green, S. & Chambon, P. (1988). Genomic organisation of the human oestrogen receptor gene. *EMBO J.* **7**, 3385–8.

Poole, S.J., Kauvar, L.M., Drees, B. & Kornberg, T. (1985). The *engrailed* locus of *Drosophila*: structural analysis of an embryonic transcript. *Cell*, **40**, 37–43.

Posakony, J.W., Flytzanis, C.N., Britten, R.J. & Davidson, E.H. (1983). Interspersed sequence organisation and development representation of cloned poly(A)⁺ RNA's from sea urchin eggs. *J. Mol. Biol.* **167**, 361–89.

Potter, H., Weir, L. & Leder, P. (1984). Enhancer dependent expression of human κ immunoglobulin genes introduced into mouse pre-B lymphocytes by electroporation. *Proc. Natl. Acad. Sci. (USA)* **81**, 7161–5.

Poulson, D.F. (1950). Histogenesis, organogenesis and differentiation in the embryo of *Drosophila melanogaster* Meiger. In *Biology of* Drosophila, ed. M. Demerec, pp. 168–93. Wiley, New York.

Preiss, A., Rosenberg, U.B., Kienlin, A., Seifert, E. & Jackle, H. (1985). Molecular genetics of *Kruppel*, a gene required for segmentation of the *Drosophila* embryo. *Nature (Lond.)* **313**, 27–30.

Preiss, A., Hartley, D.A. & Artavanis-Tsakonas, S. (1988). The molecular genetics of *Enhancer of split*, a gene required for embryonic neural development in *Drosophila*. *EMBO J.* **7**, 3917–27.

Price, J.V., Clifford, R.J. & Schupbach, T. (1989). The maternal ventralising locus *torpedo* is allelic to *faint little ball*, an embryonic lethal, and encodes the *Drosophila* EGF receptor homologue. *Cell*, **56**, 1085–92.

Priess, J.R. & Thomson, J.N. (1987). Cellular interactions in early *Caenorhabditis elegans* embryos. *Cell*, **48**, 241–50.

Priess, J.R., Schnabel, H. & Schnabel, R. (1987). The *glp-1* locus and cellular interactions in early *C. elegans* embryos. *Cell*, **51**, 601–10.

Prost, E., Deryckere, F., Roos, C., Haenlin, M., Pantesco, V. & Mohier, E. (1988). Role of the oocyte nucleus in determination of the dorsoventral polarity of *Drosophila* as revealed by molecular analysis of the K10 gene. *Genes & Dev.* **2**, 891–900.

Proudfoot, N.J. & Maniatis, T. (1980). The structure of a human α globin pseudogene and its relationship to α globin gene duplication. *Cell*, **21**, 537–44.

Proudfoot, N.J., Gill, A. & Maniatis, T. (1982). The structure of the human zeta globin gene and a closely-linked, nearly identical pseudogene. *Cell*, **31**, 553–63.

Ptashne, M. (1986). Gene regulation by proteins acting nearby and at a distance. *Nature (Lond.)* **322**, 697–701.

Ptashne, M. (1988). How eucaryotic transcriptional activators work. *Nature (Lond.)* **335**, 683–9.

Pultz, M.A., Diederich, R.J., Cribbs, D.L. & Kaufman, T.C. (1988). The *proboscipedia* locus of the Antennapedia complex: a molecular and genetic analysis. *Genes & Dev.* **2**, 901–20.

Queen, C. & Baltimore, D. (1983). Immunoglobulin gene transcription is activated by downstream sequence elements. *Cell*, **33**, 741–8.

Rabin, M., Ferguson-Smith, A., Hart, C.P. & Ruddle, F.H. (1986). Cognate homeobox loci mapped on homologous human and mouse chromosomes. *Proc. Natl. Acad. Sci. (USA)* **83**, 9104–8.

Raghavan, K.V., Crosby, M.A., Mathers, P.H. & Meyerowitz, E.M. (1986). Sequences sufficient for correct regulation of *sgs-3* lie close to or within the gene. *EMBO J.* **5**, 3321–6.

Raven, C.P. (1961). *Oogenesis: the Storage of Developmental Information*. Pergamon Press; New York, Oxford, London & Paris.

Raven, C.P. (1963). Mechanisms of determination in the development of gastropods. *Adv. Morphogen.* **3**, 1–32.

Readhead, C., Popko, B., Takahashi, N., Shine, H.D., Saavedra, R.A., Sidman, R.L. & Hood, L. (1987). Expression of a myelin basic protein gene in transgenic *shiverer* mice: correction of the dysmyelinating phenotype. *Cell*, **48**, 703–12.

Rebagliati, M.R., Weeks, D.L., Harvey, R.P. & Melton, D.A. (1985). Identification and cloning of localised maternal RNAs from *Xenopus* eggs. *Cell*, **42**, 769–77.

Redemann, N., Gaul, U. & Jackle, H. (1988). Disruption of a putative Cys–zinc interaction eliminates the biological activity of the *Kruppel* finger protein. *Nature (Lond.)* **332**, 90–2.

Reed, R., Griffith, J. & Maniatis, T. (1988). Purification and visualisation of native splicesomes. *Cell,* **53**, 949–61.

Regulski, M., Harding, K., Kostriken, R., Karch, F., Levine, M. & McGinnis, W. (1985). Homeoboxes of the Antennapedia and bithorax gene complexes of *Drosophila*. *Cell,* **43**, 71–80.

Regulski, M., McGinnis, N. Chadwick, R. & McGinnis, W. (1987). Developmental and molecular analysis of *Deformed*: a homoeotic gene controlling *Drosophila* head development. *EMBO J.* **6**, 767–77.

Reinke, R. & Zipursky, S.L. (1988). Cell-cell interaction in the *Drosophila* retina: the *bride of sevenless* gene is required in photoreceptor cell R8 for R7 cell development. *Cell,* **55**, 321–30.

Reitman, M. & Felsenfeld, G. (1988). Mutational analysis of the chicken β-globin enhancer reveals two positive-acting domains. *Proc. Natl. Acad. Sci. (USA)* **85**, 6267–71.

Render, J.A. (1983). The second polar lobe of the *Sabellaria cementarium* embryo plays an inhibitory role in apical tuft formation. *Roux' Arch. Devl. Biol.* **192**, 120–9.

Renkawitz, R., Schutz, G., Von der Ahe, D. & Beato, M. (1984). Sequences in the promoter region of the chicken lysozyme gene required for steroid regulation and receptor binding. *Cell,* **37**, 503–10.

Reth, M., Gehrmann, P., Petrac, E. & Weise, P. (1986). A novel V_H to $V_H DJ_H$ joining mechanism in heavy-chain-negative (null) pre B cells results in heavy chain production. *Nature (Lond.)* **322**, 840–2.

Richter, J.D. & Smith, L.D. (1983). Developmentally regulated RNA binding proteins during oogenesis in *Xenopus laevis. J. Biol. Chem.* **258**, 4864–9.

Richter, J.D. & Smith, L.D. (1984). Reversible inhibition of translation by *Xenopus* oocyte-specific proteins. *Nature (Lond.)* **309**, 378–80.

Richter, J.D., Evers, D.C. & Smith, L.D. (1983). The recruitment of membrane-bound mRNAs for translation in microinjected *Xenopus* oocytes. *J. Biol. Chem.* **258**, 2614–20.

Riddle, D.L. (1987). Postembryonic development in *Caenorhabditis elegans. Int. J. Parasitol.* **17**, 223–31.

Riddle, D.L. (1988). The dauer larva. In *The nematode* Caenorhabditis elegans, ed. W.B. Wood. CSH Monographs 17, 393–412.

Riggleman, B., Wieschaus, E. & Schedl, P. (1989). Molecular analysis of the *armadillo* locus: uniformly distributed transcripts and a protein with novel internal repeats are associated with a *Drosophila* segment polarity gene. *Genes & Dev.* **3**, 96–113.

Rijsewijk, F., Schuermann, M., Wagenaar, E., Parren, P., Wiegel, D. & Nusse, R. (1987). The *Drosophila* homolog of the mouse mammary oncogene *int-1* is identical to the segment polarity gene *wingless. Cell,* **50**, 649–57.

Riley, P.D., Carroll, S.B. & Scott, M.P. (1987). The expression and regulation of *Sex combs reduced* protein in *Drosophila* embryos. *Genes & Dev.* **1**, 716–30.

Rixon, M.W. & Gelinas, R.E. (1988). A foetal globin gene mutation in $^A\gamma$ nondeletion hereditary persistence of foetal haemoglobin increases promoter strength in a nonerythroid cell. *Mol. & Cell. Biol.* **8**, 713–21.

Roberts, J.M., Buck, L.B. & Axel, R. (1983). A structure for amplified DNA. *Cell*, **33**, 53–63.

Robertson, M. (1988a). Homeoboxes, POU proteins and the limits to promiscuity. *Nature (Lond.)* **336**, 522–4.

Robertson, M. (1988b). Developmental decisions and pattern formation. *Nature (Lond.)* **335**, 494–5.

Rodgers, W.H. & Gross, P.R. (1978). Inhomogeneous distribution of egg RNA sequences in the early embryo. *Cell*, **14**, 279–88.

Rogers, J., Early, P., Carter, C., Calame, K., Hood, L. & Wall, R. (1980). Two mRNAs with different 3' ends encode membrane-bound and secreted forms of immunoglobulin μ chain. *Cell*, **20**, 303–12.

Rosbash, M. & Ford, P.J. (1974). Polyadenylic acid-containing RNA in *Xenopus laevis* oocytes. *J. Mol. Biol.* **85**, 87–101.

Rose, S.J. III, Rosenberg, M.J., Britten, R.J. & Davidson, E.H. (1987). Expression of myosin heavy chain gene in the sea urchin: coregulation with muscle actin gene transcription in early development. *Devl. Biol.* **123**, 115–24.

Rosenberg, U.B., Preiss, A., Siefert, E., Jackle, H. & Knipple, D.C. (1985). Production of phenocopies by *Kruppel* antisense RNA injection into *Drosophila* embryos. *Nature (Lond.)* **313**, 703–6.

Rosenberg, U.B., Schroder, C., Preiss, A., Keinlin, A, Côté, S., Riede, I. & Jackle, H. (1986). Structural homology of the product of the *Drosophila Kruppel* gene with *Xenopus* transcription factor IIIA. *Nature (Lond.)* **319**, 336–8.

Rosenquist, T.A. & Kimble, J. (1988). Molecular cloning and transcript analysis of *fem-3*, a sex-determination gene in *Caenorhabditis elegans*. *Genes & Dev.* **2**, 606–16.

Rothe, M., Nauber, U. & Jackle, J. (1989). Three hormone receptor-like *Drosophila* genes encode an identical DNA-binding finger. *EMBO J.* **8**, 3087–94.

Rowe, A. & Akam, M. (1988). The structure and expression of a hybrid homoeotic gene. *EMBO J.* **7**, 1107–14.

Royal, A., Garapin, A., Cami, B., Perrin, F., Mandel, J.L., Le Meur, M., Brégegère, F., Gannon, R., Le Pennec, J.P., Chambon, P. & Kourilsky, P. (1979). The ovalbumin gene region: common features in the organisation of three genes expressed in chick oviduct under hormonal control. *Nature (Lond.)* **279**, 125–32.

Rubin, G.M. (1989). Development of the *Drosophila* retina: inductive events studied at single cell resolution. *Cell*, **57**, 519–20.

Rubin, M.R., Toth, L.E., Patel, M.D., D'Eustachio, P. & Nguyen-Huu, M.C. (1986). A mouse homeobox gene is expressed in spermatocytes and embryos. *Science*, **233**, 663–5.

Rubin, M.R., King, W., Toth, L.E., Sawczuk, I.S., Levine, M.S., D'Eustachio, P. & Nguyen-Huu, M.C. (1987). Murine *Hox 1.7* homeobox gene: cloning, chromosomal location and expression. *Mol. & Cell. Biol.* **7**, 3836–41.

Ruddell, A. & Jacobs-Lorena, M. (1985). Biphasic pattern of histone gene expression during *Drosophila* oogenesis. *Proc. Natl. Acad. Sci. (USA)* **83**, 3316–9.

Ruderman, J.V., Woodland, H.R. & Sturgess, E.A. (1979). Modulations of histone messenger RNA during the early development of *Xenopus laevis*. *Devl. Biol.* **71**, 71–82.

Ruiz i Atalba, A. & Melton, D.A. (1989a). Involvement of the *Xenopus* homeobox gene *Xhox 3* in pattern formation along the anterior posterior axis. *Cell*, **57**, 317–26.

Ruiz i Atalba, A. & Melton, D.A. (1989b). Bimodal and graded expression of the *Xenopus* homeobox gene *Xhox 3* during embryonic development. *Development*, **106**, 173–83.

Runnström. J. (1928). Plasmabau und Determination bei dem Ei von *Paracentrotus lividus*, LK. *Roux' Arch.* **113**, 556–81.

Rushlow, C.E. Doyle, H., Hoey, T. & Levine, M. (1987a). Molecular characterisation of the *zerknullt* region of the Antennapedia gene complex in *Drosophila*. *Genes & Dev.* **1**, 1268–79.

Rushlow, C., Frasch, M., Doyle, H. & Levine, M. (1987b). Maternal regulation of *zerknullt*: a homeobox gene controlling differentiation of dorsal tissues in *Drosophila*. *Nature (Lond.)* **330**, 583–6.

Rushlow, C.A., Hogan, A., Pinchin, S.M., Howe, K.M., Lardelli, M. & Ish-Horowicz, D. (1989). The *Drosophila hairy* protein acts in both segmentation and bristle patterning and shows homology to N-*myc*. *EMBO J.* **8**, 3095–103.

Ruskin, B., Krainer, A.R., Maniatis, T. & Green, M.R. (1984). Excision of an intact intron as a novel lariat structure during pre-mRNA splicing *in vitro*. *Cell*, **38**, 317–31.

Ruvkun, G. & Giusto, J. (1989). The *Caenorhabditis elegans* heterochronic gene *lin-14* encodes a nuclear protein that forms a temporal developmental switch. *Nature (Lond.)* **338**, 313–9.

Ryffel, G.U. & Wahli, W. (1983). Regulation and structure of the vitellogenin genes. In *Eucaryotic Genes: their Structure, Activity and Regulation*, ed. N. Maclean, S.P. Gregory & R.A. Flavell, pp. 329–41. Butterworths, London.

Ryffel, G.U., Muellener, D.B., Wyler, T., Wahli, W. & Weber, R. (1981). Transcription of single copy vitellogenin gene of *Xenopus* involves expression of middle repetitive DNA. *Nature (Lond.)* **291**, 429–30.

Ryffel, G.U., Muellner, D.B., Gerber-Huber, S., Wyler, T. & Wahli, W. (1983). Scattering of repetitive DNA sequences in the albumin and vitellogenin gene loci of *Xenopus laevis*. *Nucleic Acids Res.* **11**, 7701–16.

Rykowski, M.C., Parmelee, S.J., Agard, D.A. & Sedat, J.W. (1988). Precise determination of the molecular limits of a polytene chromosome band: regulatory sequences for the *Notch* gene are in the interband. *Cell*, **54**, 46–72.

Saari, G. & Bienz, M. (1987). The structure of the *Ultrabithorax* promoter of *Drosophila melanogaster*. *EMBO J.* **6**, 1775–9.

Sachs, L. (1987). The molecular control of blood cell development. *Science*, **238**, 1374–9.

Sagami, I., Tsai, S.Y., Wang, H., Tsai, M.-J. & O'Malley, B.W. (1986). Identification of two factors required for transcription of the ovalbumin gene. *Mol. & Cell. Biol.* **6**, 4259–67.

Saint, R., Kalionis, B., Lockett, T.J. & Elizur, A. (1988). Pattern formation in the developing eye of *Drosophila melanogaster* is regulated by the homeobox gene *rough*. *Nature (Lond.)* **334**, 151–3.

Sakai, D.D., Helms, S., Carlstedt-Duke, J., Gustafsson, J.-A., Rottman, F.M. & Yamamoto, K.R. (1988). Hormone-mediated repression: a negative glucocorticoid response element from the bovine prolactin gene. *Genes & Dev.* **2**, 1144–54.

Sakano, H., Rogers, J.H., Huppi, K., Brack, C., Traunecker, A., Maki, R., Wall, R. & Tonegawa, S. (1979). Domains and the hinge region of an immunoglobulin heavy chain are encoded in separate DNA segments. *Nature (Lond.)* **277**, 627–31.

Sakonju, S., Bogenhagen, D.F. & Brown, D.D. (1980). A control region in the centre of the 5S gene directs specific initiation of transcription: I, The 5' border of the region. *Cell*, **19**, 13–25.

Saluz, H.P., Feavers, I.M., Jiricny, J. & Jost, J.P. (1988). Genomic sequencing and *in vivo* footprinting of an expression-specific DNaseI-hypersensitive site of avian vitellogenin II promoter reveal a demethylation of a mCpG and a change in specific interactions of proteins with DNA. *Proc. Natl. Acad. Sci. (USA)* **85**, 6697–700.

Salz, H.K., Maine, E.M., Keyes, L.N., Samuels, M.E., Cline, T.W. & Schedl, P. (1989). The *Drosophila* female-specific sex-determination gene, *Sex-lethal*, has stage-, tissue- and sex-specific RNAs suggesting multiple modes of regulation. *Genes & Dev.* **3**, 708–19.

Sanchez-Herrero, E. & Akam, M. (1989). Spatially ordered transcription of regulatory DNA in the bithorax complex of *Drosophila*. *Development*, **107**, 321–9.

Sanchez-Herrero, E. & Crosby, M.A. (1988). The *Abdominal-B* gene of *Drosophila melanogaster*: overlapping transcripts exhibit two different spatial distributions. *EMBO J.* **7**, 2163–73.

Sanchez-Herrero, E. & Morata, G. (1984). The *Ubx* syndrome of *Drosophila*: the prothoracic transformation (*ppx*) is independent of *bx*, *bxd* and *pbx*. *Roux' Arch. Devl. Biol.* **193**, 263–5.

Sanchez-Herrero, E., Vernos, I., Marco, R. & Morata, G. (1985). Genetic organisation of *Drosophila* bithorax complex. *Nature (Lond.)* **313**, 108–12.

Sander, K. & Lehmann, R. (1988). *Drosophila* nurse cells produce a posterior signal required for embryonic segmentation and polarity. *Nature (Lond.)* **335**, 68–70.

Sang, J.H. (1984). *Genetics and Development*. Longman Group, London.

Santamaria, P. & Nusslein-Volhard, C. (1983). Partial rescue of *dorsal*, a maternal-effect mutation affecting the dorso-ventral polarity of the *Drosophila* embryo, by the injection of wild-type cytoplasm. *EMBO J.* **2**, 1695–9.

Santon, J.B. & Pellegrini, M. (1980). Expression of ribosomal proteins during *Drosophila* early development. *Proc. Natl. Acad. Sci. (USA)* **77**, 5649–53.

Santoro, C., Mermod, N., Andrews, P.C. & Tijan, R. (1988). A family of human CCAAT-box-binding proteins active in transcription and DNA replication: cloning and expression of multiple cDNAs. *Nature (Lond.)* **334**, 218–21.

Sassoon, D.A, Garner, I. & Buckingham, M. (1988). Transcripts of α-cardiac and α-skeletal actins are early markers for myogenesis in the mouse embryo. *Development*, **104**, 155–64.

Sassoon, D., Lyons, G., Wright, W.E., Lin, V., Lassar, A., Weintraub, H. & Buckingham, M. (1989). Expression of two myogenic factors myogenin and MyoD1 during mouse embryogenesis. *Nature (Lond.)* **341**, 303–7.

Sato, T. & Denell, R.E. (1986). Segmental identity of caudal cuticular features of *Drosophila melanogaster* larvae and its control by the bithorax complex. *Devl. Biol.* **116**, 78–91.

Satoh, N. (1979). On the 'clock' mechanism determining the time of tissue-

specific enzyme development during ascidean embryogenesis: I Acetylcholinesterase development in cleavage-arrested embryos. *J. Embryol. Exp.
Morphol.* **54**, 131–9.

Satoh, N. & Ikegami, S. (1981). A definite number of aphidicolin-sensitive cell-
cyclic events are required for acetylcholinesterase development in the presumptive muscle of the ascidean embryo. *J. Embryol. Exp. Morphol.* **61**, 1–13.

Savage, C., Hamelin, M., Culotti, J.G., Coulson, A., Albertson, D.G. & Chalfie,
M. (1989). *mec-7* is a β-tubulin gene required for the production of 15-
protofilament microtubules in *Caenorhabditis elegans. Genes & Dev.* **3**, 870–81.

Savard, P., Bates, P.B. & Brockes, J.P. (1988). Position dependent expression of
a homeobox gene transcript in relationship to amphibian limb generation.
EMBO J. **7**, 4275–82.

Sawicki, J.A., Magnuson, T. & Epstein, C.J. (1981). Evidence for expression of
the paternal genome in the two-cell mouse embryo. *Nature (Lond.)* **294**, 450–2.

Schaltmann, K. & Pongs, O. (1982). Identification and characterisation of the
ecdysterone receptor in *Drosophila melanogaster* by photoaffinity labeling.
Proc. Natl. Acad. Sci. (USA) **79**, 6–10.

Schatz, D.G. & Baltimore, D. (1988). Stable expression of immunoglobulin
gene V(D)J recombinase activity by gene transfer into 3T3 fibroblasts. *Cell,*
53, 107–15.

Scheidereit, C., Heguy, A. & Roeder, R.G. (1987). Identification and purification of a human lymphoid-specific octamer-binding protein (OTF-2) that
activates transcription of an immunoglobulin promoter *in vitro. Cell,* **51**,783–93.

Scheller, R.H., Costantini, F.D., Kozlowski, M.R., Britten, R.J. & Davidson,
E.H. (1978). Specific representation of cloned repetitive DNA sequences in
sea urchin RNAs. *Cell,* **15**, 189–203.

Schierenberg, E. (1985). Cell determination during early embryogenesis of the
nematode *Caenorhabditis elegans. Cold Spring Harbor Symp. Quant. Biol.* **50**,
59–68.

Schimke, R.T. (1984). Gene amplification in cultured animal cells. *Cell,* **37**,
705–13.

Schlissel, M.S. & Baltimore, D. (1989). Activation of immunoglobulin kappa
gene rearrangement correlates with induction of germline kappa gene transcription. *Cell,* **58**, 1001–7.

Schneuwly, S., Kuroiwa, A., Baumgartner, P. & Gehring, W.J. (1986). Structural
organisation and sequence of the homeotic gene *Antennapedia* of *Drosophila
melanogaster. EMBO J.* **5**, 733–9.

Schneuwly, S., Klemenz, R. & Gehring, W.J. (1987a). Redesigning the body
plan of *Drosophila* by ectopic expression of the homoeotic gene *Antennapedia. Nature (Lond.)* **325**, 816–9.

Schneuwly, S., Kuroiwa, A. & Gehring, W.J. (1987b). Molecular analysis of the
dominant homoeotic *Antennapedia* phenotype. *EMBO J.* **6**, 201–6.

Scholer, H.R., Hatzopoulos, A.K., Balling, R., Suzuki, N. & Gruss, P. (1989). A
family of octamer-specific proteins present during mouse embryogenesis:
evidence for germline–specific expression of an Oct factor. *EMBO J.* **8**, 2543–
50.

Scholnick, S., Morgan, B. & Hirsh, J. (1983). The cloned *dopa decarboxylase*
gene is developmentally regulated when reintegrated into the *Drosophila*
genome. *Cell,* **34**, 37–45.

Schroder, C., Tautz, D., Seifert, E. & Jackle, H. (1988). Differential regulation of the two promoters from the *Drosophila* gap segmentation gene *hunchback*. *EMBO J.* **7**, 2881–7.

Schubiger, G. (1971). Regeneration, pattern duplication and transdetermination in fragments of the leg disc of *Drosophila melanogaster*. *Devl. Biol.* **26**, 277–95.

Schuh, R., Aicher, W., Gaul, U., Côté, A., Preiss, A., Maier, D., Seifert, E., Nauber, U., Schroder, C., Kemler, R. & Jackle, H. (1986). A conserved family of nuclear proteins containing structural elements of the finger protein encoded by *Kruppel*, a *Drosophila* segmentation gene. *Cell*, **47**, 1025–32.

Schule, R., Muller, M., Otsuka-Murakami, H. & Renkawitz, R. (1988). Cooperativity of the glucocorticoid receptor and the CACCC-box binding factor. *Nature (Lond.)* **332**, 87–90.

Schupbach, T. (1982). Autosomal mutations that interfere with sex determination in somatic cells of *Drosophila* have no direct effect on the germ line. *Devl. Biol.* **89**, 117–27.

Schupbach, T. (1987). Germ line and soma cooperate during oogenesis to establish the dorso-ventral pattern of egg shell and embryo in *Drosophila melanogaster*. *Cell*, **49**, 699–707.

Schupbach, T. & Wieschaus, E. (1986). Germ line autonomy of maternal effect mutations altering the embryonic body pattern of *Drosophila*. *Devl. Biol.* **113**, 443–8.

Schupbach, T., Wieschaus, E. & Nothiger, R. (1978). A study of the female germ line in mosaics of *Drosophila*. *Roux' Arch. Devl. Biol.* **184**, 41–56.

Schutz, G., Nguyen-Huu, M.C., Giesecke, K., Hynes, N.E., Groner, B., Wurtz, T. & Sippel, A.E. (1977). Hormonal control of egg white protein messenger RNA synthesis in the chicken oviduct. *Cold Spring Harbour Symp. Quant. Biol.* **42**, 617–24.

Scott, M.P. & Carroll, S.B. (1987). The segmentation and homoeotic gene network in early *Drosophila* development. *Cell*, **51**, 689–98.

Scott, M.P. & Weiner, A.J. (1984). Structural relationships among genes that control development: sequence homology beween the *Antennapedia*, *Ultrabithorax* and *fushi tarazu* loci of *Drosophila*. *Proc. Natl. Acad. Sci. (USA)* **81**, 4115–9.

Scott, S.E.M. & Sommerville, J. (1974). Location of nuclear proteins on the chromosomes of newt oocytes. *Nature (Lond.)* **250**, 680–2.

Searle, P.F. & Tata, J.R. (1981). Vitellogenin gene expression in male *Xenopus* hepatocytes during primary and secondary stimulation with estrogen in cell culture. *Cell*, **23**, 741–6.

Seiler-Tuyns, A., Eldridge, J.D. & Paterson, B.M. (1984). Expression and regulation of chicken actin genes introduced into mouse myogenic and nonmyogenic cells. *Proc. Natl. Acad. Sci. (USA)* **81**, 2980–4.

Sen, R. & Baltimore, D. (1986a). Multiple nuclear factors interact with the immunoglobulin enhancer sequences. *Cell*, **46**, 70–16.

Sen, R. & Baltimore, D. (1986b). Inducibility of κ immunoglobulin enhancerbinding protein NF-κB by a posttranslational mechanism. *Cell*, **47**, 921–8.

Senger, D.R. & Gross, P.R. (1978). Macromolecule synthesis and determination in sea urchin blastomeres at the sixteen-cell stage. *Devl. Biol.* **65**, 404–15.

Serfling, E., Jasin, M. & Schaffner, W. (1985). Enhancers and eukaryotic gene transcription. *Trends Genet.* **1**, 224–30.

Severne, Y., Wieland, S., Schaffner, W. & Rusconi, S. (1988). Metal binding 'finger' structures in the glucocorticoid receptor defined by site-directed mutagenesis. *EMBO J.* **7**, 2503–8.

Seybold, W.D. & Sullivan, D.T. (1978). Protein synthetic patterns during differentiation of imaginal discs *in vitro. Devl. Biol.* **65**, 69–80.

Shackleford, G.M. & Varmus, H.E. (1987). Expression of the proto-oncogene *int-1* is restricted to post-meiotic male germ cells and the neural tube of mid-gestational embryos. *Cell,* **50**, 89–95.

Shani, M. (1985). Tissue-specific expression of rat myosin light-chain 2 gene in transgenic mice. *Nature (Lond.)* **314**, 283–5.

Sharp, P.A. (1987). *Trans* splicing: variation on a familiar theme. *Cell,* **50**, 147–8.

Sharp, S., de Franco, D., Dingermann, T., Farrell, P. & Soll, D. (1981). Internal control regions for transcription of eucaryotic tRNA genes. *Proc. Natl. Acad. Sci. (USA)* **78**, 6657–61.

Sharpe, C.R., Fritz, A., de Robertis, E.M. & Gurdon, J.B. (1987). A homeobox-containing marker of posterior neural differentiation shows the importance of predetermination in neural induction. *Cell,* **50**, 749–58.

Sharpe, P.T., Miller, J.R., Evans, E.P., Burtenshaw, M.D. & Gaunt, S.J. (1988). Isolation and expression of a new mouse homeobox gene. *Development,* **102**, 397–407.

Sharrock, W.J. (1984). Cleavage of two yolk proteins from a precursor in *Caenorhabditis elegans. J. Mol. Biol.* **174**, 419–31.

Shastry, B.S., Honda, B.M. & Roeder, R.G. (1984). Altered levels of a 5S gene-specific transcription factor (TFIIIA) during oogenesis and embryonic development of *Xenopus laevis. J. Biol. Chem.* **259**, 11373–82.

Shen, D.-W., Real, F.X., DeLeo, A.B., Old, L.J., Marks, P.A. & Rifkind, R.A. (1983). Protein p53 and inducer-mediated erythroleukemia cell commitment to terminal cell division. *Proc. Natl. Acad. Sci. (USA)* **80**, 5919–22.

Shen, M.M. & Hodgkin, J. (1988). *mab-3*, a gene required for sex-specific yolk protein expression and a male-specific lineage in *C. elegans. Cell,* **54**, 1019–31.

Shen, S.-H. & Smithies, O. (1982). Human globin ψ β2 is not a globin-related sequence. *Nucleic Acids Res.* **10**, 7809–7818.

Shepherd, J.C.W., McGinnis, W., Carrasco, A.E., de Robertis, E.M. & Gehring, W.J. (1984). Fly and frog homeodomains show homologies with yeast mating-type regulatory proteins. *Nature (Lond.)* **310**, 70–2.

Shermoen, A.W., Jongens, J., Barrett, S.W., Flynn, K. & Beckendorf, S.K. (1987). Developmental regulation by an enhancer from the *sgs-4* gene of *Drosophila. EMBO J.* **7**, 207–14.

Shimizu, A. & Honjo, T. (1984). Immunoglobulin class switching. *Cell,* **36**, 801–3.

Shore, E.M. & Guild, G.M. (1987). Closely linked DNA elements control the expression of the *sgs-5* glue protein gene in *Drosophila. Genes & Dev.* **1**, 829–39.

Short, N.J. (1988). Regulation of transcription: flexible interpretation. *Nature (Lond.)* **334**, 192–3.

Shott, R.J., Lee, J.J., Britten, R.J. & Davidson, E.H. (1984). Differential expression of the actin gene family of *Strongylocentrotus purpuratus. Devl. Biol.* **101**, 295–306.

Simcox, A.A., Wurst, G., Hersperger, E. & Shearn, A. (1987). The *defective dorsal discs* gene of *Drosophila* is required for the growth of specific imaginal discs. *Devl. Biol.* **122**, 559–67.

Simeone, A., Mavilio, F., Bottero, L., Giampaolo, A., Russo, G., Faiella, A., Boncinelli, E. & Peschle, A. (1986). A human homeobox gene specifically expressed in spinal cord during embryonic development. *Nature (Lond.)* **330**, 763–5.

Simeone, A., Pannese, M., Acampora, D., D'Esposito, M. & Boncinelli, E. (1988). At least three human homeoboxes on chromosome 12 belong to the same transcription unit. *Nucleic Acids Res.* **16**, 5379–87.

Simpson, P. (1983). Maternal–zygotic gene interactions during formation of the dorso-ventral pattern in *Drosophila* embryos. *Genetics*, **105**, 615–32.

Sina, B.J. & Pellegrini, M. (1982). Genomic clones coding for some of the initial genes expressed during *Drosophila* development. *Proc. Natl. Acad. Sci. (USA)* **79**, 7351–5.

Singh, H., Sen, R., Baltimore, D. & Sharp, P.A. (1986). A nuclear factor that binds to a conserved sequence motif in transcriptional control elements of immunoglobulin genes. *Nature (Lond.)* **319**, 154–7.

Sitia, R., Neuberger, M.S. & Milstein, C. (1987). Regulation of membrane IgM expression in secretory B cells: translational and post-translational events. *EMBO J.* **6**, 3969–77.

Skoultchi, A. & Gross, P.R. (1973). Maternal histone messenger RNA: detection by molecular hybridisation. *Proc. Natl. Acad. Sci. (USA)* **70**, 2840–4.

Slack, J.M.W. (1985). Peanut lectin receptors in the early amphibian embryo: regional markers for the study of embryonic induction. *Cell*, **41**, 237–47.

Slack, J.M.W. (1987). We have a morphogen! *Nature (Lond.)* **327**, 553–4.

Slightom, J.L., Blechl, A.E. & Smithies, O. (1980). Human fetal $^G\gamma$ and ^{A}globin genes: complete nucleotide sequences. *Cell*, **21**, 627–38.

Smale, S.T. & Baltimore, D. (1989). The 'initiator' as a transcriptional control element. *Cell*, **57**, 103–13.

Smith, L. & Thorogood, P. (1983). Transfilter studies on the mechanism of epithelio–mesenchymal interaction leading to chondrogenic differentiation of neural crest cells. *J. Embryol. Exp. Morphol.* **75**, 165–88.

Smithies, O., Gregg, R.G., Boggs, S.S., Koralewski, M.A. & Kucherlapati, R.S. (1985). Insertion of DNA sequences into the human chromosomal globin locus by homologous recombination. *Nature (Lond.)* **317**, 230–4.

Snape, A., Wylie, C.C., Smith, J.C. & Heasman, J. (1987). Changes in states of commitment of single animal pole blastomeres in *Xenopus laevis*. *Devl. Biol.* **119**, 503–10.

Snyder, P.B., Galanopoulos, V.K. & Kafatos, F.C. (1986). *trans*-acting amplification mutants and other eggshell mutants of the third chromsome in *Drosophila melanogaster*. *Proc. Natl. Acad. Sci. (USA)* **83**, 3341–5.

Soeller, W.C., Poole, S.J. & Kornberg, T. (1988). *In vitro* transcription of the *Drosophila engrailed* gene. *Genes & Dev.* **2**, 68–81.

Sollner-Webb, B. (1988). Surprises in polymerase III transcription. *Cell*, **52**, 153–4.

Spangrude, G.J., Heimfeld, S. & Weissman, I.L. (1988). Purificiation and characterisation of mouse haematopoietic stem cells. *Science*, **241**, 58–62.

Spemann, H. (1928). Die entwicklung seitlicher und dorso-ventraler keim-hälfter bei verzögerter kernversorgung. *Z. Wiss. Zool.* **132**, 105–34.

Spieth, J. & Blumenthal, T. (1985). The *Caenorhabditis elegans* vitellogenin gene family includes a gene encoding a distantly related protein. *Mol. & Cell. Biol.* **5**, 2495–501.

Spieth, J., Denison, K., Zucker, E. & Blumenthal, T. (1985a). The nucleotide sequence of a nematode vitellogenin gene. *Nucleic Acids Res.* **13**, 7129–38.

Spieth, J., Denison, K., Kirtland, S., Cane, J. & Blumenthal, T. (1985b). The *Caenorhabditis elegans* vitellogenin genes: short sequence repeats in the promoter regions and homology to the vertebrate genes. *Nucleic Acids Res.* **13**, 5283–95.

Spirin, A.S. (1966). On 'masked' forms of messenger RNA in early embryogenesis and in other differentiating systems. *Curr. Topics Devl. Biol.* **1**, 1–38.

Spradling, A.C. (1981). The organisation and amplification of two chromosomal domains containing *Drosophila* chorion genes. *Cell*, **27**, 193–201.

Spradling, A.C. & Mahowald, A.P. (1980). Amplification of genes for chorion proteins during oogenesis in *Drosophila melanogaster*. *Proc. Natl. Acad. Sci. (USA)* **77**, 1096–100.

Spradling, A.C. & Rubin, G.M. (1983). The effect of chromosomal position on the expression of the *Drosophila* xanthine dehydrogenase gene. *Cell*, **34**, 47–57.

Spradling, A.C., de Cicco, D.V., Wakimoto, B.T., Levine, J.F., Kalfayan, L.J. & Cooley, L. (1987). Amplification of the X-linked *Drosophila* chorion gene cluster requires a region upstream from the s38 chorion gene. *EMBO J.* **6**, 1045–53.

Sprenger, F., Stevens, L.M. & Nusslein-Volhard, C. (1989). The *Drosophila* gene *torso* encodes a putative receptor tyrosine kinase. *Nature (Lond.)* **338**, 478–82.

Spritz, R.A., de Riel, J.K., Forget, B.G. & Weissman, S.M. (1980). Complete nucleotide sequence of the human δ globin gene. *Cell*, **21**, 639–46.

Stalder, J., Groudine, M., Dodgson, J.B., Engel, J.D. & Weintraub, H. (1980). Hb switching in chickens. *Cell*, **19**, 973–80.

Stamatoyannopoulos. G., Constantoulakis, P., Bryce, M., Kurachi, S. & Papayannopoulou, T. (1987). Coexpression of embryonic, foetal and adult globins in erythroid cells of human embryos: relevance to the cell-lineage models of globin switching. *Devl. Biol.* **123**, 191–7.

Standart, N.M., Bray, S.J., George, E.L., Hunt, T. & Ruderman, J.V. (1985). The small subunit of ribonucleotide reductase is encoded by one of the most abundant translationally regulated maternal RNAs in clam and sea urchin eggs. *J. Cell Biol.* **100**, 1968–76.

Stanley, E.R., Bartocci, A., Patinkin, D., Rosendaal, M. & Bradley, T.R. (1986). Regulation of very primitive, multipotent, haemopoietic cells by haemopoietin I. *Cell*, **45**, 667–74.

Stanojevic, D., Hoey, T. & Levine, M. (1989). Sequence specific DNA-binding activities of the gap proteins encoded by *hunchback* and *Kruppel* in *Drosophila*. *Nature (Lond.)* **341**, 331–5.

Steele, R.E., Thomas, P.S. & Reeder, R.H. (1984). Anucleolate frog embryos contain ribosomal DNA sequences and a nucleolar antigen. *Devl. Biol.* **102**, 409–16.

Stein, J.P., Catterall, J.F. Kristo, P., Means, A.R. & O'Malley, B.W. (1980).

Ovomucoid intervening sequences specify functional domains and generate protein polymorphism. *Cell*, **21**, 681–7.

Stein, R., Sciaky-Gallili, N., Razin, A. & Cedar, H. (1983). Pattern of methylation of two genes coding for housekeeping functions. *Proc. Natl. Acad. Sci. (USA)* **80**, 2422–6.

Steinmann-Zwicky, M. (1988). Sex determination in *Drosophila*: the X-chromosomal gene *liz* is required for *Sxl* activity. *EMBO J.* **7**, 3889–98.

Stephenson, E.C., Erba, H.P. & Gall, J.G. (1981). Histone gene clusters of the newt *Notophthalmus* are separated by long tracts of satellite DNA. *Cell*, **24**, 639–47.

Stern, C.D. & Keynes, R.J. (1987). Interactions between somite cells: the formation and maintenance of segment boundaries in the chick embryo. *Development* **99**, 261–72.

Sternberg, P.W. (1988). Lateral inhibition during vulval induction in *Caenorhabditis elegans*. *Nature (Lond.)* **335**, 551–4.

Sternberg, P.W. & Horvitz, H.R. (1984). The genetic control of cell lineage during nematode development. *Ann. Rev. Genet.* **18**, 489–524.

Sternberg, P.W. & Horvitz, H.R. (1986). Pattern formation during vulval development in *Caenorhabditis elegans*. *Cell*, **41**, 76–72.

Sternberg, P.W. & Horvitz, H.R. (1988). *lin-17* mutations of *Caenorhabditis elegans* disrupt certain asymmetric cell divisions. *Devl. Biol.* **130**, 67–73.

Sternberg, P.W. & Horvitz, H.R. (1989). The combined action of two intercellular signalling pathways specifies three cell fates during vulval induction in *C. elegans*. *Cell*. **58**, 679–93.

Steward, F.C., Mapes, M.O., Kent, A.E. & Holsten, R.D. (1964). Growth and development of cultured plant cells. *Science*, **143**, 20–7.

Steward, R. (1987). *dorsal*, an embryonic polarity gene in *Drosophila*, is homologous to the vertebrate proto-oncogene *c-rel*. *Science*, **238**, 692–4.

Steward, R., McNally, F.J. & Schedl, P. (1984). Isolation of the *dorsal* locus of *Drosophila*. *Nature (Lond.)* **311**, 262–4.

Steward, R., Ambrosio, L. & Schedl, P. (1985). Expression of the *dorsal* gene. *Cold Spring Harbor Symp. Quant. Biol.* **50**, 223–8.

Steward, R., Zusman, S.B., Huang, L.H. & Schedl, P. (1988). The *dorsal* protein is distributed in a gradient in early *Drosophila* embryos. *Cell*, **55**, 487–95.

Stillman, D.J., Bankier, A.T., Seddon, A., Groenhout, E.G. & Nasmyth, K.A. (1988). Characterisation of a transcription factor involved in mother-cell specific transcription of the yeast HO gene. *EMBO J.* **7**, 485–94.

St Johnston, R.D. & Gelbart, W.M. (1987). *decapentaplegic* transcripts are localised along the dorsal–ventral axis of the *Drosophila* embryo. *EMBO J.* **6**, 2785–91.

Strahle, U., Klock, G. & Schutz, G. (1987). A 15bp oligomer is sufficient to mediate both glucocorticoid and progesterone induction. *Proc. Natl. Acad. Sci. (USA)* **84**, 7871–5.

Strahle, U., Boshart, M., Klock, G., Stewart, F. & Schutz, G. (1989). Glucocorticoid- and progesterone-specific effects are determined by differential expression of the respective hormone receptors. *Nature (Lond.)* **339**, 629–31.

Strecker, T.R., Kongsuwan, K., Lengyel, J.A. & Merriam, J.R. (1986). The zygotic mutant *tailless* affects the anterior and posterior ectodermal regions of the *Drosophila* embryo. *Devl. Biol.* **113**, 64–76.

Strecker, T.R., Merriam, J.R. & Lengyel, J.A. (1988). Graded requirement for the zygotic terminal gene, *tailless*, in the brain and tail region of the *Drosophila* embryo. *Development*, **102**, 721–34.

Strecker, T.R., Halsell, S.R., Fisher, W.W. & Lipshitz, H.D. (1989). Reciprocal effects of hyper- and hypoactivity mutations in the *Drosophila* pattern gene *torso*. *Science*, **243**, 1062–5.

Stroeher, V.L., Jorgensen, E.M. & Garber, R.L. (1986). Multiple transcripts from the *Antennapedia* gene of *Drosophila melanogaster*. *Mol. & Cell. Biol.* **6**, 4667–75.

Strome, S. & Wood, W.B. (1982). Immunofluorescent visualisation of germ-line-specific cytoplasmic granules in embryos, larvae and adults of *Caenorhabditis elegans*. *Proc. Natl. Acad. Sci. (USA)* **79**, 1558–62.

Strome, S. & Wood, W.B. (1983). Generation of asymmetry and segregation of germ-line granules in early *C. elegans* embryos. *Cell*, **35**, 15–25.

Strub, K., Galli, G., Busslinger, M. & Birnstiel, M.L. (1984). The cDNA sequences of the sea urchin U7 small nuclear RNA suggest specific contacts between histone mRNA precursor and U7 RNA during RNA processing. *EMBO J.* **3**, 2801–4.

Struhl, G. (1981a). A blastoderm fate map of compartments and segments of the *Drosophila* head. *Devl. Biol.* **84**, 386–96.

Struhl, G. (1981b). A gene product required for correct initiation of segmental determination in *Drosophila*. *Nature (Lond.)* **293**, 36–9.

Struhl, G. (1981c). A homoeotic mutation transforming leg to antenna in *Drosophila*. *Nature (Lond.)* **292**, 635–7.

Struhl, G. (1982). Genes controlling segmental specification in the *Drosophila* thorax. *Proc. Natl. Acad. Sci. (USA)* **79**, 7380–4.

Struhl, G. (1983). Role for the *esc*$^+$ gene product in ensuring the selective expression of segment-specific homoeotic genes in *Drosophila*. *J. Embryol. Exp. Morphol.* **76**, 297–331.

Struhl, G. (1984a). Splitting the bithorax complex of *Drosophila*. *Nature (Lond.)* **308**, 454–6.

Struhl, G. (1984b). A universal genetic key to body plan? *Nature (Lond.)* **330**, 10–11.

Struhl, G. (1985). Near-reciprocal phenotypes caused by inactivation or indiscriminate expression of the *Drosophila* segmentation gene *ftz*. *Nature (Lond.)* **318**, 677–9.

Struhl, G. (1989). Differing strategies for organising anterior and posterior body pattern in *Drosophila* embryos. *Nature (Lond.)* **338**, 741–4.

Struhl, G. & Akam, M. (1985). Altered distributions of *Ultrabithorax* gene of *Drosophila* by other *bithorax* complex genes. *Cell*, **43**, 507–19.

Struhl, G. & White, R.A.H. (1985). Regulation of the *Ultrabithorax* gene of *Drosophila* by other bithorax complex genes. *Cell*, **43**, 507–19.

Struhl, K. (1989). Helix-turn-helix, zinc-finger, and leucine-zipper motifs for eucaryotic transcriptional regulatory proteins. *Trends Biochem. Sci.* **14**, 137–40.

Sturkie, P.D. & Mueller, W.J. (1976). Reproduction in the female and egg production. In *Avian Physiology*, 3rd edn, ed. P.D. Sturkie, pp. 302–30. Springer Verlag, New York, Heidelberg, Berlin.

Sturm, R.A. & Herr, W. (1988). The POU domain is a bipartite DNA-binding structure. *Nature (Lond.)* **336**, 601–3.

Sturtevant, E.H. (1929). The claret mutant type of *Drosophila simulans*: a study of chromosome elimination and cell lineage. *Z. Wiss. Zool.* **135**, 323–6.

Stryer, L. (1981). *Biochemistry*, 2nd ed. W.H. Freeman & Co., San Francisco.

Sucov, H.M., Benson, S., Robinson, J.J., Britten, R.J., Wilt, F. & Davidson, E.H. (1987). A lineage-specific gene encoding a major matrix protein of the sea urchin embryo spicule: II structure of the gene and derived sequence of the protein. *Devl. Biol.* **120**, 507–19.

Sucov, H.M., Hough-Evans, B.R., Franks, R.R., Britten, R.J. & Davidson, E.H. (1988). A regulatory domain that directs lineage-specific expression of a skeletal matrix protein gene in the sea urchin embryo. *Genes & Dev.* **2**, 1238–50.

Sulston, J. (1976). Postembryonic development in the ventral nerve cord of *Caenorhabditis elegans*. *Phil. Trans. Roy. Soc. Lond. B*, **275**, 287–98.

Sulston, J.E. (1988). Cell lineage. In Wood, W.B. (ed.) *The Nematode* Caenorhabditis elegans, ed. W.B. Wood. CSH Monographs, **17**, 81–122.

Sulston, J. & Horvitz, H.R. (1977). Postembryonic cell lineages of the nematode *Caenorhabditis elegans*. *Devl. Biol.* **56**, 110–56.

Sulston, J.E. & Horvitz, H.R. (1981). Abnormal cell lineages in mutants of the nematode *Caenorhabditis elegans*. *Devl. Biol.* **82**, 41–55.

Sulston, J.E., Albertson, D.G. & Thomson, J.N. (1980). The *Caenorhabditis elegans* male: postembryonic development of non-gonadal structures. *Devl. Biol.* **78**, 543–76.

Sulston, J.E., Schierenberg, E., White, J.G. & Thomson, J.N. (1983). The embryonic cell lineage of the nematode *Caenorhabditis elegans*. *Devl. Biol.* **100**, 64–119.

Summers, M.C., Bedian, V. & Kaufman, S.A. (1986). An analysis of stage-specific protein synthesis in the early *Drosophila* embryo using high-resolution, two-dimensional gel electrophoresis. *Devl. Biol.* **113**, 49–63.

Superti-Furga, G., Barberis, A., Schaffner, G. & Busslinger, M. (1988). The -117 mutation in Greek HPFH affects the binding of three nuclear factors to the CCAAT region of the γ globin gene. *EMBO J.* **7**, 3099–107.

Swift, G.H., Hammer, R.E., MacDonald, R.J. & Brinster, R.L. (1984). Tissue-specific expression of the rat pancreatic elastase I gene in transgenic mice. *Cell*, **38**, 639–46.

Szabad, J. & Bryant, P.J. (1982). The mode of action of 'discless' mutations in *Drosophila melanogaster*. *Devl. Biol.* **93**, 240–56.

Talbot, D., Collis, P., Antoniou, M., Vidal, M., Grosveld, F. & Greaves, D.R. (1989). A dominant control region from the human β-globin locus conferring integration site-independent gene expression. *Nature (Lond.)* **338**, 352–4.

Tang, P., Sharpe, C.R., Mohun, T.J. & Wylie, C.C. (1988). Vimentin expression in oocytes, eggs and early embryos of *Xenopus laevis*. *Development*, **103**, 279–87.

Tapscott, S.J., Davis, R.L., Thayer, M.J., Cheng, P.-F., Weintraub, H. & Lassar, A.B. (1988). Myo D1: nuclear phosphoprotein requiring a *myc* homology region to convert fibroblasts to myoblasts. *Science*, **242**, 405–11.

Tautz, D. (1988). Regulation of the *Drosophila* segmentation gene *hunchback* by two maternal morphogenetic centres. *Nature (Lond.)* **332**, 281–4.

Tautz, D., Lehmann, R., Schnurch, H., Schuh, R., Seifert, E., Keinlin, A., Jones, K. & Jackle, H. (1987). Finger protein of novel structure encoded by *hunchback*, a second member of the gap class of *Drosophila* segmentation genes. *Nature (Lond.)* **327**, 383–7.

Taylor, W., Jackson, I.J., Siegel, N., Kumar, A. & Brown, D.D. (1986). The developmental expression of the gene for TFIIIA in *Xenopus laevis*. *Nucleic Acids Res.* **14**, 6185–94.

Technau, G.M. & Campos-Ortega, J.A. (1987). Cell autonomy of expression of neurogenic genes of *Drosophila melanogaster*. *Proc. Natl. Acad. Sci. (USA)* **84**, 4500–4.

Teugels, E. & Ghysen, A. (1983). Independence of the numbers of legs and leg ganglia in *Drosophila* bithorax mutants. *Nature (Lond.)* **304**, 440–3.

Teugels, E. & Ghysen, A. (1985). Domains of action of bithorax genes in *Drosophila* central nervous system. *Nature (Lond.)* **314**, 558–60.

Thayer, M.J., Tapscott, S.J., Davis, R.L., Wright, W.E., Lassar, A.B. & Weintraub, H. (1989). Positive autoregulation of the myogenic determination gene MyoD1. *Cell*, **58**, 241–8.

Thiesen, M. Stief, A. & Sippel, A.E. (1986). The lysozyme enhancer: cell-specific activation of the chicken lysozyme gene by a far-upstream DNA element. *EMBO J.* **5**, 719–24.

Thireos, G., Griffin-Shea, R. & Kafatos, F.C. (1980). Untranslated mRNA for a chorion protein of *Drosophila melanogaster* accumulates transiently at the onset of specific gene amplification. *Proc. Natl. Acad. Sci. (USA)* **77**, 5789–93.

Thisse, B., Stoetzel, C., El-Messal, M. & Perrin-Schmitt, F. (1987). Genes of the maternal dorsal group control the specific expression of the zygotic gene *twist* in presumptive mesodermal cells. *Genes & Dev.* **1**, 709–15.

Thisse, B., Stoetzel, C., Gorostiza-Thisse, C. & Perrin-Schmidt, F. (1988). Sequence of the *twist* gene and nuclear localisation of its protein in endomesodermal cells of early *Drosophila* embryos. *EMBO J.* **7**, 2175–83.

Thomas, J.B. & Wyman, R.J. (1984). Duplicated neural structure in bithorax mutant *Drosophila*. *Devl. Biol.* **102**, 531–3.

Thomas, J.B., Bastiani, M.J., Bate, M. & Goodman, C.S. (1984). From grasshopper to *Drosophila*: a common plan for neuronal development. *Nature (Lond.)* **310**, 203–7.

Thomas, J.B., Crews, S.T. & Goodman, C.S. (1988). Molecular genetics of the *single minded* locus: a gene involved in the development of the *Drosophila* nervous system. *Cell*, **52**, 133–41.

Tilghman, S.M., Tiemeier, D.C., Seidman, J.G., Peterlin, B.M., Sullivan, M., Maizel, J.V. & Leder, P. (1978a). Intervening sequences of DNA identified in the structural portion of a mouse β globin gene. *Proc. Natl. Acad. Sci. (USA)* **75**, 725–9.

Tilghman, S.M., Curtis, P.J., Tiemeier, D.C., Leder, P. & Weissman, C. (1978b). The intervening sequence of a mouse β globin gene is transcribed within the 15S β globin mRNA precursor. *Proc. Natl. Acad. Sci. (USA)* **75**, 1309–13.

Tiong, S.Y.K., Whittle, J.R.S. & Gribbin, M.C. (1987). Chromosomal continuity in the abdominal region of the bithorax complex is not essential for its contribution to metameric identity. *Development*, **101**, 135–42.

Togashi, S., Kobayashi, S. & Okada, M. (1986). Functions of maternal mRNA as a cytoplasmic factor responsible for pole cell formation in *Drosophila* embryos. *Devl. Biol.* **118**, 352–60.

Tomlinson, A. (1988). Cellular interactions in the developing *Drosophila* eye. *Development*, **104**, 183–93.

Tomlinson, A., Bowtell, D.D.L., Hafen, E. & Rubin, G.M. (1987). Localisation of the *sevenless* protein, a putative receptor for positional information in the eye imaginal disc of *Drosophila. Cell*, **51**, 143–50.

Tomlinson, A., Kimmel, B.E. & Rubin, G.M. (1988). *rough*, a *Drosophila* homeobox gene required in photoreceptors R2 and R5 for inductive interactions in the developing eye. *Cell*, **55**, 771–84.

Tomlinson, C.R., Beach, R.L. & Jeffery, W.R. (1987a). Differential expression of a muscle actin gene in muscle cell lineages of ascidean embryos. *Development*, **101**, 751–65.

Tomlinson, C.R., Bates, W.R. & Jeffery, W.R. (1987b). Development of a muscle actin specified by maternal and zygotic mRNA in ascidean embryos. *Devl. Biol.* **123**, 470–82.

Toth, L.E., Slawin, K.L., Pintar, J.E. & Nguyen-Huu, M.C. (1987). Region-specific expression of mouse homeobox genes in the embryonic mesoderm and central nervous system. *Proc. Natl. Acad. Sci. (USA)* **84**, 6790–4.

Tower, J., Cizewski-Culotta, V. & Sollner-Webb, B. (1986). Factors and nucleotide sequences that direct ribosomal DNA transcription and their relationship to the stable transcription complex. *Mol. & Cell. Biol.* **6**, 3451–62.

Townes, T.M., Lingrel, J.B., Chen, H.-Y., Brinster, R.L. & Palmiter, R.D. (1985). Erythroid-specific expression of human β globin genes in transgenic mice. *EMBO J.* **4**, 1715–23.

Trainor, C.D., Stamler, D.J. & Engel, J.D. (1987). Erythroid-specific transcription of the chicken histone H5 gene is directed by a 3' enhancer. *Nature (Lond.)* **328**, 827–9.

Treisman, J. & Desplan, C. (1989). The products of the *Drosophila* gap genes *hunchback* and *Kruppel* bind to the *hunchback* promoters. *Nature (Lond.)* **341**, 335–7.

Treisman, J., Gonczy, P., Vashishtha, M., Harris, E. & Desplan, C. (1989). A single amino acid can determine the DNA-binding specificity of homeodomain proteins. *Cell*, **59**, 553–62.

Treisman, R., Proudfoot, N.J., Shander, M. & Maniatis, T. (1982). A single-base change at a splice site in a β⁰-thalassaemic gene causes abnormal RNA splicing. *Cell*, **29**, 903–11.

Tremml, G. & Bienz, M. (1989a). Homeotic gene expression in the visceral mesoderm of *Drosophila* embryos. *EMBO J.* **8**, 2677–85.

Tremml, G. & Bienz, M. (1989b). An essential role of *even-skipped* for homeotic gene expression in the *Drosophila* visceral mesoderm. *EMBO J.* **8**, 2687–93.

Truman, D.E.S. (1974). *Biochemistry of Cytodifferentiation*. Blackwell Scientific Publications, Oxford.

Truman, D.E.S. (1982). Taxonomies of differentiation. In *Stability and Switching in Cellular Differentiation*, ed. R.M. Clayton & D.E.S. Truman, Adv. Exp. Med. Biol. **158**, 45–53. Plenum Press, New York & London.

Tsai, M.J., Tring, A.C., Nordstrom, J.L., Zimmer, W. & O'Malley, B.W. (1980). Processing of high molecular weight ovalbumin and ovomucoid precursor RNAs to messenger RNA. *Cell*, **22**, 219–30.

Tsai, S.-F., Martin, D.I.K., Zon, L.I., D'Andrea, A.D., Wong, G.G. & Orkin, S.H. (1989). Cloning of cDNA for the major DNA-binding protein of the erythroid lineage through expression in mammalian cells. *Nature (Lond.)* **339**, 446–50.

Tsai, S.Y., Sagami, I., Wang, H., Tsai, M.-J. & O'Malley, B.W. (1987). Inter-actions between a DNA-binding transcription factor (COUP) and a non-DNA binding factor (S300-II). *Cell*, **50**, 701–9.

Tsai, S.Y., Carlstedt-Duke, J., Weigel, N.L., Dahlman, K., Gustafsson, J.-A., Tsai, M.-J. & O'Malley, B.W. (1988). Molecular interactions of steroid hormone receptor with its enhancer element: evidence for receptor dimer formation. *Cell*, **55**, 361–9.

Tufaro, F. & Brandhorst, B.P. (1979). Similarity of proteins synthesised by iso-lated blastomeres of early sea urchin embryos. *Devl. Biol.* **72**, 390–7.

Tung, T.-C., Wu, S.-C., Yeh, Y.-F., Li, K.-S. & Hsu, M.-C. (1977). Cell differenti-ation in ascidean studied by nuclear transplantation. *Scientia Sinica* (English edition) **20**, 222–33.

Tyc, K. & Steitz, J.A. (1989). U3, U8 and U13 comprise a new class of mamma-lian snRNPs localised in the cell nucleolus. *EMBO J.* **8**, 3113–8.

Ubbels, G.A., Hara, K., Koster, C.H. & Kirschner, M.W. (1983). Evidence for a functional role of the cytoskeleton in determination of the dorso-ventral axis in *Xenopus laevis* eggs. *J. Embryol. Exp. Morphol.* **77**, 15–37.

Ueda, R. & Okada, M. (1982). Induction of pole cells in sterilised *Drosophila* embryos by injection of subcellular fraction from eggs. *Proc. Natl. Acad. Sci. (USA)* **79**, 6946–50.

Urieli-Shoval, S., Gruenbaum, Y., Sedat, J. & Razin, A. (1982). The absence of detectable methylated bases in *Drosophila melanogaster* DNA. *FEBS Lett.* **146**, 148–53.

Utset, M.F., Awgulewitsch, A., Ruddle, F.H. & McGinnis, W. (1987). Region-specific expression of two mouse homeobox genes. *Science*, **235**, 1379–82.

van Assendelft, G.B., Hanscombe, O., Grosveld, F. & Greaves, D.R. (1989). The β-globin dominant control region activates homologous and heterologous promoters in a tissue-specific manner. *Cell*, **56**, 969–77.

Van den Berg, J., van Ooyen, A., Mantei, N., Schambock, A., Grosveld, G., Flavell, R.A. & Weissman, C. (1978). Comparison of cloned rabbit and mouse β globin genes showing strong evolutionary divergence of two homol-ogous pairs of introns. *Nature (Lond.)* **276**, 37–44.

van den Biggelaar, J.A.M. & Guerrier, P. (1979). Dorsoventral polarity and mesentoblast determination as concomitant results of cellular interactions in the mollusk *Patella vulgata*. *Devl. Biol.* **68**, 462–71.

Van Dongen, W.M.A.M., Moorman, A.F.M. & Destrée, O.H.J. (1983). The accumulation of the maternal pool of histone H1A during oogenesis in *Xenopus laevis*. *Cell Differ.* **12**, 257–64.

Van Doren, K. & Hirsh, D. (1988). *Trans*-spliced leader RNA exists as small nuclear ribonucleoprotein particles in *Caenorhabditis elegans*. *Nature (Lond.)* **335**, 536–9.

Vanin, E.F., Goldberg, G.I., Tucker, P.N. & Smithies, O. (1980). A mouse α-globin related pseudogene lacking intervening sequences. *Nature (Lond.)* **286**, 222–6.

Varley, J.M., Macgregor, H.C. & Erba, H.P. (1980). Satellite DNA is transcribed on lampbrush chromosomes. *Nature (Lond.)* **283**, 686–8.

Vassin, H., Bremer, K.A., Knust, E. & Campos-Ortega, J.A. (1987). The neurogenic gene *Delta* of *Drosophila melanogaster* is expressed in neurogenic territories and encodes a putative transmembrane protein with EGF-like repeats. *EMBO J.* 6, 3431–40.

Verdonk, N.H. (1968). The effect of removing the polar lobe in centrifuged eggs of *Dentalium. J. Embryol, Exp. Morphol.* 19, 33–42.

Villeneuve, A.M. & Meyer, B.J. (1987). *sdc-1*: a link between sex determination and dosage compensation in *C. elegans. Cell*, 48, 25–37.

Vincent, A., O'Connell, P., Gray, M.R. & Rosbash, M. (1984). *Drosophila* maternal and embryo mRNAs transcribed from a single transcription unit use alternate combinations of exons. *EMBO J.* 3, 1003–13.

Vitelli, L., Kemler, I., Lauber, B., Birnstiel, M.L. & Busslinger, M. (1988). Developmental regulation of micro-injected histone genes in sea urchin embryos. *Devl. Biol.* 127, 54–63.

von Mende, N., Bird, D.M., Albert, P.S. & Riddle, D.L. (1988). *dpy-13*: a nematode collagen gene that affects body shape. *Cell*, 55, 567–76.

Wahli, W. & Dawid, I.B. (1979). Vitellogenin in *Xenopus laevis* is encoded by a small family of genes. *Cell*, 16, 535–49.

Wahli, W., Dawid, I.B., Wyler, T., Weber, R. & Ryffel, G.U. (1980). Comparative analysis of the structural organisation of two closely related vitellogenin genes in *X. laevis. Cell*, 20, 107–17.

Wahli, W., Dawid, I.B., Ryffel, G.U. & Weber, R. (1981). Vitellogenesis and the vitellogenin gene family. *Science*, 212, 298–306.

Wahli, W., Germond, G.E., ten Heggeler, B. & May, F.E.B. (1982). Vitellogenin genes A1 and B1 are linked in the *Xenopus laevis* genome. *Proc. Natl. Acad. Sci. (USA)*, 79, 6832–6.

Walker, F., Nicola, N.A., Metcalf, D. & Burgess, A.W. (1985). Hierarchical down-modulation of haemopoietic growth factor receptors. *Cell*, 43, 269–76.

Walker, P., Brown-Leudi, M., Germond, J.E., Wahli, W., Meijlink, F.C.P., van het Schip, A.D., Roelink, H., Gruber, M. & AB, G. (1983). Sequence homologies within the 5' end region of the oestrogen-controlled vitellogenin gene in *Xenopus* and chicken. *EMBO J.* 2, 2271–9.

Walker, V.K. & Ashburner, M. (1981). The control of ecdysterone-regulated puffs in *Drosophila* salivary glands. *Cell*, 26, 269–77.

Wall, L., de Boer, E. & Grosveld, F. (1988). The human β-globin gene 3' enhancer contains multiple binding sites for an erythroid-specific protein. *Genes & Dev.* 2, 1089–1100.

Wang, L.H., Tsai, S.Y., Sagami, I., Tsai, M.-J. & O'Malley, B.W. (1987). Purificiation and characterisation of chicken ovalbumin upstream promoter transcription factor from Hela cells. *J. Biol. Chem.* 260, 16080–6.

Wang, L.-H., Tsai, S.Y., Cook, R.G., Beattie, W.G., Tsai, M.-J. & O'Malley, B.W. (1989). COUP transcription factor is a member of the steroid receptor superfamily. *Nature (Lond.)* 340, 163–5.

Ward, S., Burke, D.J., Sulston, J.E., Coulson, A.R., Albertson, D.E., Ammons, D., Klass, M. & Hogan, E. (1988). The genomic organisation of the transcribed major sperm protein genes and pseudogenes in the nematode *Caenorhabditis elegans. J. Mol. Biol.* 199, 1–13.

Waring, G.L. & Mahowald, A.P. (1979). Identification and time of synthesis of chorion proteins in *Drosophila melanogaster. Cell*, 16, 599–607.

Waring, G.L., Allis, C.D. & Mahowald, A.P. (1978). Isolation of polar granules and the identification of polar granule specific protein. *Devl. Biol.* **66**, 197–206.

Wassarman, P.M. (1987). The biology and chemistry of fertilisation. *Science,* **235**, 553–60.

Wasylyk, C. & Wasylyk, B. (1986). The immunoglobulin heavy-chain B-lymphocyte enhancer efficiently stimulates transcription in non-lymphoid cells. *EMBO J.* **5**, 553–60.

Waterston, R.H. (1988). Muscle. In *The Nematode* Caenorhabditis elegans, ed. W.B.Wood. CSH Monographs 17, 281–335.

Watson, J.D. (1987). *Molecular Biology of the Gene,* 4th ed. Benjamin, New York.

Way, J.C. & Chalfie, M. (1988). *mec-3,* a homeobox-containing gene that specifies differentiation of the touch receptor neurons in *C. elegans. Cell,* **54**, 5–16.

Weatherall, D.J. & Clegg, J.B. (1979). Recent developments in the molecular genetics of human haemoglobins. *Cell,* **16**, 467–79.

Wedeen, C., Harding, K. & Levine, M. (1986). Spatial regulation of Antennapedia and bithorax gene expression by the *Polycomb* locus in *Drosophila. Cell,* **44**, 739–48.

Weeks, D.L. & Melton, D.A. (1987). A maternal mRNA localised to the vegetal hemisphere in *Xenopus* eggs codes for a growth factor related to TGF-β. *Cell,* **51**, 861–7.

Weigel, D., Jurgens, G., Kuttner, F., Seifert, E. & Jackle, H. (1989). The homeotic gene *fork head* encodes a nuclear protein and is expressed in the terminal regions of the *Drosophila* embryo. *Cell,* **57**, 645–58.

Weigert, M., Perry, R., Kelley, D., Hunkapiller, T., Schilling, J.S. & Hood, L. (1980). The joining of V and J gene segments creates antibody diversity. *Nature (Lond.)* **283**, 497–9.

Weinberger, C., Thompson, C.C., Ong, E.S., Lebo, R., Gruol, D.J. & Evans, R.M. (1986a). The *c-erb-A* gene encodes a thyroid hormone receptor. *Nature (Lond.)* **324**, 641–6.

Weinberger, J., Baltimore, D. & Sharp, P.A. (1986b). Distinct factors bind to apparently homologous sequences in the immunoglobulin heavy-chain enhancer. *Nature (Lond.)* **322**, 846–8.

Weiner, A.J., Scott, M.P. & Kaufman, T.C. (1984). A molecular analysis of *fushi tarazu,* a gene in *Drosophila melanogaster* that encodes a product affecting embryonic segment number and cell fate. *Cell,* **37**, 843–51.

Weintraub, H. (1985). Assembly and propagation of repressed and derepressed chromosomal states. *Cell,* **42**, 705–11.

Weintraub, H., Larsen, A. & Groudine, M. (1981). α globin switching during the development of chicken embryos: expression and chromatin structure. *Cell,* **24**, 333–44.

Weintraub, H., Tapscott, S.J., Davis, R.L., Thayer, M.J., Adam, M.A., Lassar, A.B. & Miller, A.D. (1989). Activation of muscle-specific genes in pigment, nerve, fat, liver and fibroblast cell lines by forced expression of MyoD. *Proc. Natl. Acad. Sci. (USA)* **86**, 5434–8.

Weir, M.P. & Kornberg, T. (1985). Patterns of *engrailed* and *fushi tarazu* transcripts reveal novel intermediate stages in *Drosophila* segmentation. *Nature (Lond.)* **318**, 433–6.

Weir, M.P., Edgar, B.A., Kornberg, T. & Schubiger, G. (1988). Spatial regulation of *engrailed* expression in the *Drosophila* embryo. *Genes & Dev.* 2, 1194–203.

Weisbrod, S. (1982). Active chromatin. *Nature (Lond.)* 297, 289–95.

Weisinger, G. & Sachs, L. (1983). DNA-binding protein that induces cell differentiation. *EMBO J.* 2, 2103–7.

Weissmann, C. (1984). Excision of introns in lariat form. *Nature (Lond.)* 311, 103–4.

Wells, D.E., Showman, R.M., Klein, W.H. & Raff, R.A. (1981). Delayed recruitment of maternal H3 mRNA in sea urchin embryos. *Nature (Lond.)* 292, 477–9.

Welshons, W.V., Lieberman, M.E. & Gorski, J. (1984). Nuclear localisation of unoccupied oestrogen receptors. *Nature (Lond.)* 307, 747–9.

Wessel, G.M. & McClay, D.R. (1985). Sequential expression of germ-layer specific molecules in the sea urchin embryo. *Devl. Biol.* 111, 451–63.

Wharton, K.A., Yedvobnick, B., Finnerty, V.G. & Artavanis-Tsakonas, S. (1985a). *opa*: a novel family of transcribed repeats shared by the *Notch* locus and other developmentally regulated loci in *D. melanogaster. Cell*, 40, 55–62.

Wharton, K.A., Johansen, K.M., Xu, T. & Artavanis-Tsakonas, S. (1985b). Nucleotide sequence from the neurogenic locus *Notch* implies a gene product that shares homology with proteins containing EGF-like repeats. *Cell*, 43, 567–81.

White, R.A.H., & Akam, M.E. (1985). *Contrabithorax* mutations cause inappropriate expression of *Ultrabithorax* products in *Drosophila*. *Nature (Lond.)* 318, 567–70.

White, R.A.H. & Lehmann, R. (1986). A gap gene, *hunchback*, regulates the spatial expression of *Ultrabithorax*. *Cell*, 47, 311–21.

White, R.A.H. & Wilcox, M. (1984). Protein products of the bithorax complex in *Drosophila. Cell*, 39, 163–171.

White, R.A.H. & Wilcox, M. (1985a). Distribution of *Ultrabithorax* proteins in *Drosophila. EMBO J.* 4, 2035–43.

White, R.A.H. & Wilcox, M. (1985b). Regulation of the distribution of *Ultrabithorax* proteins in *Drosophila. Nature (Lond.)* 318, 563–6.

Whitfield, W.G.F., Gonzalez, C., Sanchez-Herrero, E. & Glover, D.M. (1989). Transcripts of one of two *Drosophila* cyclin genes become localised in pole cells during embryogenesis. *Nature (Lond.)* 338, 337–9.

Whittaker, J.R. (1980). Acetylcholinesterase development in extra cells caused by changing the distribution of myoplasm in ascidean embryos. *J. Embryol. Exp. Morphol.* 55, 343–54.

Wienzierl, R., Axton, J.M., Ghysen, A. & Akam, M. (1987). *Ultrabithorax* mutations in constant and variable regions of the protein coding sequence. *Genes & Dev.* 1, 386–97.

Wieslander, L. & Daneholt, B. (1977). Demonstration of Balbiani ring RNA sequences in polysomes. *J. cell Biol.* 73, 260–4.

Wilde, C.D. & Akam, M.E. (1987). Conserved sequence elements in the 5' region of the *Ultrabithorax* transcription unit. *EMBO J.* 6, 1393–1401.

Wiley, H.S. & Wallace, R.A. (1981). The structure of vitellogenin. *J. Biol. Chem.* 256, 8626–34.

Wilkins, A. (1986). *Genetic Analysis of Animal Development.* Wiley, New York.

Wilkinson, D.G., Bailes, J.A. & McMahon, A.P. (1987). Expression of the proto-oncogene *int-1* is restricted to specific neural cells in the developing mouse embryo. *Cell*, **50**, 79–88.

Wilkinson, D.G., Bhatt, S., Chavrier, P., Bravo, R. & Charnay, P. (1989a). Segment-specific expression of a zinc-finger gene in the developing nervous system of the mouse. *Nature (Lond.)* **337**, 461–4.

Wilkinson, D.G., Bhatt, S., Cook, M., Boncinelli, E. & Krumlauf, R. (1989). Segmental expression of *Hox-2* homeobox-containing genes in the developing mouse hindbrain. *Nature (Lond.)* **341**, 405–9.

Willusz, J. & Shenk, T. (1988). A 64 kd nuclear protein binds to RNA segments that include the AAUAAA polyadenylation motif. *Cell*, **52**, 221–8.

Wilson, E.B. (1904a). Experimental studies in germinal localisation: I The germ-regions in the egg of *Dentalium*. *J. Exp. Zool.* **1**, 1–72.

Wilson, E.B. (1904b). Experimental studies in germinal localisation: II Experiments on the cleavage-mosaic in *Patella* and *Dentalium*. *J. Exp. Zool.* **1**, 197–268.

Winkler, M.M., Nelson, E.M., Lashbrook, C. & Hershey, J.W.B. (1985). Multiple levels of regulation of protein synthesis at fertilisation in sea urchin eggs. *Devl. Biol.* **107**, 290–300.

Winslow, G.M., Hayashi, S., Krasnow, M., Hogness, D.S. & Scott, M.P. (1989). Transcriptional activation by the *Antennapedia* and *fushi tarazu* proteins in cultured *Drosophila* cells. *Cell*, **57**, 1017–30.

Wirth, T. & Baltimore, D. (1988). Nuclear factor NF-κB can interact functionally with its cognate binding site to provide lymphoid-specific promoter function. *EMBO J.* **7**, 3109–13.

Wirz, J., Fessler, L.I. & Gehring, W.J. (1986). Localisation of the *Antennapedia* protein in *Drosophila* embryos and imaginal discs. *EMBO J.* **5**, 3327–34.

Wiskocil, R., Bensky, P., Dower, W., Goldberger, R.F., Gordon, J.I. & Deeley, R.G. (1980). Coordinate regulation of two estrogen-dependent genes in avian liver. *Proc. Natl. Acad. Sci. (USA)* **77**, 4474–8.

Wold, B.J., Klein, W.H., Hough-Evans, B.R., Britten, R.J. & Davidson, E.H. (1978). Sea urchin embryo mRNA sequences expressed in the nuclear RNA of adult tissues. *Cell*, **14**, 941–50.

Wolff, E. (1968). Specific interactions between tissues during embryogenesis. *Curr. Topics Devl. Biol.* **3**, 65–94.

Wolffe, A.P. (1988). Transcription fraction TFIIIC can regulate differential 5S RNA gene transcription *in vitro*. *EMBO J.* **7**, 1071–9.

Wolffe, A.P. & Brown, D.D. (1988). Developmental regulation of two 5S ribosomal RNA genes. *Science*, **241**, 1626–32.

Wolfner, M.F. (1988). Sex-specific gene expression in somatic tissues of *Drosophila melanogaster*. *Trends Genet.* **4**, 333–7.

Wolgemuth, D.J., Engelmeyer, E., Duggal, R.N., Gizang-Ginsberg, E., Mutter, C.-I., Ponzetto, C., Viviano, C. & Zakeri, Z.F. (1986). Isolation of a mouse cDNA coding for a developmentally regulated, testis-specific transcript containing homeobox homology. *EMBO J.* **5**, 1229–35.

Wolgemuth, D.J., Viviano, C.M., Gizang-Ginsberg, E, Froham, M.A. Joyner, A.L. & Martin, G.R. (1987). Differential expression of the mouse homeobox-containing gene *Hox 1.4* during male germ cell differentiation and embryonic development. *Proc. Natl. Acad. Sci. (USA)* **84**, 5813–7.

Wolgemuth, D.J., Behringer, R.R., Mostoller, M.P., Brinster, R.L. & Palmiter, R.D. (1989). Transgenic mice overexpressing the mouse homeobox-containing gene *Hox-1.4* exhibit abnormal gut development. *Nature (Lond.)* **337**, 464–6.

Wolpert, L. (1969). Positional information and the spatial pattern of cellular differentiation. *J. Theor. Biol.* **25**, 1–47.

Wood, W.B. (ed.) (1988a). *The Nematode* Caenorhabditis elegans. CSH Monographs **17**, Cold Spring Harbor Press.

Wood, W.B. (1988b). Embryology. In *The Nematode* Caenorhabditis elegans. ed. W.B. Wood. CSH Monographs **17**, 215–41.

Wood, W.B. Hecht, R., Carr, S., Vanderslice, R., Wolf, N. & Hirsh, D. (1980). Parental effects of mutations that affect early development in *Caenorhabditis elegans*. *Devl. Biol.* **74**, 446–69.

Wood, W.B., Schierenberg, E. & Strome, S. (1984). Localisation and determination in embryos of *Caenorhabditis elegans*. *UCLA Symp. Mol. Cell. Biol.* **19**, 37–49.

Wood, W.G., Old, J.M., Roberts, A.V.S., Clegg, J.B., Weatherall, D.J. & Quattrin, N. (1978). Human globin gene expression; control of β, δ and $\delta\beta$ chain production. *Cell*, **15**, 437–46.

Woodland, H.R. & Jones, E.A. (1986). Unscrambling egg structure. *Nature (Lond.)* **329**, 261–2.

Woodland, H.R., Flynn, J.M. & Wyllie, A.J. (1979). Utilisation of stored mRNA in *Xenopus* embryos and its replacement by newly synthesised transcripts: histone H1 synthesis using interspecies hybrids. *Cell*, **18**, 165–71.

Woodland, H.R. Old, R.W., Sturgess, E.A., Ballantine, J.E.M., Aldridge, T.C. & Turner, P.C. (1983). The strategy of histone gene expression in the development of *Xenopus*. In *Current Problems in Germ Cell Determination*, ed. A. McLaren & C.C. Wylie, pp. 353–75. Cambridge University Press, Cambridge.

Woychik, R.P., Stewart, T.A., Davis, L.G., D'Eustachio, P. & Leder, P. (1985). An inherited limb deformity created by insertional mutagenesis in a transgenic mouse. *Nature (Lond.)* **318**, 36–40.

Wright, C.V.E., Schnegelsberg, P. & de Robertis, E.M. (1988). *XlHbox 8*: a novel *Xenopus* homeo protein restricted to a narrow band of endoderm. *Development*, **104**, 787–94.

Wright, C.V.E., Cho, K.W.Y., Hardwicke, J., Collins, R.H. & de Robertis, E.M. (1989). Interference with function of a homeobox gene in *Xenopus* embryos produces malformations of the anterior spinal cord. *Cell*, **59**, 81–93.

Wright, S., de Boer, E., Grosveld, F.G. & Flavell, R.A. (1983). Regulated expression of the human β globin gene family in murine erythroleukaemia cells. *Nature (Lond.)* **305**, 333–5.

Wright, S., Rosenthal, A., Flavell, R. & Grosveld, F. (1984). DNA sequences required for regulated expression of β globin genes in murine erythroleukaemia cells. *Cell*, **38**, 265–73.

Wright, W.E., Sassoon, D.A. & Lin, V.K. (1989). Myogenin, a factor regulating myogenesis, has a domain homologous to Myo D. *Cell*, **56**, 607–17.

Wylie, C.C., Snape, A., Heasman, J. & Smith, J.C. (1987). Vegetal pole cells and commitment to form endoderm in *Xenopus laevis*. *Devl. Biol.* **119**, 496–502.

Yablonka-Revueni, Z. & Hille, M.B. (1983). Isolation and distribution of elongation factor 2 in eggs and embryos of sea urchins. *Biochemistry*, **22**, 5205–12.

Yamada, T. (1977). Control mechanisms in cell-type conversations in newt lens regeneration. *Monographs in Developmental Biology* Vol. **13**, ed. A. Wolsky, S. Karger, Basel.

Yamada, T. & McDevitt, D.S. (1984). Conversion of iris epithelial cells as a model of differentiation control. *Differentiation*, **27**, 1–12.

Yamamoto, K.R. (1985). Steroid receptor regulated transcription of specific genes and gene networks. *Ann. Rev. Genet.* **19**, 209–52.

Yedvobnick, B., Muskavitch, M.A.T., Wharton, K.A., Halpern, M.E., Paul, E., Grimwade, B.G. & Artavanis-Tsakonas, S. (1985). Molecular analysis of *Drosophila* neurogenesis. *Cold Spring Harbor Symp. Quant. Biol.* **50**, 841–54.

Yisraeli, J.K. & Melton, D.A. (1988). The maternal mRNA Vg1 is correctly localised following injection into *Xenopus* oocytes. *Nature (Lond.)* **336**, 592–4.

Yisraeli, J., Adelstein, R.S., Melloul, D., Nudel, U., Yaffe, D. & Cedar, H. (1986). Muscle-specific activation of a methylated chimaeric actin gene. *Cell*, **46**, 409–16.

Yochem, J. & Greenwald, I. (1989). *glp-1* and *lin-12*, genes implicated in distinct cell–cell interactions in *C. elegans*, encode similar transmembrane proteins. *Cell*, **58**, 553–63.

Yochem, J., Weston, K. & Greenwald, I. (1988). The *Caenorhabditis elegans lin-12* gene encodes a transmembrane protein with overall similarity to *Drosophila Notch. Nature (Lond.)* **335**, 547–50.

Zanjani, E.D., McGlave, P.B., Bhakthavathsalan, A. & Stamatoyannopoulos, G. (1979). Sheep foetal haematopoietic cells produce adult haemoglobin when transplanted in the adult animal. *Nature (Lond.)* **280**, 495–6.

Zelenka, P. & Piatigorsky, J. (1976). Reiteration frequency of δ-crystallin DNA in lens and non-lens tissues of chick embryos: δ-crystallin gene is not amplified during lens cell differentiation. *J. Biol. Chem.* **251**, 4294–8.

Zink, B. & Paro, R. (1989). *In vivo* binding pattern of a *trans*-regulator of homoeotic genes in *Drosophila melanogaster. Nature (Lond.)* **337**, 468–70.

Ziomek, C.A., Johnson, M.H. & Handyside, A.H. (1982). The developmental potential of mouse 16-cell blastomeres. *J. Exp. Zool.* **221**, 345–53.

References added in proof

Blanco, J., Millstein, L., Razik, M., Dilworth, S., Cote, C. & Gottesfeld, J. (1989). Two TFIIIA activities regulate expression of the *Xenopus* 5S RNA gene families. *Gene & Dev.* **3**, 1602–12.

Carroll, S.B. (1990). Zebra patterns in fly embryos: activation of stripes or repression of interstripes? *Cell*, **60**, 9–16.

Casanova, J. & Struhl, G. (1989). Localised surface activity of *torso*, a receptor tyrosine kinase, specifies terminal body pattern in *Drosophila. Genes & Dev.* **3**, 2025–38.

Davidson, E.H. (1990). How embryos work: a comparative view of diverse modes of cell fate specification. *Development*, **108**, 365–89.

Dollé, P., Izpisua-Belmonte, J.-C., Falkenstein, H., Renucci, A. & Duboule, D.

(1989a). Coordinate expression of the murine *Hox-5* complex homoeobox-containing genes during limb pattern formation. *Nature (Lond.)* **342**, 767–72.

Dollé, P., Ruberte, E., Kastner, P., Petkovitch, M., Stoner, C.M., Gudas, L.J. & Chambon, P. (1989). Differential expression of genes encoding α, β and γ retinoic acid receptors and CRABP in the developing limbs of the mouse. *Nature (Lond.)* **342** 702–4.

Enver. T., Raich, N., Ebens, A.J., Papayannopulou, T., Costantini, F. & Stamatoyannopoulos, G. (1990). Developmental regulation of human foetal-to-adult globin gene switching in transgenic mice. *Nature (Lond.)* **344**, 309–14.

Fraser, S., Keynes, R. & Lumsden, A. (1990). Segmentation in the chick embryo hindbrain is defined by cell lineage restrictions. *Nature (Lond.)* **344**, 431–5.

Gallarda, J.L., Foley, K.P., Yang, Z. & Engel, J.D. (1989). The β-globin stage selector element factor is erythroid-specific promoter/enhancer binding protein NF-E4, *Genes & Dev.* **3**, 1845–59.

Gonzalez-Reyes, A., Urquia, N., Gehring, W.J., Struhl, G. & Morata, G. (1990). Are cross-regulatory interactions between homoeotic genes functionally significant? *Nature (Lond.)* **344**, 78–80.

Hopwood, N.D., Pluck, A. & Gurdon, J.B. (1989). A *Xenopus* mRNA related to *Drosophila twist* is expressed in response to induction in the mesoderm and neural crest. *Cell*, **59**, 893–903.

Inoue, K., Hoshijima, K., Sakamoto, H. & Shimura, Y. (1990). Binding of the *Drosophila Sex-lethal* gene product to the alternative splice site of *transformer* primary transcript. *Nature (Lond.)* **344**, 461–4.

Kopczynski, C.C. & Muskavitch, M.A.T. (1989). Complex spatio-temporal accumulation of alternative transcripts from the neurogenic gene *Delta* during *Drosophila* embryogenesis. *Development*, **107**, 623–36.

Lawrence, P.A. (1990). Compartments in vertebrates? *Nature (Lond.)* **344**, 382–3.

Lewis, J. & Martin, P. (1989). Limbs: a pattern emerges. *Nature (Lond.)* **342**, 734–5.

Liu, Z. & Ambros, V. (1989). Heterochronic genes control the stage-specific initiation and expression of the dauer larva developmental program in *Caenorhabditis elegans*. *Genes & Dev.* **3**, 2039–49.

Robertson, M. (1990). More to muscle than MyoD. *Nature (Lond.)* **344**, 378–9.

Rosa, F.M. (1989). *Mix-1*, a homeobox mRNA inducible by mesoderm inducers, is expressed mostly in the presumptive endodermal cells of *Xenopus* embryos. *Cell*, **57**, 965–74.

Roth, S., Stein, D. & Nusslein-Volhard, C. (1989). A gradient of nuclear localisation of the *dorsal* protein determines dorsoventral pattern in the *Drosophila* embryo. *Cell*, **59**, 1189–202.

Rushlow, C.A., Han, K., Manley, J.L. & Levine, M. (1989). The graded distribution of the *dorsal* morphogen is initiated by selective nuclear transport in *Drosophila*. *Cell*, **59**, 1165–77.

Schatz, D.G., Oettinger, M.A. & Baltimore, D. (1989). The V(D)J recombination activating gene, RAG-1. *Cell*, **59**, 1035–48.

Schnabel, R. & Schnabel, H. (1990). Early determination in the *C. elegans* embryo: a gene, *cib-1*, required to specify a set of stem-cell-like blastomeres. *Development*, **108**, 107–19.

Steward, R. (1989). Relocalisation of the *dorsal* protein from the cytoplasm to the nucleus correlates with its function. *Cell*, **59**, 1179–88.

Suter, B., Romberg, L.M. & Steward, R. (1989). *Bicaudal-D*, a *Drosophila* gene involved in developmental asymmetry: localised transcript accumulation in ovaries and sequence similarity to myosin heavy chain tail domains. *Genes & Dev.* **3**, 1957–68.

Takagaki, Y., Ryner, L.C. & Manley, J.L. (1989). Four factors are required for 3'-end cleavage of pre-mRNAs. *Genes & Dev.* **3**, 1711–24.

Way, J.C. & Chalfie, M. (1989). The *mec-3* gene of *Caenorhabditis elegans* requires its own product for maintained expression and is expressed in three neuronal cell types. *Genes & Dev.* **3**, 1823–33.

Wharton, R.P. & Struhl, G. (1989). Structure of the *Drosophila Bicaudal-D* protein and its role in localising the posterior determinant *nanos*. *Cell*, **59**, 881–92.

Woodland, H.R. (1989). Mesoderm formation in *Xenopus*. *Cell*, **59**, 767–70.

Index